New Insights into Archaeoastronomy

New Insights into Archaeoastronomy

Edited by Hudson Pearce

MURPHY & MOORE
www.murphy-moorepublishing.com

Murphy & Moore Publishing,
1 Rockefeller Plaza,
New York City, NY 10020, USA

ISBN: 978-1-63987-390-6

Cataloging-in-Publication Data

New insights into archaeoastronomy / edited by Hudson Pearce.
p. cm.
Includes bibliographical references and index.
ISBN 978-1-63987-390-6
1. Archaeoastronomy. 2. Astronomy. 3. Astronomy--History.
4. Archaeology. I. Pearce, Hudson.
GN799.A8 C87 2022
520.93--dc23

For information on all Murphy & Moore Publications
visit our website at www.murphy-moorepublishing.com

MURPHY & MOORE

Contents

Preface

The multidisciplinary study that focuses on how the people in the past understood the phenomena in the sky, the usage of this knowledge, and the roles the sky played in their cultures is referred to as archaeoastronomy. It uses a variety of methodologies to uncover evidence of past practices such as archaeology, anthropology, astronomy, statistics and probability, and history. There are two approaches to do archaeoastronomy, namely, green archaeoastronomy and brown archaeoastronomy. Some of the major topics of archaeoastronomical research include the use of calendars, myth, and cosmology, and displays of power. This book discusses the fundamentals as well as modern approaches to archaeoastronomy. Some of the diverse topics covered herein address the varied branches that fall under this category. This book is a vital tool for all researching or studying archaeastronomy as it gives incredible insights into emerging trends and concepts.

This book unites the global concepts and researches in an organized manner for a comprehensive understanding of the subject. It is a ripe text for all researchers, students, scientists or anyone else who is interested in acquiring a better knowledge of this dynamic field.

I extend my sincere thanks to the contributors for such eloquent research chapters. Finally, I thank my family for being a source of support and help.

Editor

1

AN ARCHAEOASTRONOMICAL APPROACH TO THE MEGALITHIC SITES OF SAUDI ARABIA

Munirah A. Almushawh

King Saud University, Department of Archaeology, Riyadh, Saudi Arabia

(munira.a.a@hotmail.com)

ABSTRACT

This study attempts to decipher the cultural significance of standing stones left by the past civilizations in late Stone Ages. It provides a statistical list of 16 megalithic sites within Saudi Arabia with more focus on the site of Rajajil in northwest Saudi Arabia. Speculation as to their purposes differ, some were used for religious activities, burial sites, and astronomical observatories for celestial bodies.

In this study a field survey was conducted to the late 5th millennium B.C.E. site of Rajajil in Al-Jawf Region (Arabic: الجوف al-Ğawf), and some of the well preserved groups of megaliths in the site were measured in an attempt to shed some light on the bearings to know if astronomical interpretation is possible. Measurment indicated that there is a clear alignment along the north-south axis In one of the megalithic groups on the site. The general layout of the site applies to this description to a high extent too. The site belong to the Neolithic period, where societies began to think in a more complex way than before. The majority of megalithic sites including Rajajil belong to the era of agricultural revolution where agricultural activities demand astronomical knowledge, therefore its logical to see traces of astronomy dating back to that stage. Petroglyphs of celestial objects carved on some standing stones e.g. the figures carved in Kubat Altamathil megalithic site in Tabuk (Arabic: تبوك Tabūk) indicate that ancients observed the sky and recorded what they perceived. To ensure that the functional purposes of ancient archaeological sites might affect the type of inscriptions in the surroundings, I traced examples of ancient natural observatories still used to the present time e.g. (Majarda observatory, Taif) and (Hilat Alshams observatory, ABaha) and the result was positive.

The aim of this studey is to give an overview of existing possibilities of astronomical indications to stimulate further archaeoastronomical studies in saudi arabia.

KEYWORDS: Archaeoastronomy, Megaliths, Ancient Observatories, Rajajil site

1. INTRODUCTION

Megalithic sites are found in various continents, such as Asia, Africa, Europe, America and Australia. This raises an important question, what are these sites? And was this similarity a result of ancient cultural connection? Or is it simply a similar human thought? These sites reveal many cultural aspects of the life of our ancestors, they reflect advanced mentality, strong memory, and very sophisticated ability to plan. Megalithic structures are often associated with Neolithic agricultural revolution that originated from the Fertile Crescent, "what is now called Syria, Iraq, and Turkey", and then spread over subsequent periods to neighboring continents (Patton, 1993).

Studies have emerged that attempt to interpret these sites from an archaeoastronomical perspective. Unfortunately, they are few in the Arabian Peninsula although the archaeological material suitable for this type of study is available and of great value. In the twenty- first century, archaeoastronomy is starting to be considered one of the "key concepts" in the growth and progress of archaeological thinking and method (Renfrew and Bahn, 2005). More solid theoretical studies appeared, unlike the studies that spread in the 1960s and 1970s. This has resulted in the broad acceptance of cultural astronomy globally. But some disciplines still find it tough to garner the attention of Middle East scholars. Only recently, some serious studies like (Hoskin, 2001; Belmonte, González-García, and Polcaro 2013; Belmonte and González-García, 2014; González-García, Belmonte and Polcaro 2016, Liritzis et al., 2015) have emerged and aroused the curiosity of those interested in this type of archaeological studies in the area. Authoritative contributions like the encyclopedia of cosmology and myth (Ruggles, 2005) has been a great scientific value to serious scholars regardless of their disciplines, covering broad definitions and case studies from five continents, presenting deferent methodologies, and giving a complete representation of this field. It is only nowadays that this interdisciplinary field is starting to be recognized in Saudi Arabia, and the hope is that with an overarching conceptual framework, it will be possible to understand more about megalithic civilizations worldwide once comparative studies are presented from this area in line with the increasing research topics and case studies in the world.

The earliest known megalithic sites were set up about 9,000 years B.C.E. in Asia (Reingruber, 2005), then sites were built in Africa, and according to (Rosen & Brophy, 2005) the oldest site in Africa is probably Napata dating about 6800 B.C.E., then sites occurred in Europe during the Late Neolithic period to the Copper Age (4500-1500 B.C.E.). This study sheds light on the site of Rajajil, a 5th millennium B.C.E. site (Gebel, 2016). The purpose of this study is revealing the cultural indications of such sites in Arabia.

2. MEGALITHIC SITES OF SAUDI ARABIA

Megalithic sites spread throughout the Arabian Peninsula some in the central regions and others in the northern Arabian Peninsula, such as Jordan, as well as in the far south, such as Yemen (as shown in Figure 1).

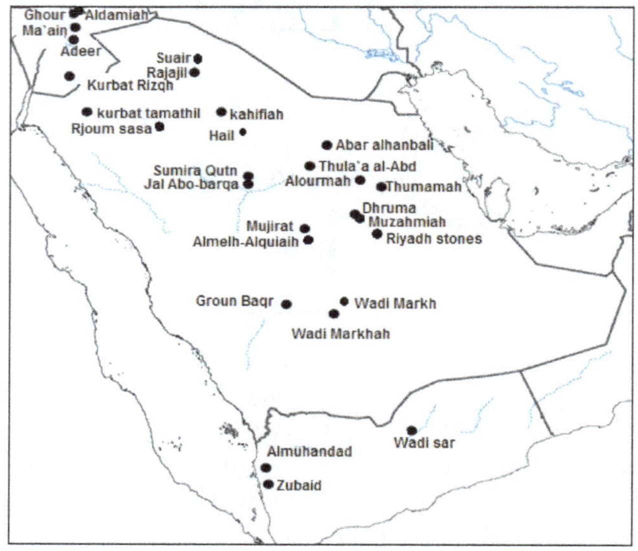

Figure 1. A map of megalithic sites within the Arabian Peninsula

The size of these megaliths varies from one site to another, and the layout of the site also differs. Standing stones are found either in the form of an alignment, a stone circle or a monolith.

2.1. *MEGALITHIC SITES IN NORTHWEST SAUDI ARABIA*

The north-western region of the Kingdom is considered an area of great importance due to its unique location in the ancient world. It is a region rich in geological and geographic aspects. It encompasses a large area of the Arabian Peninsula. It is also characterized by an environment suitable for life and stability. It was an active settlement area, and this is evident from the large number of archaeological finds in various sites. Excavations enabled scientists to develop an historical sequence based on the archaeological material discovered, from the Stone Age to the present time. One of the first settlement sites in the region is the site of Shuweihtia dating back to more than a million years ago, making it the oldest settlement site in the Arabian Peninsula to date (Whalen et al., 1989). This region also contains two sites registered on the UNESCO World Heritage List, the multi-temporal site of Al-Hajar in Al-Ula, which includes major Nabataean remains, and the Hail rock art site. As for megalithic sites, this part of Saudi Arabia has one the most significant sites in terms of preservation, complexity and size. See Table 1 for a list of sites from Hail, Tabouk and Al-Jouf.

Table 1. Megalithic sites in north-west Saudi Arabia.

Site	Region	longitude	Latitude	Type
1- Rjajil	Jouf	E 40° 13′ 13″	N 29° 48′ 45″	Standing stones
2- Souier	Jouf	E 40° 23′ 30″	N 30° 02′ 45″	Standing stones
3- Hail stones	Hail	E 42° 14′ 46″	N 27° 20′ 09″	Standing stones
4- Alkhifia	Hail	Near hail city E 42° 14′ 46″	Near hail city N 27° 20′ 09″	Standing stones
5- Kubat Tamathil	Tabuk	Near Al Athem center E 37° 33′ 44″	Near Al Athem center N 27° 39′ 19″	Standing stones
6-Rjoum Sasa	Tabuk	Near "Taima" E 38° 32′ 0. 11″	Near "Taima" N 27° 37′ 0. 9″	Standing stones

2.2. *MEGALITHIC SITES OF RIYADH*

Riyadh is the capital city of Saudi Arabia. It is situated in the centre of the Arabian Peninsula. There are many megalithic sites Riyadh such as the Riyadh standing stones located in the north of the wadi Al-Dawasir, as well as those of Mount Al-Arma, Muzahmiya, Mujirat south of Dawadmi, Abar Hanbali in the province of Al-Majma`ah, Dhruma and the site of Quaiya. See Table 2 for Riyadh sites.

Table 2. Megalithic sites in Riyadh.

Site	Region	longitude	Latitude	Type
1- Riyadh stones	Riyadh	E 44° 20′ 53″	N 21° 00′ 25″	Standing stones
2- Mount Al-Arma	Riyadh	E 46° 16′ 32″	N 25° 51′ 14″	Standing stones
3- Muzahmiya	Riyadh	E 46° 18′ 53″	N 24° 21′ 20″	Standing stones
4- Mujirat	Riyadh	E 44° 26′ 18″	N 24° 14′ 49″	Standing stones
5- Abar Hanbali	Riyadh	E 45° 04′ 37″	N 27° 19′ 16″	Standing stones
6- Dhruma	Riyadh	E 46° 7′60″	N 24° 35′ 60″	Standing stones
7- Quaiya	Riyadh	E 45° 16′ 50″	N 24° 3′ 56″	Standing stones

Table 3. Megalithic sites in Qassim.

Site	Region	longitude	Latitude	Type
1- Jal Abu Baraka	Qassim	N 44° 23′ 34″	N 26° 17′ 30″	Standing stones
2- Thula`a al-Abd	Qassim	Near Thulah E 43° 57′ 46. 84″	Near Thulah N 25° 58′ 23. 62″	Standing stones
3- Samira Cotton	Qassim	Near Cotton E 42° 20′ 4. 77″	Near Cotton N 25° 59′ 75″	Standing stones

2.3. *MEGALITHIC SITES OF QASSIM*

Qassim is one of the thirteen administrative regions of Saudi Arabia. Located at the heart of the kingdom. There are several megalithic sites in Qassim area, such as: the site of Jal Abu Baraka (Al-shaya, 2009), the site of Thula`a al-Abd and the site of Samira Cotton demonstrated. See Table 3 for Qassim sites.

3. FIELD STUDY (RAJAJIL MEGALITHIC SITE)

Rajajil site is located in Sakaka / Al-Jouf, it is one of the most important megalithic sites in Saudi Arabia. This site was excavated during the comprehensive survey of the lands of Saudi Arabia under the supervision of the Department of Antiquities and Museums in 1965.

The site was described by the two western historians, Fred Winnet and Willem Reed, in their journey through Wadi Sarhan in 1962, they referred to it as a large stone circle, where standing stones were erected in separate collections (Winnett & Reed, 1970). The Archaeology Department sent teams to survey the site from 1965, during which time the researcher were able to study the site better. The site dates bac to the end of the fifth millennium B.C.E., accordi to the Results of Rajajil Joint Archaeological Proj (Gebel, 2016).

In the study reviled in this paper more focus made in the fenced area (as shown in Figure 2). area contains standing stones and many stone arrowheads, and fragments of other tools, whi found within the fence and also without.

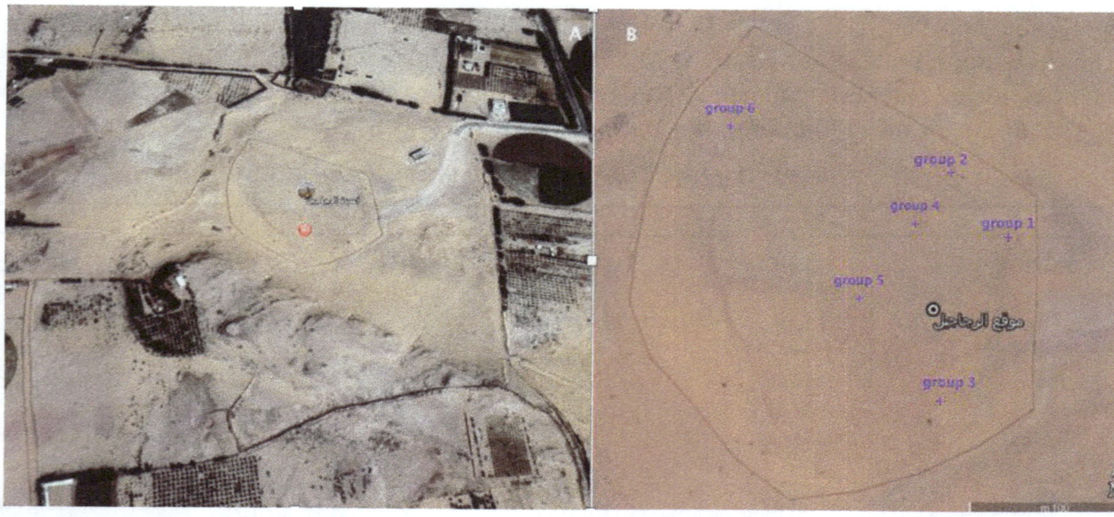

Figure 2. A. General view of the site; (B) The fenced area is the boundary of the field work.

3.1. SITE DESCRIPTION

The site is situated between a hill on the south, and a low slope on the north, with an altitude of 600m above sea level. A panoramic photo with a wide-angle view of the site is shown in Figure 3. A 360° view can be found online in at Rajajil Archaeological Site (2011).

Figure 3. panoramic photo of the site location.

'n the conducted survey of the site in 2016 only six 'ps could still possibly be measured. The existing 's have lost some of the stones through time in 'ison to the first archaeological plan of the site 1977 (as shown in Figure 4). The rest of the e disturbed by either vandalism or natural 's clear that all standing stones in the six erected according to a certain plan and lomly placed. Most of the stone rows gned towards the North-South Axis. sting standing stone is about 4.20 allest is around 50cm. This highest oup located on the southern side of the fence. Other standing stones in the site are within the range of two to three and four meters high. As for the measurement procedure, the alignment was calculated by measuring the azimuth, the angle from north, of the structure and the altitude of the horizon it faces. The azimuth was measured using a compass with a precision of a degree. The declination was calculated using the same formulae as "GETDEC", the program referred to in Astronomy in Prehistoric Britain and Ireland (Ruggles, 2005, p.169) and can be found online in (Ruggles, 2018). See Table 4 for details of the approximate measurements.

Table 4. details of the groups.

alignment	Latitude	Longitude	Azimuth	Apparent declination
NS	+29 81 30	+39 29 32	0°	+80 19 36
NS	+29 48 47	+39 00 00	3°	+79 46 58
NS	+29 48 48	+36 16 27	8°	+80 00 32
W-SSE	+29 48 44	+30 09 37	348°	+79 36 18
W	+29 48 49	+38 25 11	354°	+79 23 35
NW	+29 48 42	+37 29 44	352°	+79 50 48

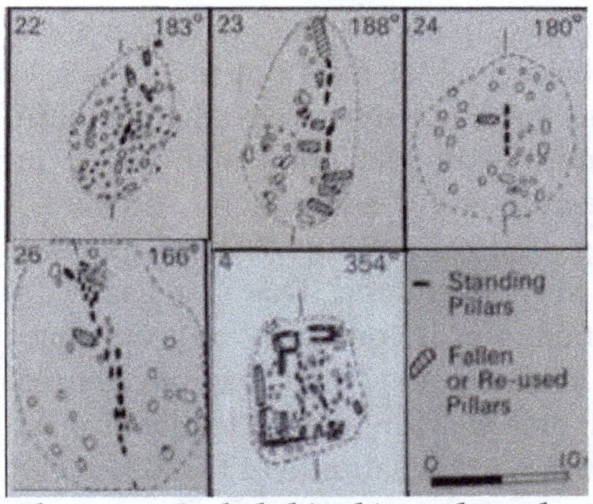

Figure 4. The groups included in this study as drawn in the 1st archeological plan of the site (Parr et al,1978)

3.1.1 Group 1

This is the most distinguished group in the site because it is still in a very good condition. The row of standing stones is precisely aligned upon the North-South Axis.

The group is composed of four standing stones, fixed and strong, with a North-South Axis, and a slight slope to the north. The group is surrounded by broken parts of the stones on the ground. The most important of which is a 1.60 m stone, which seems to have been broken in the past. It contains five deeply engraved circles, which are not a result of the erosion processes. There are also some inscriptions and signs on the surface. One highly representing an arrow directed up high, maybe to the sky. The measurements of the stones are as follows: in order from north to south: First - height: 3.10 m, second - height: 2.12 m, third - height: 2.75 m, fourth - height: 2.10 m (as shown in both Figure 5A and 5C). Extending side by side with a width of 2.68 meters (as shown in Figure 5 D).

Figure 5 A. GROUP 1; (B) Facing north; (C) Facing east; (D) Facing west

3.1.2 Group 2

Two standing stones almost oriented toward the geographical north with an axis of North-South, it seems as if they are about to fall, leaning towards the west, and around them there are parts of stones on the ground. The heights of these are: 3.75 m and 2.50 m.

3.1.3 Group 3

The group is composed of four vertical standing stones. Although they seem unaligned in their current condition, they actually were aligned in the past but have since tilted over in time and ended up in this arrangement: Their heights are between 4.20 meters to 2.60 meter (Figure 6)

Figure 6. Group 3: stones probably tilted over time

3.2. *Samples of archaeological remains*

Many archaeological remains were found, some were sent to the laboratories for dating and analysis, and others were placed in AL-Jouf Museum. We must bear in mind that some of the findings belong to the civilization that built Rajajil site, but other items may have been added to it later. Therefore, the accuracy of collecting evidence is mandatory in order to help reveal the functional purposes of the site. See Figure 7 for some Samples collected from the site.

Figure 7. Lithics collected from the surface of the site of Rajajil

4. RESULTS

Measurements indicate that all the six groups measured in this study (five different stone rows and one horse shoe structure) are set along a particular axis, the North-South Axis (as shown in Figure 8). Group 1 located in the northern part of the fence is at an angle of 360 ° due north.

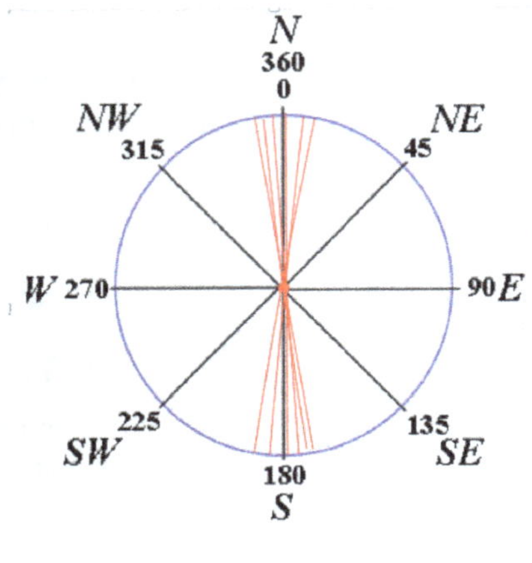

Figure 8. Orientation of 6 well-preserved groups in Rajajil

This rises an important question; how did the prehistoric inhabitants of the site know where the cardinal directions were? It seems that the builders of this site in addition to their high physical ability to transport and build these huge megalithic monuments, they had the ability to choose significant directions for it too. Due to the absence of written records of a site dating back to the fifth millennium B.C.E like Rajajil it is difficult to prove or reject this theory. Researchers in the prehistoric era are left with the option of reviewing the evidence and measuring its relevance to reality. Therefore, confirmation or disconfirmation by extra pragmatic data will then immediately strengthen or weaken the hypotheses. There are other sites with similar axis alignment like Nabta Playa, one of the most important monumental monuments in the African continent. The north-south axis in the site of Nabta is slightly lower in two degrees. This is fair enough if we consider that during the time Nabta functioned (6100-5600 B.C.E) the star that indicated the north was not as bright as the star we see today (Malville et al., 2008). Due to the precession of the equinoxes and the stars' motions, the role of the north star changes from one star to another through long periods of time, and the star closest to the Earth's north celestial pole is considered the north star. It is a point in the sky around which the stars seem to rotate, it also constitutes true north therefore it will be easy to determine where the south is, it's easy to figure it out in the northern hemisphere. In our days it is called Polaris, it appears due north in the sky to a precision better than one degree and if you find it, you've found true north. Therefore, it becomes possible to know the rest of the directions; this has made the north star very important in ancient times. In the case of Rajajil, if the site is deliberately aligned to the north than the possibility that they knew the starry skies well increases. They might have recognized Thuban; the Pole Star Around 5000 years ago Thuban (α Draconis) laying a mere 0°.1 from the pole (Norton & Ridpath, 2004). During the course of the night, the north star does not rise or set, it remains nearly in the same spot above the northern horizon year-round while the other stars circle around it. Using astronomical simulations software, the four standing stones in group 1 seem in relation with Polaris in present time. This star has a declination +89°. Therefore, it constantly appears due north in the sky to a precision better than one degree, and the angle it makes with respect to the true horizon is almost equal to the latitude of the observer. This star is in the constellation of small dipper (as shown in Figure 9).

Figure 9. Simulation of the alignment found in group 1, using the software Skyview.

However, again the author is quite aware that stars change their locations in the sky over thousands of years and investigations for the era of Rajajil reveal that the star of the north was in the Dragon constellation and the sky would have appeared as we see (in Figure 10). Importantly, the movement of stars does not change the fact that the site is aligned to the north

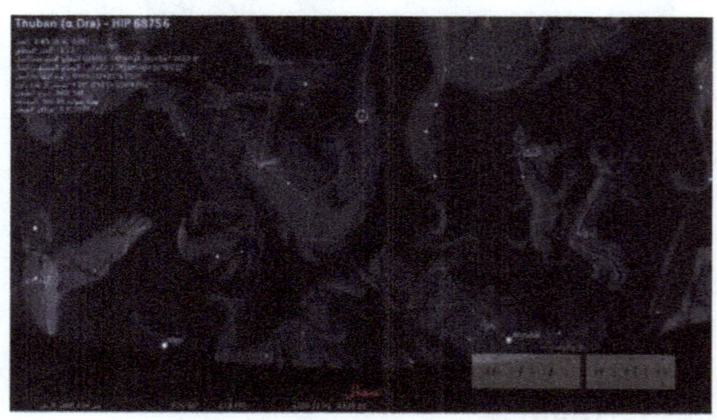

Figure 10. The sky as seen when standing at the site in Al-Jawf in the fifth millennium B.C.E., using the public software Stellarium.

two sites registered on the UNESCO World Heritage List, the multi-temporal site of Al-Hajar in Al-Ula, which includes major Nabataean remains, and the Hail rock art site. As for megalithic sites, this part of Saudi Arabia has one the most significant sites in terms of preservation, complexity and size. See Table 1 for a list of sites from Hail, Tabouk and Al-Jouf.

Table 1. Megalithic sites in north-west Saudi Arabia.

Site	Region	longitude	Latitude	Type
1- Rjajil	Jouf	E 40° 13′ 13″	N 29° 48′ 45″	Standing stones
2- Souier	Jouf	E 40° 23′ 30″	N 30° 02′ 45″	Standing stones
3- Hail stones	Hail	E 42° 14′ 46″	N 27° 20′ 09″	Standing stones
4- Alkhifia	Hail	Near hail city E 42° 14′ 46″	Near hail city N 27° 20′ 09″	Standing stones
5- Kubat Tamathil	Tabuk	Near Al Athem center E 37° 33′ 44″	Near Al Athem center N 27° 39′ 19″	Standing stones
6-Rjoum Sasa	Tabuk	Near "Taima" E 38° 32′ 0. 11″	Near "Taima" N 27° 37′ 0. 9″	Standing stones

2.2. *MEGALITHIC SITES OF RIYADH*

Riyadh is the capital city of Saudi Arabia. It is situated in the centre of the Arabian Peninsula. There are many megalithic sites Riyadh such as the Riyadh standing stones located in the north of the wadi Al-Dawasir, as well as those of Mount Al-Arma, Muzahmiya, Mujirat south of Dawadmi, Abar Hanbali in the province of Al-Majma`ah, Dhruma and the site of Quaiya. See Table 2 for Riyadh sites.

Table 2. Megalithic sites in Riyadh.

Site	Region	longitude	Latitude	Type
1- Riyadh stones	Riyadh	E 44° 20′ 53″	N 21° 00′ 25″	Standing stones
2- Mount Al-Arma	Riyadh	E 46° 16′ 32″	N 25° 51′ 14″	Standing stones
3- Mu-zahmiya	Riyadh	E 46° 18′ 53″	N 24° 21′ 20″	Standing stones
4- Mujirat	Riyadh	E 44° 26′ 18″	N 24° 14′ 49″	Standing stones
5- Abar Hanbali	Riyadh	E 45° 04′ 37″	N 27° 19′ 16″	Standing stones
6- Dhruma	Riyadh	E 46° 7′60″	N 24° 35′ 60″	Standing stones
7- Quaiya	Riyadh	E 45° 16′ 50″	N 24° 3′ 56″	Standing stones

Table 3. Megalithic sites in Qassim.

Site	Region	longitude	Latitude	Type
1- Jal Abu Baraka	Qassim	N 44° 23′ 34″	N 26° 17′ 30″	Standing stones
2- Thula`a al-Abd	Qassim	Near Thu-lah E 43° 57′ 46. 84″	Near Thulah N 25° 58′ 23. 62″	Standing stones
3- Samira Cotton	Qassim	Near Cot-ton E 42° 20′ 4. 77″	Near Cotton N 25° 59′ 75″	Standing stones

2.3. *MEGALITHIC SITES OF QASSIM*

Qassim is one of the thirteen administrative regions of Saudi Arabia. Located at the heart of the kingdom. There are several megalithic sites in Qassim area, such as: the site of Jal Abu Baraka (Al-shaya, 2009), the site of Thula`a al-Abd and the site of Samira Cotton demonstrated. See Table 3 for Qassim sites.

3. FIELD STUDY (RAJAJIL MEGALITHIC SITE)

Rajajil site is located in Sakaka / Al-Jouf, it is one of the most important megalithic sites in Saudi Arabia. This site was excavated during the comprehensive survey of the lands of Saudi Arabia under the supervision of the Department of Antiquities and Museums in 1965.

The site was described by the two western historians, Fred Winnet and Willem Reed, in their journey through Wadi Sarhan in 1962, they referred to it as a large stone circle, where standing stones were erected in separate collections (Winnett & Reed, 1970). The Archaeology Department sent teams to survey the site from 1965, during which time the researchers were able to study the site better. The site dates back to the end of the fifth millennium B.C.E., according to the Results of Rajajil Joint Archaeological Project (Gebel, 2016).

In the study reviled in this paper more focus was made in the fenced area (as shown in Figure 2). This area contains standing stones and many stone tools, arrowheads, and fragments of other tools, which are found within the fence and also without.

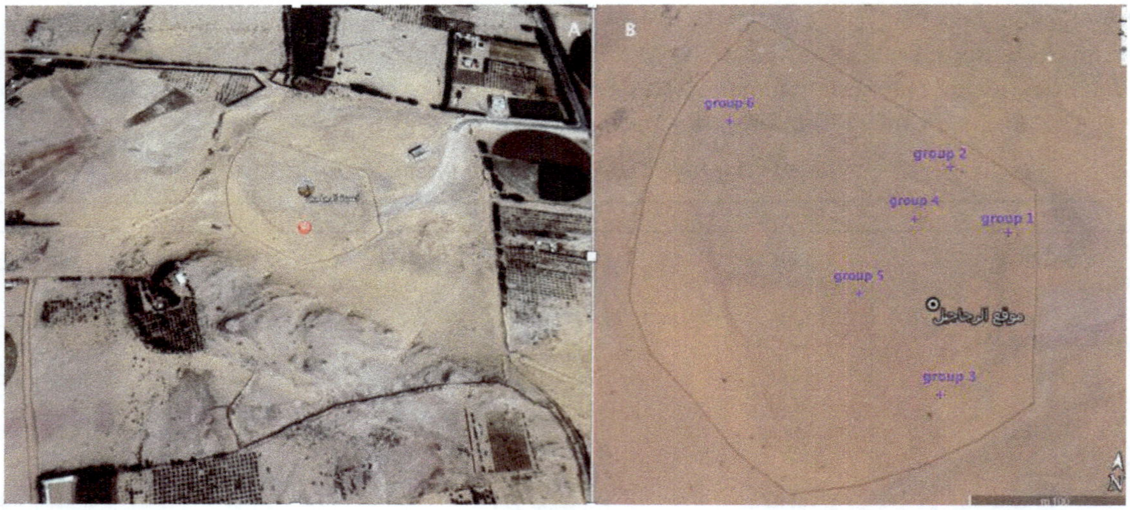

Figure 2. A. General view of the site; (B) The fenced area is the boundary of the field work.

3.1. SITE DESCRIPTION

The site is situated between a hill on the south, and a low slope on the north, with an altitude of 600m above sea level. A panoramic photo with a wide-angle view of the site is shown in Figure 3. A 360° view can be found online in at Rajajil Archaeological Site (2011).

Figure 3. panoramic photo of the site location.

In the conducted survey of the site in 2016 only six groups could still possibly be measured. The existing groups have lost some of the stones through time in comparison to the first archaeological plan of the site done in 1977 (as shown in Figure 4). The rest of the groups are disturbed by either vandalism or natural erosion. It is clear that all standing stones in the six groups were erected according to a certain plan and were not randomly placed. Most of the stone rows seem to be aligned towards the North-South Axis. The highest existing standing stone is about 4.20 meters and the smallest is around 50cm. This highest stone is part of a group located on the southern side of the fence. Other standing stones in the site are within the range of two to three and four meters high. As for the measurement procedure, the alignment was calculated by measuring the azimuth, the angle from north, of the structure and the altitude of the horizon it faces. The azimuth was measured using a compass with a precision of a degree. The declination was calculated using the same formulae as "GETDEC", the program referred to in Astronomy in Prehistoric Britain and Ireland (Ruggles, 2005, p.169) and can be found online in (Ruggles, 2018). See Table 4 for details of the approximate measurements.

Table 4. details of the groups.

Group number	alignment	Latitude	Longitude	Azimuth	Apparent declination
1	NS	+29 81 30	+39 29 32	0°	+80 19 36
2	NS	+29 48 47	+39 00 00	3°	+79 46 58
3	NS	+29 48 48	+36 16 27	8°	+80 00 32
4	NNW-SSE	+29 48 44	+30 09 37	348°	+79 36 18
5	NW	+29 48 49	+38 25 11	354°	+79 23 35
6	NW	+29 48 42	+37 29 44	352°	+79 50 48

The date and time used for this image is midnight in 4500 B.C.E, an average period of the site occupation and function. The angle of view here is a base-line viewshed of a human view of about 60 degrees when looking in the north direction. The star in the centre of the screen is Thuban; the Pole Star (α Draconis).

5. DISCUSSION AND CONCLUSION

The hypothesis of the presence of astronomical indications in archaeological sites emerged as early as 1678 when John Aubrey recorded numerous megalithic monuments in southern England and then came Henry Chauncy in 1700 with similar astronomical principles to understand the orientation of churches (Hoskin, 2001). Arguably, Heinrich Nissen, working in the mid-nineteenth century was the first archeoastronomer (Ruggles, 2005).

Since then, statistical studies have been conducted and accurate measurements have resulted in the existence of precise astronomical alignments in megalithic sites that are difficult to ignore or to think of as a coincidence, especially that these sites are associated with the period of the agricultural revolution which requires high astronomical knowledge. In Britain highly accurate alignments to 0.01/0.1 degree have been disputed, the alignments within a degree and a few degrees have been accepted and supported (Ruggles and Appleton, 1984). This study shows that all the six groups measured (five different stone rows and one horse shoe structure) are set along a particular axis, the North-South Axis. The significant North- South alignments found at the Rajajil site in this preliminary study is an attempt to predict an appropriate sample for a future full-scale project that could improve various aspects of this hypothesis in Saudi Arabian megalithic sites. Monumental alignments pointing due north are found in other archaeological sites in the world, according to (Ruggles, 2005) the North- South alignments are astronomical in a broad sense, because they follow the axis of the daily motions of the stars in the sky as they move around the celestial poles. Ruggles wrote "north-south alignments relate to the daily motion of the heavens as a whole rather than to any particular celestial body or event, three different examples illustrate the reasons for keeping architecture "in tone with the cosmos" in this way: The Great North Road at Chaco Canyon; the Forbidden City in Beijing; and the pyramids of Giza" (Ruggles, 2005, p.67).

Celestial symbols in Rock art on some megaliths, such as the case of Kurbat Tamathil, Tabouk (as shown in Figure 11) increases the validity of indications supporting this theory.

Figure 11. Crescent carvings in two standing stones found in kubat attamathil

A drawing of the sun was also found alongside a star in Sakaka, where Rajajil is located (as shown in Figure 12; Al-Kabawi, 1986 p.88). Al-Kabawi (1986) hypothesized that the circles drawn around one centre with radial lines represented the sun disk. The general composition of the sun and star, may be intended to locate a certain star (or perhaps Earth) for the sun. From this perspective we are faced with the fact that there is astronomical knowledge in the prehistoric era, and in front of a new window that enables us to see the sophisticated ideas that embraced the mind of the ancient astronomers of Arabia.

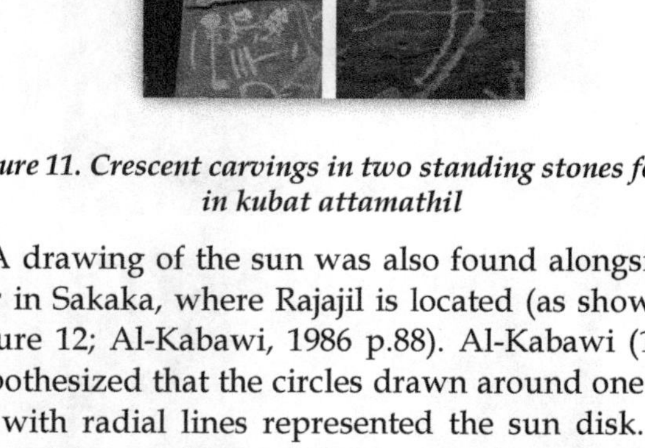

Figure 12. sun and a star in Sakaka close Rajajil site

To make sure that the functional purposes of ancient archaeological sites affect the kind of petroglyphs in its surroundings, I traced ancient natural observatories in Saudi Arabia that are still used today, for example: The Observatory of Helat Alshams "Sun Set Observatory" in Abha (as shown in Figure 13). People follow the path of the sun at certain times of the year through a hole in the mountain as it appears in section (A). The locals track the sun to see it reach its southernmost point in the sky in its journey across the horizon, in what is astronomically called the winter solstice (Gashash, 2001). Not surprisingly, Petroglyphs of celestial bodies are found at the site. In section (B) of the figure below, an image of sun carved next to geometric symbols is very evident. The aim of it is probably to illustrate the concept of the site (articles in MAA, 2017).

Figure 13. Geometric shape next to a carving of the sun in Helat alshams (Gashash, 2001)

There are many contemporary practices related to starwatching, especially among modern indigenous societies in Saudi Arabia that enriches our understanding of early astronomies. One of the most important observatories accredited to monitor the Arab months and the seasons of agriculture used by the ancient and still used by contemporary peoples to the present time is the astronomical observatory "khohila rock" in the village of Al-Majarada in Taif (Alyousfy, 2015). The people of the village consider it as a natural calendar for monitoring the seasons of agriculture and the beginning and end of lunar months. The external outline of the observatory is similar to standing stone sites. It is made up of "stone piles" 1.5 to 2 meters high and the distance between them is about 30 meters. Agriculture was practiced by most people in Taif, adding more evidence to support the hypothesis linking natural observatories and megalithic monuments in agricultural societies to astronomical aspects.

The main goal of this study is to start filling existing gaps in the field of cultural astronomy in the Kingdom of Saudi Arabia and to presents a number of megalithic sites that had not been exposed considerably, to give an idea of the type of Neolithic monuments found in this part of the world. Future investigations are necessary to validate the kinds of conclusions that can be drawn from this study. This research will contribute to the formation of a comprehensive view of astronomical knowledge in prehistoric stone ages and assist us in maintaining evidence about indigenous ways of life connected to astronomy for future generations.

ACKNOWLEDGEMENTS

The author would like to thank Deanship of scientific research for funding and supporting this research through the initiative of DSR Graduate Students Research Support (GSR). Also I would like to express my deepest appreciation to the organizers of "Road to the Stars " conference in Spain 2017, in which this paper was presented. I also thank Dr. Ali Almushawh, Vice Dean of Faculty Affairs at King Khalid College, Mrs. Noura Al Swailem, Dr. Alaa Al Sharida, Mrs. Sahar Ali, Prof. Azhari Mustafa Sadig, Professor of Archaeology, king Saud university, and Mr. Ahmed Al-Qaid, Director of the of General Authority for Tourism and National Heritage in Al-Jouf region for the help and support. I also thank the anonymous reviewer and Dr. Gail Higginbottom whose comments and suggestions helped improve and clarify this manuscript.

REFERENCES

Alshaya, A. (2009). أطلس الشواهد الأثرية على مسارات طرق القوافل القديمة في شبه الجزيرة العربية [*Atlas of archaeological evidence of ancient caravan trails routes in the Arabian Peninsula*], King Abdulaziz Foundation, KSA

Al-kabawi, A. (1986) Preliminary report on the second phase of the comprehensive survey of inscriptions and rock drawings in the northern region for the year 1405 H-1985. *Atlal*, Vol. 10, pp. 85-114.

Alyousfy, T. (2015) Restoration of the Astronomical Observatory in the village of Al-Majarada, Retrived from http://thqeef.org/26095.

Belmonte, J. A.; González-García, A. C. (2014) On The Orientation of Early Bronze Age Tombs in Ancient Magan, *Mediterranean Archaeology & Archaeometry*, Vol. 14 Issue 3, p233-246.

Belmonte, J.A., González-García, A.C. and Polcaro A. (2013) On the orientation of megalithic monuments of the Transjordan Plateau: new clues for an astronomical in- terpretation, *Journal for the History of Astronomy*, vol. 64, 429-55.

Gashash, A. (2001) The drawings and inscriptions of prehistoric rock in the Baha area and their historical and cultural significance. Retrieved from. albahatoday.cc/articles.php?action=show&id=597.

Gebel, H. (2016). The Socio-Hydraulic Foundations of Oasis Life in NW Arabia: The 5th Millennium BCE Shepherd Environs of Rajajil, Rasif and Qulban Beni Murra. *Proceedings of the International Congress held at the University, Vienna*, December 2013, pp. 79-114

González-García, A.C. Belmonte, J.A. and Polcaro A. (2016) A Diachronic Analysis of Orientation of Sacred Precincts Across Jordan, Mediterranean Archaeology and Archaeometry, vol. 16 (4), 133-41.

Hoskin, M. (2001) *Tombs, Temples and their Orientations, New Perspectives in Mediterranean Prehistory*, Ocarina Books, Bognor Regis, UK.

Liritzis, I, Al-Otaibi, F.M, Castro, B, Drivaliari, A (2015) Nabatean tombs orientation by remote sensing: Provisional results. *Mediterranean Archaeology and Archaeometry*, Vol. 15, No 3, pp. 289-299 (DOI: 10.5281/zenodo.33835)

MAA (2017) Mediterranean Archaeology and Archaeometry, Vol.17, No.4, 1-102 (http://www.maajournal.com/Issues2017d.php)

Malville J.M., Schild R., Wendorf F., Brenmer R. (2008) Astronomy of Nabta Playa, African Cultural Astronomy. *Astrophysics and Space Science Proceedings*. Springer, Dordrecht.

Norton, A. P., Ridpath, I. (2004). *Norton's star atlas and reference handbook: Epoch 2000*, Pi Press, New York.

Patton, M. (1993) *Statements in stone: Monuments and society in Neolithic Brittany*. London, Routledge.

Parr, P; Zarins, J; Ibrahim, M; Waechter, D; Garrard, A; Clarke, C; Bidmead, M; (1978) Preliminary report on the second phase of the Northern Province Survey 1397/1977. *Atlal*. Journal of Saudi Arabian Archaeology, Vol. 2, pp. 29-49.

Rajajeel Archeological Site. (2011). Retrieved from https://www.360cities.net/image/052-al-rajajeel-archeological-site-sakakah?utm_source=google_earth&utm_medium=all_images

Reingruber, T. A. (2005) 14C database for the Aegean catchment (eastern Greece, southern Balkans and western Turkey) 10,000–5500 cal BC. In: Lichter C (ed) How did farming reach Europe? Anatolian-European relations from the second half of the seventh through the first half of the sixth millennium cal. *International workshop at Istanbul*, Vol. 2, pp. 295–323.

Renfrew, C., & Bahn, P. G. (2005). *Archaeology: The key concepts*. London: Routledge.

Rosen, P &. Brophy, T. (2005). Satellite Imagery Measures of The Astronomically Aligned Megaliths at Nabta Playa. *Mediterranean Archaeology and Archaeometry*, Vol. 5, pp. 14-25.

Ruggles, C. L. N., & Appleton, P. N. (1984). *Megalithic astronomy: A new archaeological and statistical study of 300 western Scottish sites*. Oxford: B.A.R.

Ruggles, C. L. N. (1999). Astronomy in prehistoric Britain and Ireland. New Haven, Conn: Yale University Press.

Ruggels, C. L. N. (2005) Ancient Astronomy. *An Encyclopedia of Cosmologies and Myth*, ABC-CLIO,Santa Barbara, CA.

Ruggels, C. L. N. (2018) Declination calculator. *Retrieved from https://www3.cliveruggles.com/index.php/tools/declination-calculator.*

Whalen, N., Davis, W., and Pease, D. (1989) Early Pleistocene migrations into Saudi Arabia. *Atlal*, Vol. 12, pp. 59-75.

Winnett, F. V., & Reed, W. L. F. (1970). *Ancient records from North Arabia. Toronto*, University of Toronto Press.

2

THE CASE OF THE TWO CHURCHES OF SANT'APOLLINARE IN PIEDMONT (ITALY): CAN ARCHAEOASTRONOMY HELP TO IDENTIFY WHICH OF THEM IS THE TEMPLAR ONE?

Silvia Motta, Adriano Gaspani

I.N.A.F. Istituto Nazionale di Astrofisica Osservatorio Astronomico di Brera- Milano Italy
silvia.motta78@gmail.com; adriano.gaspani@brera.inaf.it

ABSTRACT

In this paper we analyze the strange case of two churches located not far from each other, identified with the same name, Sant'Apollinare, situated one in Carpignano Sesia (No) and the other in Fisrengo-Casalbeltrame (No), at about ten kilometers each other, in the region of Piedmont, Italy. In a deed dated 1174 A.D it is mentioned the Templar Mansione Sanctum Apollinarem, in the area of Novara, where the Count Guido of Biandrate donates to the Templars everything he owned in the region Ruspalia, but there is not any indication or land registry map to identify the exact position of the mansio. In these two places the Order's presence is indicated by the archive historical documents, mostly, related to acts of buying and selling, but despite this there is a diatribe for the identification of the " Templar Church". Additional means of identification can be offered by an archaeoastronomical analysis of the temples within the context of their surrounding landscape and skyscape. A twofold approach was chosen, consisting of an archaeoastronomical examination of the temple's orientations, and an analysis of placenames and documents which. The two churches have been measured "in situ. Subsequently an appropriate statistical study was carried out in order to infer the distribution function of the astronomical orientations with the aim to perform an appropriate archaeoastronomical analysis. A set of appropriate statistical tests, based on artificial Neural Network, were designed and applied in order to check the possible solutions. The result is that the church that best meets the features of the Templar churches built in Piedmont is the church Sant'Apollinare in Carpignano Sesia.

KEYWORDS: Archaeoastronomy, Templar Knights, Templar church, Astronomy

1. INTRODUCTION

1.1. *A brief history of the Templar Order*

The Poor Knights of Christ and of the Temple of Solomon, (Pauperes commilitones Christi templique Salomonis) a religious military order of knighthood was originally founded to protect Christian pilgrims to the Holy Land and subsequently the order assumed greater military duties during the 12th century.

In 1119, the French knight Hugues de Payens approached King Baldwin II of Jerusalem and Warmund, Patriarch of Jerusalem, and proposed creating a monastic order for the protection of these pilgrims. Thanks to the significant contribution of Bernard of Clairvaux, at the Council of Troyes, 1129, the Templars were given a proper Rule, which was followed by the issue of three key bulls establishing the Templar as a privileged Order under Rome. In 1139, Pope Innocent II papal bull *Omne Datum Optimum* exempted the Order from obedience to local laws..Templars began to accumulate a substantial landed base in the West, not only in France, Provence, Iberia and England, where they were first known, but also in Italy, Germany, Dalmatia and Cyprus .By the late 13th century they may have had as many as 870 castles, preceptories and subsidiary houses spread across Latin Christendom. During the 12th and 13th centuries these properties were built into a network of support which provided men, horses, money and supplies for the Templars in the East.

On Friday, 13 October 1307 King Philip IV ordered de Molay and scores of other French Templars to be simultaneously arrested.At the Council of Vienne in 1312, Pope Clement issued a series of papal bulls, including *Vox in excelso*, which officially suppressed the Order, and *"Ad providam"*, which turned over most Templar assets to the Hospitallers. As for the leaders of the Order, the elderly Grand Master Jacques de Molay and Geoffroi de Charney, Preceptor of Normandy, were declared guilty of being relapsed heretics, and they were sentenced to burn alive at the stake in Paris on 18 March 1314.

1.2. *Templar Mansio in Piedmont*

A strange destiny occurred to all the Templar mansions, churches and properties formerly existing in the bishopric of Novara, Piedmont, Italy, condemned by time and unlucky events to undergo the uncertainty of the identification of the location site. The limited number of documents that cite the places is often characterized by nebulous and inaccurate references. Moreover, the buildings have been systematically destroyed and the residual traces, due to the anthropization, have been hidden. The scarce presence of the Templars in this area had to deal with political problems due to their evident support to the imperial cause and culminated with the close relationship with the Biandrate family, enemy of Novara. That's why following the annihilation of the political power of this family the domus were abandoned. A serious problem of identification has arisen for the Templar Mansio of Sant'Apollinare. It is mentioned in a Templar notary deed of 1164, a document that has generated many misunderstandings.

In the Templar Mansio of Sant'Apollinare was drawn up the famous donation of the real estate of Santa Maria di Ruspaglia, in San Giorgio Canavese, to the Templar Knights Order.

In this act, dated September 25, 1164, we read that "the Count "Guido il Grande" of Biandrate in the presence of his sons and friars Alchenius and Portonarius, at the mansio, donates to the Templar Knights Order everything he owned in San Giorgio Canavese, Ruspalia". The question is that there is no geographical indication of the location of the Mansio of Sant'Apollinare.

Two academic scholars who are experts in local history have provided two different locations for the Mansio. The Vercelli historian Avonto, in a work of 1982, identifies the Templar Mansio with the Church of Sant'Apollinare in Fisrengo, citing the following reasons: the observation of what is still visible today, the strategic location of the alleged mansio in relation to the road network of the time, the surrounding territory rich in fountains and springs and he dedicatory title of the church of Fisrengo which would remind the Biandrate family (Avonto,L. *I Templari a Vercelli*, Vercelli 1977)

Some buildings of the farmhouse that incorporate the church, however, although they are ancient, are placed at most to date back to the sixteenth century. Regarding the strategic location, close to the major roads of communication, it has ensured that the distance of Fisrengo from the Via Francisca was 10km, from the Via Blandrena 5 km, 6 km from the Via Novarese, and at the time it was not right next to each. Regarding the dedication "Sant'Apollinare", it is possible to connect with the election to Bishop of Ravenna, in 1159, of Guido son of "Guido il Grande" of Biandrate. Analyzing the documents, however, it does not appear that Fisrengo ever belonged to the counts of Biandrate, but it turns out to have always belonged to the counts De Fisrenghi. The Novara historian Fiori, after having carefully studied the documents, identifies the church of Sant'Apollinare in the locality of Carpignano Sesia as the one mentioned in the deed. It is documented that Carpignano was a possession of the counts as early as 1070 A.D. East of Carpi runs the via Blandrena, a fundamental axis for power, economy and commerce. This ancient road from Biandrate went up the Valsesia, on the left bank of the river, up to the slopes of Monte Rosa. The village is at the crossroads of this street and another very important route that came from Ticino and flows from east to west, it passed through Carpignano and crossed the Sesia. The Knights Templar built their suburban mansio on the streets of great commercial communication and pilgrimage, so to be closer to their land properties and better serve the pilgrims on their journey with hospital and refreshment facilities.

2. MATERIALS AND METHODS

Here we present the results of our analysis of the architectural alignments of the two Churches, whose data are based on measurements made by us in the field in different stages, from spring 2016 until March 2017. The complete mapping of the area was carried out using the GPS satellite tracking system with a portable Garmin Etrex set to the WGS84 coordinate system; the azimuth of orientation were measured with two compasses, a Sunto and a Silva; we also quite had a clinometer for detecting directly the heights of the mountains located on the opposite site.The error of a single measurement, detected by the profile of the scale of the instrument, is about ½° for the azimuth and ½° for the local horizon height. The azimuth measured evaluated, being detected with a magnetic compass, were corrected by the magnetic declination obtained from the website *http://www.ngdc.noaa.gov* and mathematical calculations.

2.1. *Sant'Apollinare in Fisrengo (45°25′00″ N; 8°29′35″ E)*

The farmhouse of Sant'Apollinare in Fisrengo rises about 1 km east of the municipality of Fisrengo, a hamlet of Casalbeltrame, province of Novara. The complex consists of four main parts dating back to different periods: the ancient church; an eighteenth-century building leaning against the left side of the church looking at the façade; an ancient building leaning against the church, but hidden from view from the outside and various successive rural buildings. The church has clear signs of restoration work, mostly from the 17th century, but also later. The most interesting part is the construction, hidden from the outside, which bears traces of ancient origin. The historian Avonto proposes that this building rises on the foundations of the ancient Templar domus annexed to the church of Sant'Apollinare, mentioned in the document of 1164 (Figure.1).

Figure 1: Sant'Apollinare in Fisrengo.

The church is composed of a main nave flanked by two altars housed in a sort of side aisle separated by a pillar with two round arches. All the walls and the ceiling are painted with architectural and floral motifs in shades of yellow ocher and with various figures in the panels. I found no trace of the sixteenth-century fresco with the Madonna delle Grazie. To the left of the entrance there is what remains of the marble high altar, while the barrel vault overhangs with the adoration of the monstrance. At the center of the nave there is an amazing sarcophagus of gilded wood placed on a wooden base surrounded by statues also wooden. These are the relics of a saint: Santa Esuberanza, whose body simulacrum, covered, is located inside with the vase containing the true relics. Next to the second altar, on the wall to the right of the presbytery, there is a large, rather battered canvas representing the Virgin with Child surrounded by various figures of saints: St. Joseph, St. Anthony of Padua, St. Charles Borromeo and at the foot two holy bishops. On the back wall of the church, what had to have the now walled entrance door, there is a large crucifix. On the side we see an old door that allowed a time to probably access a choir, a sort of wooden balcony whose signs can still be seen in the walls. Behind the high altar a small door leads into the bell tower.

The measurements were carried out on 5/10/2016 and 15/03/2017. The main side of the church is 19 meters, while the smaller side is 5.50 meters.The perimeter walls of the church, dating to the first building, are oriented with a measured azimuth of (93.9° ± 0.5°) and (273.9° ± 0.5°), the same orientation of the axis of the nave was detected. The churches were often oriented to the rising sun at the equinox, they were mostly equinoctial, that is, with an azimuth of orientation of the nave of 90°. When the sky was observed the day of the September 23, 1164, the Autumn Equinox, the sun arose at the local horizon with an azimuth of orientation which value was (93.9° ± 0.5°). We have calculated the real date of the Autumn equinox in the year 1164 A.D. and it was September 16th. and for about a century it did not change. The solar year related to the astronomical year, over the centuries had accumulated a small delay each year until you get to about 7 days in the twelfth century, because of the error in the Julian calendar. In this way, the azimuth of the rising sun measured to plan the construction of the church is different from the correct one, because the equinox day was September 16, 1164, that is seven days before and in this date the true value of the astronomical azimuth of the rising sun on the local horizon was 90.0°degrees. Figure 2

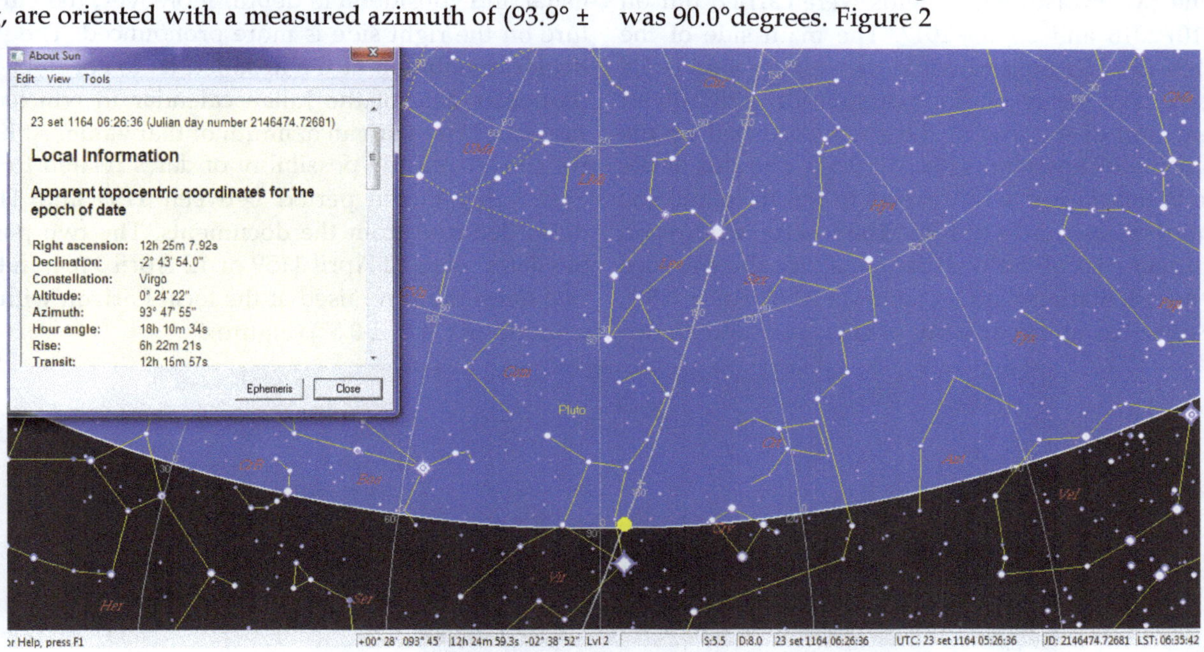

Figure 2: Sun rising on the Day of Equinox,23th September 1164.

2.2 *Sant'Apollinare in Carpignano Sesia (45°31'44" N ; 8°25'48" E)*

The chapel is located south of the town in the direction of Sillavengo, close to the route of the ancient via Blandratina (or Biandrina), a strategic and economic pillar of the road network of the possessions of the counts and secondary branch of the "via pellegrinale alta" of the" Francigena "for the steps towards France and Santiago (Fiori, S.,2015). The large part of the building, rebuilt at the beginning of the 16th century, is rebuilt on ancient walls, proof of which are the fragments of foundations that emerge from the inner corners of the south-east and south-west. It is formed by a single nave and an apse with an irregularity of curvature. The church consists of a single-apse inside a single-chamber chapel, probably built between 1159 (date of election of Guido di Biandrate, son of Guido il Grande, as a bishop of Ravenna) and 1164 (date of the donation of Ruspalia to the Templars, drawn up to his interior). The original apse presents a construction technique typical of the second half of the 12th century (Figure 3). It is located in the territory of Carpignano (Calpinianum), belonging, since 1070, to the ancestral nucleus of the feudal possessions of the de Blanderate family, called "Comitato della Biandrina".

Figure 2: Carpignano Sesia: S. Apollinare, details of the apse.

The two measurement fields were carried out on 25/10/2016 and 25/03/2017. The main side of the church is 12.35 meters, while the smaller side is 6.8 meters. The value of the astronomical azimuth of orientation of the axis of the nave, corrected by the magnetic declination, is (68.8° ± 0.5°). The axis of the apse is misaligned with respect to the nave and the azimuth measures 74.8° ± 0.5. The rosette window on the upper part of the west facade above the entrance has an orientation with an azimuth of (248.8° ± 0.5°). The apse has an irregularity of curvature and an unusual and considerable depth; moreover, the curvature on the right side is more pronounced. The particular orientation of the apse led us to consider the particular date of the Julian calendar in which the sun could rise with an azimuth of that value. At first, we considered the possibility of dates related to the Easter day in the period between 1159 and 1164, dates deduced from the documents. The two possible dates were 12 April 1159 or 12 April 1164. In this two dates the sun raised at the local horizon with an azimuth of (74.6° ± 0.5°) (Figure 4).

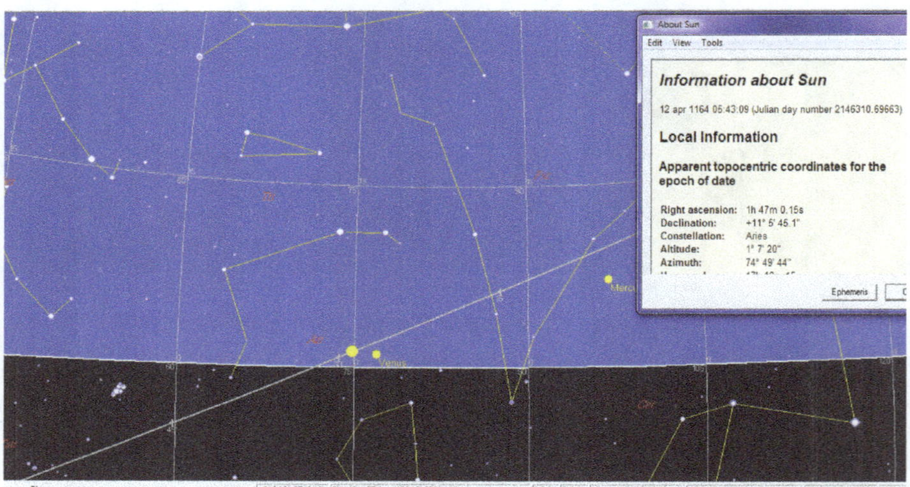

Figure 3: Sun rising on Easter Sunday April 12, 1164.

3. ORIENTATIONS OF TEMPLAR CHURCHES

We must bear in mind that in the Late Middle Ages, despite the indications and the Papal Bull of Nicea Council in 323 A.D. issued by the Church that prohibited their use, the astrological rules were widely used in building construction as churches and castles. The Council Fathers affirmed in AD 325: "ecclesiarum situs plerimque talis erat, ut fideles facie altar versa orantes orientem solem, symbolum Christi qui est sol iustitia et lux mundi interentur" (Carolus Kozma "De Papi", 1861). During the Middle Ages the construction of a church had to undergo very precise rules of orientation of its entrance-apse axis, but also in establishing the period in which the foundation rite was to be celebrated. Guido Bonatti da Forlì, mathematician, astronomer and astrologer active in Paris during the thirteenth century, in his "Decem continens tractatus astronomiae", of which there is an edition published in Venice in 1506, highlights that the churches, being centers of divine power, they had to be raised according to scrupulous ritual rules following the course of the heavens and which had to be built up when certain favorable astral conjunctions occur. In particular, the era of foundation of the churches was chosen in accordance with the rising on the horizon, for the first time during the year, of the stars of the constellation of Aries, so the period chosen was shortly after the spring equinox and it was in accord with the astronomical rules of the celebration of the Christian Easter. The reason was not only mystical, but also corresponded to two very specific practical needs: first, this was the period when the frost and winter rains ceased and the ground became softer allowing the workers to work; second, there was sufficient time, until the following winter, to complete the construction work.

In 1406, Jean Ganivet wrote: "Si velis aedificare aedificium duraturum, considers fundas in primal et conferas eis planetas benevolos (Jean Ganivet," Coeli enarrant ", Lyon 1406)" If you want to build a durable building, in the foundation observes primarily the fixed stars and compare them with the benevolent planets ". Therefore not only the rising heliacal rising of the stars of Aries defined the most favorable seasonal period, but the planetary positions, especially those of Mars and Jupiter, in the zodiacal constellations established the most suitable years for the building of sacred buildings, especially those of relevant importance. The position of the moon with its phases within a certain constellation was held in high esteem; it was not to be found absolutely in Pisces, Capricorn, Scorpio and never in conjunction with Saturn. This direction could be correlated with the date of Easter which, as is known, is celebrated on the Sunday closest to the first full moon after the spring equinox. Being, however, the date of the Passover Easter with respect to the date of the equinox because of the lunar constraints, the orientation according to the position of the rising sun at Easter could not be codified in a fixed way. Since the date of Easter can oscillate roughly 30 days past the spring equinox, i.e. 1 lunar synodic month (29.5306 days), the difference in orientation with respect to the equinoctial line can reach up to about 18° north of the East. This means that orientations between 72° and 90° could be correlated with the position of the sunrise on Easter Sunday of the church's foundation year. Taking into account these considerations, given that on 12 April 1159 the moon was passing through Pisces, an unfavorable constellation, while on 12 April 1164 it was in Aquarius, a not unfavorable constellation, it is more probable that the year of construction was 1164.

The Templar churches considered can be divided into two main groups, those acquired as a result of donations and that preserve the original geometry and orientation, and those built ex-novo by the Templars, thus with geometric Templar criteria and Templar orientations. The astronomical orientation of the churches built by the Templar Order (Figure.5), in a sample of about 50 Templar churches built in Europe, provided that the axis alignment of the aisle on sight was aligned towards one of these targets: the point of the sunrise on calendar day of 21th March (Equinox); the point of the sunrise on calendar day of 25th March (Annunciation); the point of the sunrise on the calendar day of Easter; along the equinoctial line obtained by gnomonic methods.

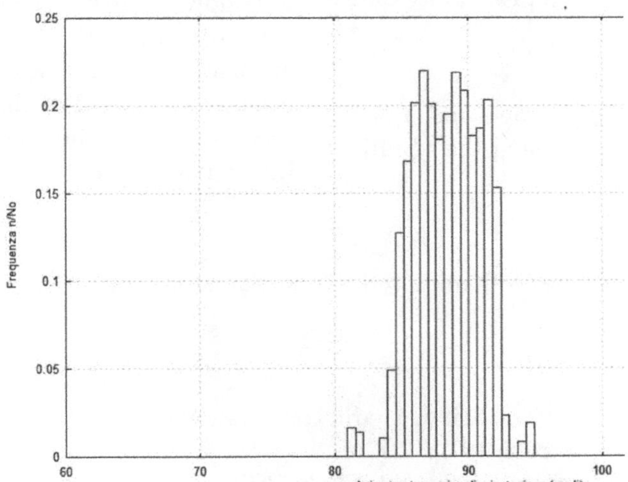

Figure 4: Astronomical azimuths of the Templar churches.

4. RESULTS

We processed the data making use of the statistics for circular and axial data. In order to estimate the cross correlation between the orientation of the axis of the nave of the church and the astronomical target. we applied some techniques (Fisher 1995), by the relation (Proakis 1989):

$$R = \cos(\theta),\ R = |1|$$

For $\theta = 0°$ is the case of perfect correlation and $R = 0$; if $\theta = \pm 90°$ there isn't any correlation between the processed directions; if $\theta = \pm 180°$ we have $R = -1$ and the vectors are unrelated. P is the probability of randomness that the two vectors are randomly correlated with each other having that particular correlation R value.

To estimate the cross correlation between the orientation of the axis of the nave of the church and the astronomical target, we have calculated the value for R in Sant'Apollinare in Carpignano Sesia and we obtained a minimum value R = 0.9999. This is the correlation between the two vectors within the range of an angle 0.013°. That means that the two orientations are well correlated. Taking into account the correlation coefficient R obtained, we have estimated a probability P(R)= 0.000228 to obtain a correlation coefficient R=0.9999 only by chance, so in this case there is a good correlation between the direction of the axis of the nave of the church and the astronomical target.

We operated in the same way with the data of Sant'Apollinare in Fisrengo and we obtained a minimum value R = 0.9999. This is the correlation between the two vectors within the range of an angle 0.1353° and with a probability P(R)= 0.0002361 that they are randomly correlated. Now using the calculation of the probability and statistical test we want to establish which of the two churches Sant'Apollinare is more likely to be the Templar one mentioned in the document of 1164 A.D. The probability density function is:

$$\boldsymbol{P(x)} = \boldsymbol{e}^{-0.5\left(\frac{x-xm}{s}\right)^2}$$

where x is our value, xm the average and s is the standard deviation, all obtained using the t-Student distribution.

The probability density function related to the length (a) for the Templar churches located in the Piedmont area is:

$$\boldsymbol{P(a)} = \boldsymbol{e}^{-0.5\left(\frac{a-11.6}{1.6}\right)^2}$$

The probability density function related to the ratio $r = a/b$, where b is the short side of the church

$$\boldsymbol{P(r)} = \boldsymbol{e}^{-0.5\left(\frac{r-1.75}{0.09}\right)^2}$$

The probability density function related to the azimuth

$$\boldsymbol{P(Az)} = \boldsymbol{e}^{-0.5\left(\frac{Az-84.53}{11}\right)^2}$$

When a deliberate astronomical alignment is proposed, it is of course fundamental to investigate on the possibility of a mere coincidence.

We assumed that the "null hypothesis" H_0 is the random orientation, that is the church is not a templar church and the alternative hypothesis H_1 is the deliberate orientation along the inferred directions, in this case it is a templar church. The decision criterion H_0 has to be rejected if

$$P_n(\text{random}) < P(\sigma)$$

where P_n(random) is the probability that the direction of the orientation is chance, $P(\sigma)$ is the tabulated integral of the Gaussian function.

Adopting a 97.0% confidence level the value is $P(\sigma) = 0.03$, the probability that a church had been built by the Templar Knight Order is:

$$P(\text{Az,r.a}) = 1 - [(1-P(\text{Az}))(1-P(\text{r})(1-P(\text{a}))]$$

For the church Sant'Apollinare in Carpignano Sesia:

a = 15.35 m b= 6.80 m $r = a/b$ = 2.5

and a value of the probability to be Templar is:

P(Az,r.a) = 98.7%

For the church Sant'Apollinare in Fisrengo:

a = 24.50 m b = 5.50 m $r = a/b$ = 4.45

the value of the probability to be Templar is

P(Az,r.a) = 69.4%

5. CONCLUSION

Following the archaeoastronomical analysis, we can suggest that the church that best meets the average features such as size, geometry and astronomical orientation typical of the Templar churches built in Piedmont is the church Sant'Apollinare in Carpignano Sesia.

ACKNOWLEDGEMENTS

The authors would especially like to thank Arch. FIORI Salvatore and the Ing. DANILO Alberto for their support and technical cooperation.

REFERENCES

A.A.V.V., *Guida all'Italia dei Templari*, Edizioni Mediterranee, Roma, 1989.

Avonto,L. *I Templari a Vercelli*, Vercelli, 1977.

Bellomo, E. 2008. *The Templar Order in North-west of Italy*, Brill.

Fiori, S., I *Templari nel territorio Novarese, Ediz. Interlinea, 2015.*

Fisher,N.I.1995, *Statistical Analysis of Circular Data*, Cambridge University Press.
Jean Ganivet, "*Caeli enarrant*", Lione 1406.
Gaspani, A., 2016, *Il Codice Astronomico dei Cavalieri Templari*, Connla Editrice - Ivrea.
D. Gauzolino P., *Storia antica di Livorno Ferraris*, Crescentino, 1989.
Giuliano G.F., *Santa Maria d'Isana*, Santhià, 2006.
Guidonis Bonati. *Foroliviensis Mathematici de Astronomia Tractatus X Universum quod ad iudiciariam rationem nativitatum, Aëris, Tempestatum attinet, comprehendentes. Adiectus est Cl. Ptolemaei liber Fructus, cum Commentarijs* Georgij Trapezuntij, Basileae, Anno MDL.
Le carte dell'archivio arcivescovile di Vercelli, ed. D. Arnoldi, 2 vols., BSSS 85/2+appendix, (Pinerolo, 1917–1932).
Le carte dell'archivio capitolare di Vercelli, eds. D. Arnoldi, G.C. Faccio, F. Gabotto, G. Rocchi, 2 vols., BSSS 70–71 (Pinerolo, 1912–1914).
Meeus, J., 1978. *Astronomical Formulae for Calculators*, William Bell.
Ruggles, C. L. N. 2005, *Archaeoastronomy*. In C. Renfrew & P. Bahn (eds), Archaeology: the Key Concepts, Routledge, Abingdon.
Regione Piemonte, atti del convegno: *I Templari in Piemonte*, Torino, 1994.
Regione Piemonte, *I Templari tra storia, mito e iconografia*, Stupinigi, 1994.

3

EVOLUTION OF ASTRONOMICAL FACILITIES AND PRACTICES IN ANCIENT THRACE

Alexey Stoev[1], Penka Maglova[1] and Mina Spasova[2]

[1]*Space Research and Technology Institute, Bulgarian Academy of Sciences, Stara Zagora Department*
[2]*Institute for the Study of Societies and Knowledge, Bulgarian Academy of Sciences, Sofia*

**Corresponding author: Penka Maglova (penm@abv.bg)*

ABSTRACT

Evolution of megalithic and rock-cut monuments used for astronomical observations and practices in ancient Thrace are presented in this work. Classificaton is made after independent archaeoastronomical research of the rock-cut monuments and cave sanctuaries.

The discovery and the accumulation of "in situ" material allow to identify the signs of the formation of a number of their parameters - territorial, chronological, functional. One of the promising areas of research, for example, is to distinguish different rock shrines and megalithic complexes associated with long term astronomical observations and astronomical practices. Another area is the specification of the affiliation of individual objects to synchronously existed settlements and tribal communities. This way, reasons for their appearance, function and place in everyday life of citizens and society in Ancient Thrace can be justified.

Here we consider examples of positional astronomical observations, projective systems, and monochrome paintings with calendar records. We present the concept of "dynamic structure of the rock-cut monument", which means that the site is characterized not only with space but also with time parameters as cyclicity and duration of existence connected with various cults and cult practices.

The evolution of astronomical facilities and practices is shown by the suggested typology of the rock-cut monuments according to their locality, landscape of the environment, functional astronomical elements, and accuracy.

KEYWORDS: rock-cut monuments, cave sanctuaries, shapes, spatial relations, meridional and horizon observations, calendar, typology.

1. INTRODUCTION

Copper Age, the Eneolithic age (5000-3000 BC) is one of the brightest periods in Prehistory in the Bulgarian lands. According to experts, in this era very ancient ethnicity and complex culture was born (Fol A., 2009, Popov, 2014). Hundreds Eneolithic settlements were found in Bulgaria - they are well built, with streets oriented according to the cardinal directions, protected by wooden or clay fence. Further, during the Bronze Age (3100-1200 BC) a process of gradually improving of the economy and evolution of the socio-economic life takes place in Bulgarian lands. Mastering the production of metal tools completely changed the lives of native people. The tools of stone and bone were abandoned and replaced by more sophisticated tools and weapons including bronze swords, knives, arrows, spears with metal tip and others (Bailey 1998, Todorova and Vaisov 1993, Stoev and Maglova, 2014). At that time creation of megalithic and rock-cut monuments blossomed. The rock-cut monuments - sanctuaries and tombs are the only almost completely preserved representatives of monumental religious architecture of the ancient people from the Eneolithic to the end of the Iron Age in Ancient Thrace.

Data from archaeological excavations allow us to determine the character of these monuments functioning as sanctuaries, where different ritual activities were practiced. Rituals were connected with the positions of specific heavenly objects in accordance with the natural-geographical environment, economy specificity and social structure of the population. The rock-cave sanctuaries were a unique phenomenon for these civilisations - a material expression of a complicated ritual system naturally added to their philosophical doctrine (Naydenova 1986, Stoev et al. 1990; Dermendjiev et al. 1984; Stoev and Varbanova 1994; Nikolov et al. 1988; Stoev et al. 2001; Stoev et al. 2003; Radoslavova and Stoev 1991).

2. CLASSIFICATION OF ARCHAEOASTRONOMICAL OBJECTS STUDIED IN BULGARIA

Archeoastronomical monuments on the territory of rock-cut complexes arise when the society moves to a sedentary lifestyle. Rock-cut monument or cave sanctuary with astronomical purposes can be defined as a spatial structure of specific relief, rock shapes and cuts connected in a certain functional dependence with the horizon, the sky, the heavenly objects and related phenomena. Such a description gives an idea of the appearance of the monument itself, but does not affect factors associated with its emergence and development over time, which are particularly relevant to its archaeoastronomical context.

The study of the spatial organisation of the sanctuaries and the existence of obvious linear and circular structures offer sufficient and serious data about the existence of orientations, targets and solar-cult architecture. That is why the rock-cut monuments and cave sanctuaries became an object of independent archaeoastronomical research (Polcaro and Polcaro, 2009). It is based on our knowledge about the place of astronomy in the culture of ancient archaic societies who lived on these lands, as well as on registration, analysis and interpretation of their structural elements.

In general, archaeoastronomical objects can be classified as folows:

1. Objects showing ordinary interest in heaven, heavenly bodies and phenomena and their cyclical appearance.
2. Objects testifying to astronomical knowledge and skills and their practical use:
    - Megalithic and rock-cut monuments;
    - Ancient sanctuaries, masonry and rock tombs;
    - Astronomical images and calendar ornaments;
    - Ancient calendars and calendar records.

Archaeoastronomical objects allow us to learn about astronomical knowledge, practices, art, belief, custom, understandings, and values of the societies who lived on Bulgarian lands.

3. FORM AND SPACE OF THE MEGALITHIC ROCK-CUT MONUMENTS

Many authors study the problem of recreation of the main features and ways of exploitation of the megalithic and rock-cut monuments and systems that existed in prehistoric times. They, in general, tend to accept that the model should be understood as a system of interactions between cultural relationships, cult activities and astronomical observations, which in turn determine the stable existence and development of human societies inhabiting a given region (Christov , 1999, Fol A., 1986, 1990, Fol V., 2000, Maglova and Stoev, 2014, Stoev and Maglova, 2014, Stoev et al., 2006 a, b, c).

In essence, the different models aim to clarify the ways, in which individual societies use the resources of their area of life and the landscape to ensure their long-term existence. Therefore, the development of a sustainable model of the rock-cut monument must be based on the balance between the available natural resources in the region and the number of observed objects that can be used during the relevant period.

For this reason, it is necessary to carry out a complex study of the "observational purposes" of the determined number of rock-cut objects and the comparison of these purposes with the available landscape and **relief** in the given region. Assumptions about the size of the effective range of a rock-cut object are based on the distance of the main landscape factors from the structurally identified forms of the rock relief of the monument.

The definition of "form" and "space" categories for the area of megalithic and rock-cut monuments requires a complex approach. Structure of the rock-cut site includes a centrally located body and a platform to which radially oriented chutes and paths are directed.

There are additional elements to every rock-cut monument. The focus of the composition is a visual (or projective) device that allows observation of the Sun in astronomically significant points of the celestial sphere and the horizon. The autochthonous relief structure of the rock - cut monument "Harman Kaya" was used in the Eneolithic Age for positional astronomical observations of the Sun during the summer and winter solstice, and equinoxes (Maglova and Stoev, 2014) through the methods of the so - called vector - horizon astronomy (as shown in Figure 1).

Figure 1. Rock sanctuary "Harman Kaya" with three viewfinders - 1-summer solstice, 2- winter solstice, and 3- equinox.

As another example we can consider is the rock - cut monument Cabile, where the sunrise during the summer solstice can be observed (Stoev and Stoeva, 2016), (as shown in Figure 2). The bas-relief image of the Goddess Cybela is located at the west side of the sanctuary, next to the trench East - West. There are also foundations of an edifice made from large levers stones, which form a square. To the south, at a distance of about 2 km, following the North - South trench, one can see two mounds, almost one behind the other. During archaeological excavations, specialists discovered large quantities of wood ashes. This shows that mounds, which define the meridian line, were probably used as light markers at night observations of bright stars.

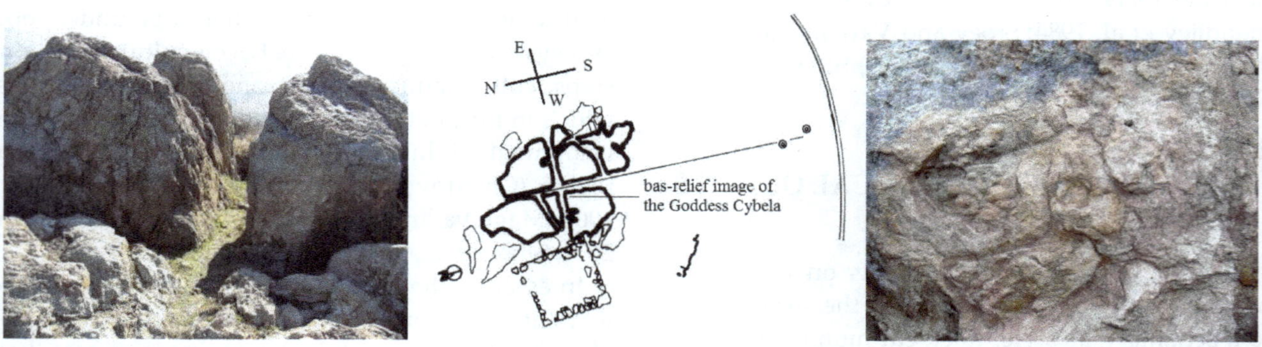

Figure 2. Cabile - trenches oriented East–West and North–South, plan of the sanctuary, bas-relief image of the Goddess Cybela.

Tangarak Kaya rock cave sanctuary can be considered as a projective system of observation of the extreme Sun culminations (as shown in Figure 3). During the winter solstice, once per year solar rays penetrate into the cave and reach the altar, embodying the sacred marriage between the Goddes-mother and the God-Sun (Stoev et al., 2001).

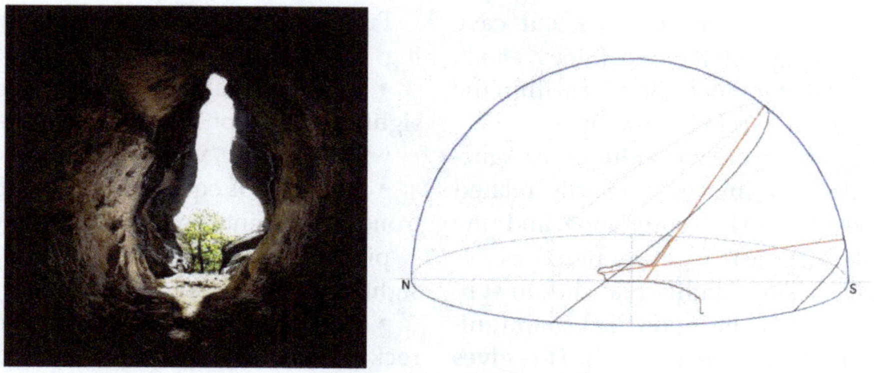

Figure 3. Tangarduk Kaya cave sanctuary. Vertical plan of the sanctuary - solar beam penetrating in the cave during the summer and winter solstice (in red).

The basic rules of shape creation made by ancient builders of megalithic and rock-cut objects are probably based on their knowledge of the organization of forms in living and non-living nature connected with the cult of the Heaven and the Sun.

Last but not least is the process of creating and maintaining a calendar, especially needed in the social organization and religious practices of the society. As an example we can consider the earliest paintings in caves on our lands - the monochrome paintings with astronomical meaning from the Magura cave (about 3000 BC), (Stoev and Maglova, 1999), (as shown in Figure 4).

Figure 4. The Magura cave - monochrome paintings composing solar calendar from the late Eneolithic.

Thus, we can suggest the term "dynamic structure of the rock-cut monument", which means that it is characterized not only with space but also with time parameters - cyclicity and duration of existence of rock-cut monuments connected with various cults and cult practices.

Systematic observations of the positions of the sun's projections during the daily culminations, as well as sunrises at extreme points on the horizon, allowed the days between the winter and the summer solstice to be counted. For those megaliths and rock-cut monuments associated with specific astronomical practices, seasonal occurrences of the observed celestial objects (sunrises and sunsets, culminations, compounds and other astronomical phenomena and events), should be also included in time parameters.

These procedures have greatly facilitated the creation and use of a primitive calendar related to the economic, religious, and domestic needs of the society in this era.

Figure 5. The megalithic complex with caves near the village of Bailovo - monochrome paintings - part of symbolic solar calendar and earlier Lunar images carved into the rock.

Monochrome paintings from the rock cut cave sanctuary near the village of Bailovo (Stoev et al., 1989) represent a finished annual calendar within the tropical solar year, (as shown in Figure 5).

The distinction between the individual rock sanctuaries and megalithic complexes is mainly related to the specific astronomical observations and the existence of long-standing astronomical practices.

Clarifying the affiliation of individual sites to synchronously existing settlements and tribal communities is very important during the research. This gives the opportunity to explore the objects of this type by regions and by chronological boundaries. In this way more reliable conclusions can be argued about the reasons for their appearance, functioning and place in the everyday life of the inhabitants and society in Ancient Thrace.

People of the then society obviously have created and continuously used the powerful solar-chthonic cult united with the cult of the Sun and this to Heaven and Stone. The rock is the personification of the Goddess Mother, who bore and bred the Sun itself. Serving this cult also requires creating and maintaining a precise calendar. This means exact determining of the initial moment of the calendar cycle and its division to equal time intervals connected with seasons. That is why, we suppose that positions of the sunrises and sunsets (during solstice or equinox), which coincide with characteristic points on the local visible horizon were observed.

4. TYPOLOGY OF THE MEGALITHIC AND ROCK-CUT MONUMENTS

Typology of the rock-cut monuments in Ancient Thrace shows the evolution of astronomical facilities and practices.

The typology of rock-cut monuments with an archaeoastronomical meaning is made according to their locality, landscape of the environment, functional astronomical elements, and accuracy of the obtained observational results (Maglova et al., 2007):

Locality:

- On high rocky peaks;
- In natural or artificial caves;
- Accidentally chosen sites with good view towards the local horizon.

Landscape of the environment:

- It almost always include rock, cave and water (springs or water reservoirs);
- Visited by a great many people or only by initiated people;
- It can be used for - astronomical observations, cult practices, time reckoning and time measuring.

Functional elements with an astronomical meaning:

- Astronomically significant azimuth of the basic sight direction (points of the summer and winter solstice, vernal or autumn equinox);
- Presence of equipment for sighting towards astronomical points on the horizon (by using of pillars, grooves, trenches, cave galleries, special sight equipment, near and far relief marks);
- Presence of astronomical objects depicted on the rock (caves, niches and open rock surfaces).

Accuracy of the obtained observational results (on angular dimensions and time) in accordance with the place and manner of observation:

- Observational grounds with a wide scope of vision without requirements for any accuracy (symbolic astronomy weaved into cult and religious practices);
- Equipment for Sun observations with an accuracy of about 5 arc degrees (±10 days - symbolic astronomy with low accuracy);
- Equipment for Sun observations with an accuracy of about 0.5÷1 arc degree (±1 ÷ ±2 days - positional astronomical observations with high accuracy).

5. CONCLUSION

In this work we investigate the evolution of astronomical facilities and practices in ancient Thrace.

Positional astronomical observations, projective systems, monochrome paintings with calendar records are shown as examples. The concept of "dynamic structure of the rock-cut monument" is introduced. This means that the site is characterized not only with space but also with time parameters as cyclicity and duration of existence connected with various cults and cult practices.

Structure of the rock-cut site includes a centrally located body and a platform to which radially oriented chutes and paths are directed. Also, there are additional elements to every rock-cut monument. The focus of the composition is a visual (or projective) device that allows observation of the Sun in astronomically significant points of the celestial sphere and the horizon.

The presented archeoastronomical objects show the mechanisms for observing extreme sunrises and culminations. The accuracy of the solstice and equinox determination is ±1 day and this allows creating and maintaining a calendar with the duration of the solar tropical year. As cave monochrome paintings show, typical holidays related to specific days are also marked in the calendar. The typology of archeoastronomical monuments shows the evolution of astronomical observational technologies and the ways of presenting astronomically significant moments by pictograms.

REFERENCES

Bailey D. (1998) Expanding the Dimensions of Early agricultural Tells: The Podgoritsa Archaeological Project , Bulgaria'. *Journal of Field Archaeology*, Vol. 25:4, pp. 373-396.

Christov I, (1999) *Mountain Thrace, population, culture, and religion in antiquity*. Faber edition, Velko Turnovo.

Dermendjiev V., Dermendjieva S., Menkov L. (1984) Rocky sites in Western Rodopes with a possible astronomic purpose. *Comptes rendus de l'Academie Bulgarie des Sciences*, Vol. 37.5, pp. 545–548.

Fol A. (1986) *Thracian orphism*. Kliment Ohridski University press, Sofia.

Fol A. (1990) *Politics and culture in ancient Thrace*. Nauka I Izkustvo, Sofia.

Fol V. (2000) *Megalithic and rock-cut monuments in Ancient Thrace*, St. Kliment Ohridsky University press.

Fol A. (2009) *The ancient culture of South-East Europe*, TanNakRa, Sofia.

Maglova P. and Stoev A. (2014) Thracian sanctuaries. In *Handbook of Archaeoastronomy and Ethnoastronomy*, part VIII , Springer Reference, Springer New York, C.L.N. Ruggles (Ed.), pp. 1385-1394.

Maglova P., Stoev A., Stoeva M., Stefanov P. (2007) Extreme rises and meridional sun culminations in the cult practice of the Thracian civilisation. *Proceedings of the 10th International Congress of Thracology*, Komotini - Alexandroupolis, Greece, 18 - 23 October 2005, Athens, pp. 407-415.

Marazov I. (1994) *Thracian mythology*. IK Sekor Press, Sofia.

Naydenova V. (1986) The rocky sanctuaries in Thracia. *Proceedings of the II symposium "Settlement life in Thracia"* Yambol, Bulgaria, pp. 15-29.

Nikolov N. Zlatev S. Vasileva K. (1988) The astronomical meaning of Tatul archaeoastronomical monument. *Archeology* 2, pp. 28-32.

Polcaro A., Polcaro V. F. (2009) Man and sky: problems and methods of Archaeoasuonomy. *In Archeologia e Calcolatori* , Vol. 20, pp. 223-245.

Popov D. (2014) *Thracian religion*, South-West Press.

Radoslavova T. Stoev A. (1991) Astronomical traces in ancient rock monuments in Bulgarian lands. *Colloquia Internationale Archaeologia a Astronomia Rome*, pp. 176 -179.

Stoev A. and Maglova P. (2014) Astronomy in the Bulgarian Neolithic. In *Handbook of Archaeoastronomy and Ethnoastronomy*, part VIII, Springer Reference, Springer New York, C.L.N. Ruggles (Ed.), pp. 1377-1384.

Stoev A. and Stoeva M. (2016) Early Bronze Age Megalithic monuments situated near large ancient settlements in Central Thrace, Bulgaria. In *Proceedings of the International Conference Astronomy and world heritage - across time and continents*, 19 - 24 August 2009, Kazan, Republic of Tatarstan, Russian Federation, pp. 118-125.

Stoev A., Muglova P., Gumarov M., Stoeva M., Tashev Y., Videnov B. (2001) Spatial orientation and acoustic reverberation in the Tahgarduk kaya cave sanctuary, near the Ilinitsa village, Kurdjali district. In *Perperek I, Perperek and the microregion around it, complex investigation of the thousand years multireligious centre in East Rhodopes*, Sofia: New Bulgarian University, pp. 226-235.

Stoev A., Gerassimova - Tomova V., Stoytchev T. (1989) Complexe de grottes pres du village Bailovo, Region de Sofia (Bulgarie) et son utilisation par l'Homme. *Proceedings, II, 10-th International Congress of Speleology*, Budapest, pp. 331-333.

Stoev A. and Muglova P. (1999) Rationality, Validity and Reliability of Astronomical Observations during Neolithic and Eneolithic Ages. In *Ancient times, modern methods - Proceedings of the Third SEAC Annual Meeting*, 1 - 3 September 1995, Sibiu, Romania, Stanescu F. (Ed.), Sibiu, Publisher: "Lucian Blaga" University from Sibiu.

Stoev A., Muglova P., Radoslavova T. (1990) Thracian rock sanctuary "Belintash" near Mostovo village, Plovdiv district - a model of an ancient solar observatory. *Interdisciplinary studies*, Vol. XVII, pp. 156-167.

Stoev A. Varbanova Y. (1994) Archaeoastronomical investigation of the Thracian rocky sanctuary "Zaichi vrah". *Proceedings of the IIIrd International Symposium "Cabyle", Settlement life in ancient Thrace* , Yambol, pp. 426-434.

Stoev A. Muglova P. Gumarov M. Stoeva M. Tashev Y. Videnov B. (2001) Spatial orientation and acoustic reverberation in the Tangarduk Kaya cave sanctuary, near the Ilinitsa village, Kurdjali district. *Perperek I: Perperek and the micro region around it, complex investigation of the thousand years multireligious centre in East Rhodopes*. New Bulgarian University, pp. 226-235.

Stoev A. Muglova P. Stoeva M. Tashev Y. Videnov B. Velkov R. Velkov V. (2003) An archaeoastronomical investigation of the Harman Kaya rock-cut mega complex in the region of Dolna Chobanka village, Momchilgrad municipality. *Thracia 15*. Tangra TanNakRa Publishing house, pp. 323-334.

Stoev, A, Maglova,P, Spasova, M Maglova,P Spasova, M (2006a) Late Chalcolithic solar-chthonic rock-cut structures for time measuring in the Eastern Rhodopes, Bulgaria. *Mediterranean Archaeology and Archaeometry*, Vol. 16, No. 4, pp.401-406.

Stoev,A, Maglova, P (2006b) Research project "Astronomical orientations and geophysical anomalies of trapezoidal niches in the Eastern Rhodopes, Bulgaria": in search of an international partnership. *Mediterranean Archaeology and Archaeometry*, Vol. 16, No. 4, pp.407-412.

Stoev,A, Maglova,P, Markov,V, Spasova, M (2006c)Prehistoric rock sanctuary with arch near the village of Kovachevitsa, Bulgaria: spatial orientation and solar projections. *Mediterranean Archaeology and Archaeometry*, Vol. 16, No 4, pp. 413-419.

Todorova H. and Vaysov I. (1993) *New Stone Age in Bulgaria*, Sofia.

4

STONE ROWS OF THE PRESELI HILLS

David A. Fisher, Ph.D.

Independent Researcher

**Corresponding author: (docfishers@gmail.com)*

ABSTRACT

It has been proven that many of the rhyolite and dolerite bluestones that reside within the Sarsen circle of Stonehenge originate from the Preseli Hills of Wales (Bevins, Ixer; 2013, 2014). Yet, no evidence can be found that the rows of stones that are circumjacent to the Preseli Hills have ever been investigated to determine if any orientations to features within the Hills exists, or, whether they may possess any astronomical orientation as occurs with Scottish stone rows (Thom, 1967; Fisher, 2013). References to these sites, are few (Burl, 1993; Williams,1988) and information is somewhat anecdotal. To expand the knowledge base, five sites, each comprising of a pair of stones, were investigated by means of site visits and 3-dimensional computerisation in order to envisage any potential orientations, to either the Hills themselves or celestial events, during the period of the suspected construction of the Mid-Neolithic to early Bronze age time frame. The resultant investigation determined that not only do the sites have potential orientations toward the lunar minor and major limits, but concur with the concept of duality as proposed by Sims (2016).

KEYWORDS: megalith, lunar limits, solar, orientation, Solstice, standstill

1. INTRODUCTION

This research is part of an ongoing investigation into the lesser investigated Neolithic stone monuments of the British Isles; in particular the monuments comprising of 2 or 3 megaliths considered to be stone rows. Earlier research work investigated sites within the Argyll area of Scotland opening a completely new perspective on how the stones were used to view the phenomena of the rising and setting of the Sun and Moon by the Neolithic people who constructed the sites (Fisher 2013).

The stone rows under this investigation are those that surround the Preseli Hills of Wales, the hills that have been proven to be the source of dolerite and rhyolite stones situated within the Stonehenge monument. The specific sites investigated are: 1) Tafarn-y-Bwlch (aka Tre-Bwlch), 2) Penparke (aka Penlan Stones) to the north east of the hills, 3) Cwm Gawr (aka Cerrig Meibion Arthur) to the South, and 4) Dolau Maen (aka Waun Lwyd) with 5) Gors Fawr (aka Mynachlog-Ddu) to the south east, see Figure 1. There is a paucity of information regarding Welsh stone rows (Anthony, 1973; Barber, 1982; 1986; Barber & Williams, 1989; Houlder, 1974; Houlder, Mannig, 1966). Burl's (1993) descriptive catalogue was the source to identify the stone rows and verification was made against the RCAHMW (2016) online database.

Earlier reviews of Welsh sites, with two exceptions, a site known as Harold's Stone (Burl, p188) and Gors Fawr (Thom, p 101), have no mention or consideration of celestial association. Branwell (1875) describes most stones as menhirs, stones of reverence or foci for gatherings. George Williams (1988) mentions the varying shapes of the Welsh stones but provides no specific description of the shapes. Lewis (1966, p9) does note that amongst stone pairs a characteristic dimorphism occurs, one stone being larger and square topped, the other being slighter and often tapered. Lynch (2000, p110) does mention the bluestone of the Preseli Hills but only in the context of polished axe heads. As a result, this investigation applies a fresh perspective as to whether celestial orientations exist or do not exist.

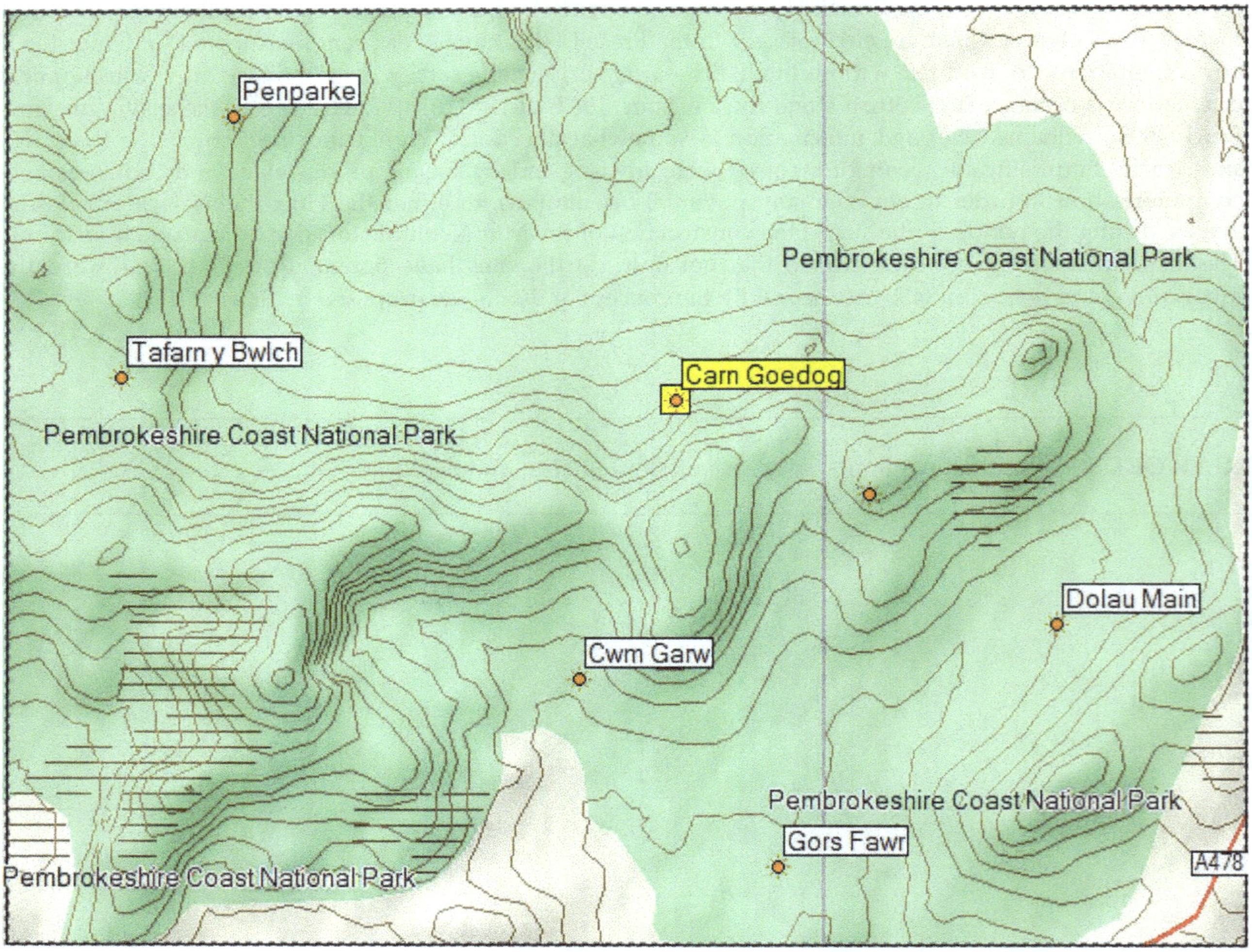

Figure 1. Location of Sites Surrounding the Preseli Hills

2. RESEARCH METHOD EMPLOYED

Each site listed above was visited and the physical dimensions, averaged GPS location readings, true bearings of each face and photographs of all faces of the stones were taken. The data collected was then utilised in the construction of 3 dimensional models of each stone. Ordnance Survey of Great Britain DTM tiles of the Preseli Hills and surrounding areas were converted into a 3 dimensional topography. Each stone model was then incorporated into 3-D landscape to their correct GPS location and set to their correct bearings at their respective locations. The completed site models were then interrogated for the days upon which the rising or setting of lunar minor and major limits (incorrectly termed a lunar standstill, see Fisher,2017) occurred in order to determine whether or not an orientation between, a bearing of a face of the stone or perpendicular to a face (i.e. viewing across the stone not along a face) and the Moon, as the Moon attained the horizon. A value of within ±1 degree was considered to be of sufficient accuracy. The same approach was employed for the Sun, as to whether an orientation to a solstice or equinox is indicated by the stones. This interrogation was conducted via personally developed celestial software set to an arbitrary date of 2200 BCE. No subjective horizon point was chosen only the objective approach of testing the bearing of each stone face in possible association with a solar or lunar event at the horizon.

The plot plans within the figures that follow contain abbreviations for the orientations disclosed by the simulations, and two values for the bearings of an orientation. The first value is the simulation computed bearing, the second is the bearing measured on site. With regards to the abbreviations, the centre letters are 'S', 'M' and 'm' which indicate solstice, major and minor respectively.

3. INTERROGATION RESULTS

All but one site have the surrounding horizon varying in height in all directions, therefore it was not anticipated that if there were any solar or lunar orientations that complimentary opposites (e.g. solstice rising and setting) would be forthcoming. An initial review of the data gathered for the sites was conducted and one feature appeared to be consistent. Each site, with the exception of Cwm Gawr, contained one stone that was triangular in shape that has two faces meeting at an acute angle rather than within ±10 degrees of a right angle. This will be evident in the site plots shown below in the individual site results that follow.

3.1. Tafarn-Y-Bwlch Pillar Stones,

N 51° 58′ 5.4 ″ W 4° 47′ 35.9″

These two stones are heavily leaning toward the northwest and the buildup of peat over the centuries around the stones has probably been worn away by the sheep that roam the hills. The peat to the southeast side of the stones helps retain the stones in a somewhat upright position. As can be seen in Figure 2 an attempt appears to have been made to prevent further erosion of the peat and the possibility of the stones falling over by the placement of boulders at their base.

Within the simulation model the stones were set upright, but no change in orientation was made. Two potential orientations were extracted from the modeling, specifically the rising and setting of the Moon at its northern minor limit. No solar or specific horizon features were made evident.

Figure 2a Tafarn Y Bwlch Stones

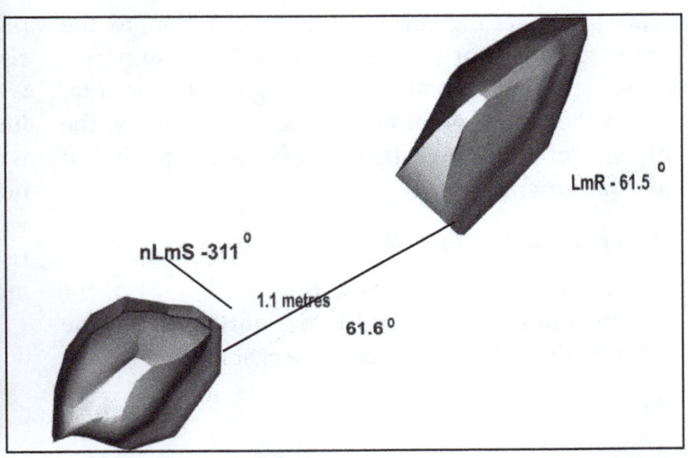

Figure 2b Tafarn Y Bwlch Plan

3.2. *Gors Fawr*

N 51° 55′ 56.9″ W 4° 42′ 46″

Gors Fawr is one of two sites examined which are in close proximity to a stone circle and situated within marshy wetlands. The stone circle is approximately 20 metres in diameter with most stones no taller than knee height. As mentioned in the introduction Thom (1967, p101) considered an orientation to the Sun, but makes no specific statement as to which stone or which solar event. Burl (1993, p188) expands the discussion on orientation stating that an initial consideration was given to a midwinter sunset and a midsummer sunrise, but this was revised after horizon altitudes were reflected upon to an orientation toward a November Samhain sunset.

Figure 3a Gors Fawr Stones

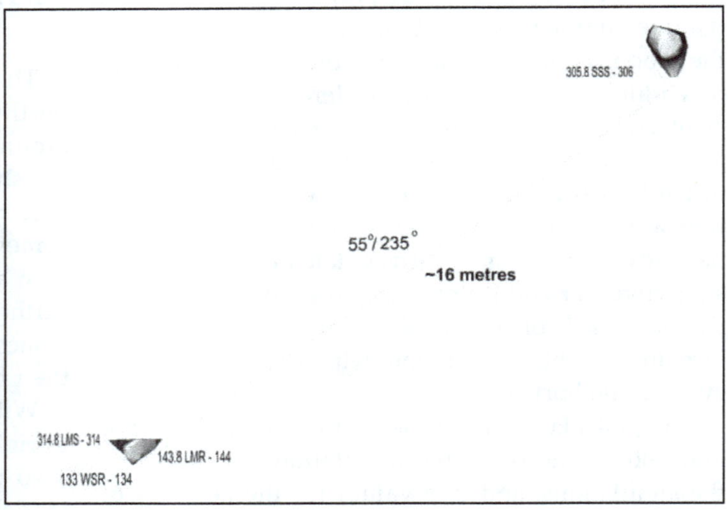

Figure 3b Gors Fawr Plan

The pronounced triangular stone, as shown in Figure 3a, during the simulations suggests multiple orientations, both a northern major limit rise and a northern major limit setting of the Moon (Figure 3b). It is interesting that due to the variable horizon that the southwest face of this triangular stone has both a major lunar setting and a solar orientation to the winter solstice sunrise of 134°. With the more northern stone only suggestive of a single solar orientation to the setting summer solstice Sun. This is the only site of the 5 investigated that is suggestive of any solar orientation.

3.3. *PenParke (PenLan)*

N 51° 59′ 12.5″ W 4° 46′ 43.6″

These stones are situated to the north west of the Preseli Hills and positioned to the east of the brow of a hill, which obscures any view of the wide open view of the Cardigan Bay's watery horizon to the west. A stony outcrop is situated toward the north. Here again is a triangular stone with the potential of multiple orientations. This triangular stone is the one exception where the horizon altitudes permit a complimentary orientation toward the southern minor rise and the northern minor setting of the Moon, as well as an orientation toward the major limit to the lunar setting. The stone to the northeast has 2 faces with orientations, the northwest face toward the northern minor moon setting and the southeast face with an orientation toward the southern minor lunar rise. Again the emphasis is on the minor limits of the moon not the major limits.

Figure 4a PenParke Stones

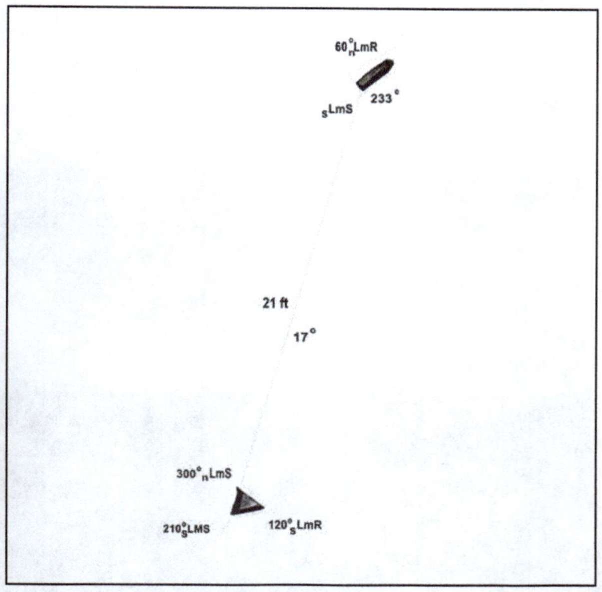

Figure 4b PenParke Plan

3.4. *Cwm Gawr*

N 5° 1 56′ 43.6″ W 4° 44′ 17.5″

This is the second site situated within marshy wetlands and in close proximity to a stone circle, the circle itself being approximately 8 metres in diameter and the stones within the circle similar in size to those at Gors Fawr; in this instance the circle is situated next to a brook. The stones at this site are aligned east west, both to each other and with regards to their individual geographic orientation. First thoughts are to equinoctial events but with an inclination to the horizon varying from +2 degrees in the east to over +3 degrees to the west, any orientation toward the horizon at either 90 or 270 degrees could not be demonstrated by the simulation. The orientations could not equate to the Sun's position on the day of a vernal or autumnal equinox.

This site also differed from the other four under investigation, as both stones were rectangular in shape; neither stone possesses an edge that approaches an acute angle.

Figure 5a Cwm Gawr Stones

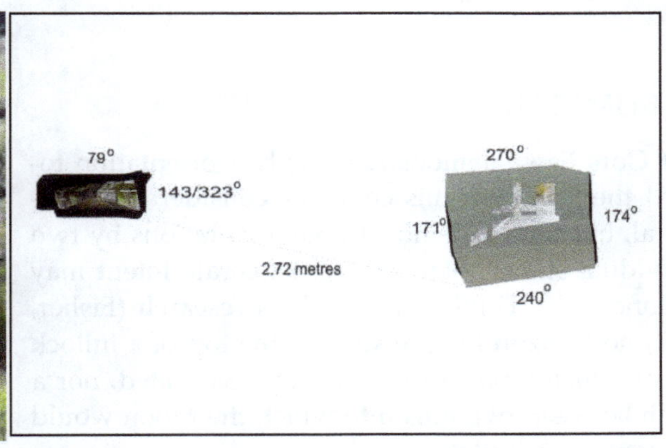

Figure 5b Cwn Gawr Plan

No orientations could be associated with this site.

3.5. *Dolau Maen,*

N 51° 56′ 56.3″ W 4° 40′ 50.9″

Dolau Maen is situated to the south east of the Preseli Hills almost diametrically opposite Tafarn y Bwlch. This arrangement of stones contains orientations that emphasise the major rising and setting of the Moon with only the southern minor setting of the Moon being suggested by the southern stone.

Figure 6a Dolau Maen

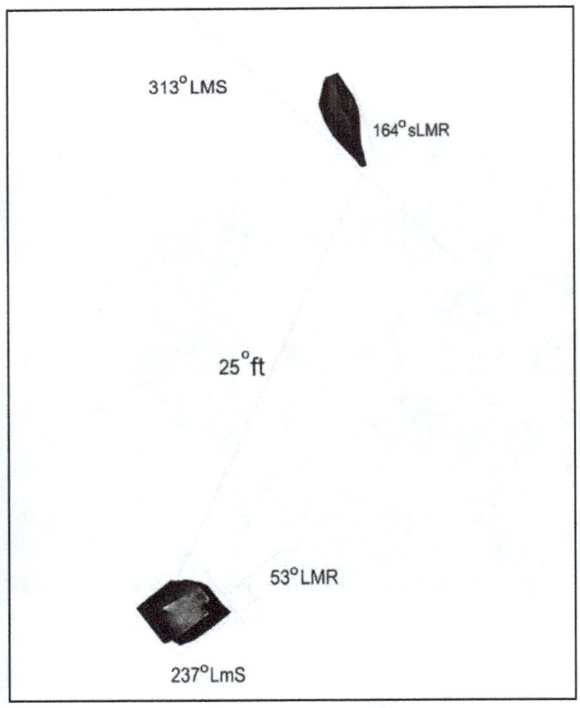

Figure 6b Dolau Maen

There is one aspect of this stone arrangement worthy of mention. The northernmost stone has a face indicating the setting point of the Moon at its northern major limit. Over the hill beyond this setting point is the quarry at Carn Goedog (highlighted in yellow in Figure 1) which has been identified (Bevin, 2014) as a major source of dolerite bluestones at Stonehenge (see Figure 7).

4. SUMMARY

If Gors Fawr demonstrated only 1 orientation toward the Sun then this could be considered coincidental, but with 2 demonstrable orientations by two individual stones then a more deliberate intent may be concluded. Unlike the previous research (Fisher, 2013) no horizon feature such as the top of a hillock out of which a Sun rise or set was associated, nor a notch between two hills into which the Moon would set, could be simulated.

Six orientations to the major lunar limits are informative but seven orientations toward the minor lunar limits across 3 sites is a revelation and quite enlightening. It is the almost isosceles triangular shaped stones at PenParke and Gors Fawr that provide these primary orientations to the minor limits which suggests deliberate shaping and not serendipitous selection. The sites to the south and east of the Preseli Hills, Gors Fawr and Dolau Maen, show a preference to the major lunar rising and setting; whereas, those to the west, northwest, PenParke and Tafarn-Y-Bwlch, have a prevalence toward the minor rising and setting of the Moon.

If we consider that it is these triangular shaped stones at PenParke and Gors Fawr that provide the opposites in orientations and that those triangular stones reside on opposing sides of the Preseli Hills then this leads to two hypotheses. The first being, that the community across the region was connected and the second being that of duality. Duality as described in Sims (2018, in press) where "each monument horizon is paired with another monument's matching but reversed horizon, so horizon alignments are also combined asymmetrically with their reverse equivalent in paired monuments whose design categories overlap each other." Thus this research appears to concur with Sims' postulations.

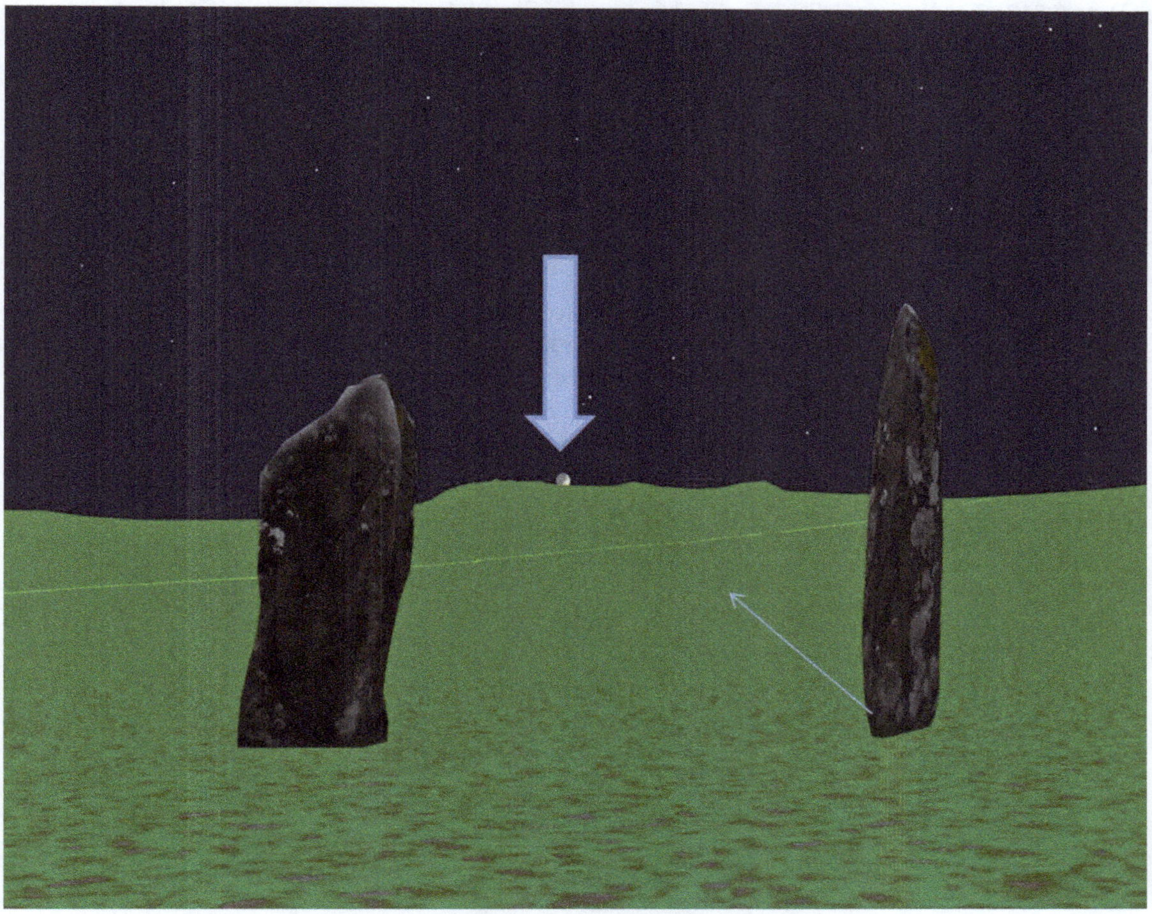

Figure 7 Lunar Major Setting at Dolau Maen

Table 1. Results Summary.

	Lunar								Solar			
Site	nMjrR	nMjrS	sMjrR	SMjrS	nMnrR	nMnrS	sMnrR	sMnrS	SSrise	SSset	WSrise	WSset
Gors Fawr	X	X	-	-	-	-	-	-	-	X	X	-
Dolau Maen	X	X	X	-	-	X	-	-	-	-	-	-
PenLan	-	-	-	X	X	X	X	X	-	-	-	-
Tafarn Bylch	-	-	-	-	X	X	-	-	-	-	-	-
Cwm Gawr	-	-	-	-	-	-	-	-	-	-	-	-

REFERENCES

Anthony, I. (1973) *Discovering regional Archaeology*. Shire Publications, Tring.

Barber, C. (1982) *Mysterious Wales*. Granada, London.

Barber, C. (1986) *More Mysterious Wales*. Granada, London.

Barber, C & Williams, G. (1989) *The Ancient Stones of Wales*. Blorenge Books, Abervenny.

Bevins, R.E., and Ixer R A. (2013) ***Carn Alw as a source of the rhyolitic component of the Stonehenge bluestones: a critical re-appraisal of the petrographical account of H.H. Thomas*** Journal of Archaeological Science Volume 40, Issue 8 - pp 3293-3301

Bevins, R.E., and Ixer R A. (2014) ***Carn Goedog is the likely major source of Stonehenge doleritic bluestones: evidence based on compatible element geochemistry and Principal Component Analysis***. Journal of Archaeological Science Volume 42 -pp 179-193

Branwell, E.L. 1875 *On Pillar-Stones of Wales*, Archaeologia Cambrensis v6, Cambrian Archaeological Association, Wales.

Burl, A. (1993) *From Carnac to Callanish*. Yale University Press. New Haven.

Fisher, D.A. (2013). Employing 3-Dimensional Computer Simulation to Examine the Celestial Dating of Scottish Megalithic Sites. *Anthropological Notebooks*. Slovene Anthropological Society, Ljubljana. P143-155.

Fisher, D.A. and Sims, L. (2017). Modelling Lunar Extremes. *Journal of Skyscape Archaeology* Vol 3.2. p207-216.

Houlder, C. (1974) Wales: an Archaeological Guide Faber and Faber, London.

Houlder, C. & Mannig. (1966) South Wales. Cory, Adams & Mackay, London.

Lynch, F., Aldhouse-Green, S. and Davies, J L. (2000) *Prehistoric Wales*. Sutton Publishing, Stroud.

Lewis, J.M. (1966) The standing stones of Pembrokeshire. *Pembrokeshire Historian* 2, 7-18. Pembrokeshire Community Council, Haverfordwest.

RCAHMW (august 2016) https://rcahmw.gov.uk/discover/coflein/

Sims, L and Fisher, D. (2018 in press). Through the gloomy vale: underworld alignments at Stonehenge. *Journal of Skyscape Archaeology*. Equinox Publishing, Sheffield.

Thom, A. (1967) *Megalithic Sites in Britain*. Oxford University Press, Oxford.

Williams, G. (1988) *The standing stones of Wales and South west England*. BAR British Series 197, Oxford

5

THE OBSERVATIONS OF THE MOON AT NARANJO – NEW FACTS AND INTERPRETATIONS

Stanisław Iwaniszewski

Posgrado en Arqueología, Escuela Nacional de Antropología e Historia, Instituto Nacional de Antropología e Historia, Mexico City

**Corresponding author: (siwanisz@yahoo.com)*

ABSTRACT

According to the Maya records, in 682 CE the Dos Pilas ruler Bahlaj Chan K'awiil, sent his daughter, "Lady Six Sky", to Naranjo (today's Guatemala) to re-establish the local dynasty. Some years later, her arrival to the city was described by a glyph that was widely associated with the age of the moon in the Maya Lunar Series, while she was portrayed in the guise of the Moon Goddess. At Naranjo, the Lunar Series attached to the date of "Lady Six Sky" arrival is associated with the start of Teeple's (1931) "Period of Uniformity". However, for some time on, another system of counting the moon run in parallel.

The name of the ruler's consort, "Lady Six Sky" appears on other sites where the title of a "weaver" is added reinforcing associations with the Moon Goddess, commonly regarded as a weaver. The number six in the name usually identifies various Maize Gods, so when used to denote the Moon Goddess emphasizes the links between the moon and planting of maize.

The paper aims to examine the Lunar Series at Naranjo and to compare them with those from La Corona, Yaxchilan and Oxpemul.

KEYWORDS: Maya Astronomy, Lunar Theory, Lunar Series, Xultun Lunar Table, Naranjo, La Corona, Oxpemul, Yaxchilan.

1. INTRODUCTION

The archaeological remains of Naranjo are situated in the northeastern part of the Guatemalan Department of Peten, not far from the border of Belize. The ancient city of Naranjo was the capital of the Classic Maya kingdom called Sa'al and flourished between late 5th and early 9th centuries CE. Located between Tikal and Caracol, and politically dominated by Calakmul, Naranjo was often defeated, even suffering the extinction of its ruling dynasties (Martin and Grube 2008:69-83).

After the defeat that Naranjo suffered at the hands of Caracol's ruler K'an II in 631 CE, the site disappeared from the political map for about 40 years. Though the reasons for this chronological gap are far from settled, at the time of this writing, there are no inscriptions that would tell us the story of what happened in this period. Be as it may be, in 682 CE reached Naranjo, a woman that would change the history of the city: her name has long been recognized as Lady Wak Chanil Ajaw (Lady Six Sky Queen), the daughter of the Dos Pilar ruler B'alaj Chan K'awiil. Her arrival brought a revival for the dynasty of Naranjo. Her son K'ak' Tiliw Chan Chaak became a new Naranjo ruler, the 38th in the line, and a series of military campaigns allowed the kingdom to reclaim the political control over the neighboring area.

The texts that record Lady Six Sky Queen "arrival" to Naranjo also register the adoption of the Lunar Uniformity. As is known, the Maya scribes used a sequence of three to six glyphs, usually displayed after the Long Count, *tzolk'in* and Glyphs G and F, to offer information regarding the synodic period of the Moon. Their so-called Lunar Series embodied a type of a lunar calendar. Glyphs E and D determined the age of the Moon, Glyphs C grouped lunar periods into sets of 6 or 18 differentiated months, Glyphs X provided individual names for lunar months and Glyphs A defined the length of the current lunar month to be either 29 or 30 days.

In his discussion of lunar reckoning among the Maya Teeple (1931) argued that most Late Classic Maya polities adopted the uniform system of reckoning moons. Within Teeple's Period of Uniformity (9.12.15.0.0 - 9.16.5.0.0, or 687 - 756 CE) all Glyphs C bore the same values. As will be shown in this paper, the Maya scribes of Naranjo kept the same count until the end of the 8th century.

2. MAYA MOON RECKONING

The Mayan lunar month began, in principle, on the evening of the first sighting of the crescent moon, although on occasions the new month was counted from the astronomical new moon. The evolution of the Lunar Series towards greater predictability and regularity led to the installation of a fixed cycle based on a standardized alternation of 29- or 30-day lunar months grouped into the sets of 6 months (of 177 days each). The six-month groupings were organized in 3 groups of 6 months (determined by both head variants and numerical coefficients of Glyph C) for a fixed total of 18 months or 531 days (9 x 30 + 9 x 29). In such a way, succeeding lunar months were arranged in a series of alternating fixed units, forming a regular and easily predictable sequence that also simplified any further calculations.

However, to accord the predicted moon to the observational one, the Maya scribes practiced intercalation through adding an extra day to a standard lunar month of 29 days. The Lunar Table from the wall paintings found in Structure 10K-2, in a residential sector of Xultun, a Classic-period (200-900 AD) city in Petén, Guatemala, shows such intercalations were made at pretty regular intervals determining the cycles of 30 lunar months (equaling 886 days, roughly 2.4 years). The text was first recognized as a Lunar Table by David Stuart (Saturno et al. 2012b: S5-S6) and dated to the early ninth century. The table consists of 27 columns, each containing six lunar months containing either 30 or 29 days, giving a total of 4784 days, because five extra days were added to 29-day months. In other words, the use of a 4784-day cycle comprising five intercalations over 162 lunar months (86 x 30 + 76 x 29 = 4784) determined the pattern of regular intercalations. This pattern defines the average lunar month to be 29.5308642 days long (Aveni et al. 2013: 2; Iwaniszewski 2012, 2015). One 4784-day cycle equals two periods of 2392-days, known as the Palenque formula (81 moons = 2392 days, or 43 x 30 + 38 x 29 days, Teeple 1931). Furthermore, one 4784-day cycle covers 2/5th parts of a 11960-day period, known from the Eclipse Table in the Dresden Codex (Dresden 51-58). However, in this respect, the Eclipse Table introduces nine groups consisting of 5 lunar months which is not the case of the Xultun table. Another difference between the Xultun and Eclipse Table is that the latter does not mention Glyphs C (also consult Aveni et al. 2013: 3; Zender and Skidmore 2012: 13). Be as it can be the Lunar Table from Xultun is the only known Late Classic record of lunar computations proving conceptual continuity with the lunar tables of the Postclassic period Dresden Codex. Nevertheless, it should be mentioned that though the Dresden Codex was manufactured during the Late Postclassic period, its content includes tables computed in earlier centuries, for example, the Eclipse Table is dated to 755 CE (Bricker and Bricker 2011: 254).

Epigraphers also found out that the wall computations were made by the *taaj* scribal-priests, "who celebrated its members' achievements in consulting and producing work for their sovereign's reign" (Saturno et al. 2015: 134). The texts, carefully painted on the walls already received many interpretations (Aveni et al. 2013; Bricker et al. 2014; Iwaniszewski 2012; Zender and Skidmore 2012) providing direct evidence of the Maya involvements with calendrical and astronomical computations previously known only from Late Postclassic (1250-1530) codices.

3. THE LUNAR SERIES AT NARANJO

The Lunar Table from Xultun is dated to the early ninth century CE (Saturno et al. 2012: 717). This period roughly corresponds to the reign of Waxaklajuun Ub'aah K'awiil, the last known ruler of Naranjo who acceded in 814 CE. The last known and dated Lunar Series comes from Stela 8 (800 CE) commissioned by Itzamnaaj K'awiil, his predecessor, who reigned between 790 and 810 CE. My computations of lunar cycles at Naranjo suggest that the process of employing the rigid 4784-day period to the Lunar Series began during the reign of Lady Six Sky Queen. In Table 1, I present the lunar data at Naranjo. Glyphs C displayed on Stelae 24 and 29, the earliest known monuments erected by Lady Six Sky Queen and her son K'ahk' Tiliw Chan Chaak, belong to Teeple's (1931: 54-61) and Aldana's (2006: 240, 248) periods of Uniformity. In both cases, the lunar data are counted backward from the same Dedicatory Date at 9.13.10.0.0 (702 CE) using the fixed 4784-day intervals. Interestingly, to perform backward computations, Stela 24 utilized a 7088-day period (8 x 886 days), perhaps signaling the knowledge of an 886-day cycle cited above (Barthel 1951: 234; Iwaniszewski 2013: 264, 2015: 25-27). However, the Xultun 4784-day cycle is composed of 354 + 4430 days, so we should deal with the alternating sequence of 354 + 5 x 886 + 354 + 5 x 886 + 354 +..., rather than simply with the rigid sequence of 8 x 886. This problem can be resolved by computing the lunar data between Stela 24 and Stela 30. Both IS dates are separated by 11768 days. The application of 886 cycles calls for 13 intercalations (11768 = 13 x 886 + 250 days), while the use of the 4784-day cycle (or of the 9568-day cycle) calls for 12 intercalations. From Table 2 we learn that between both dates were 210 months of 30 days and 188 months with 29 days, implying there were only 11 intercalations made. Therefore, our reconstruction using the 4784-day cycle better approximates the lunar data recorded on Stela 30 than the alternative interval scheme basing entirely on the 886-day period. From Table 2 we learn that the best solution is to count 210 months of 30 days and 188 months of 29 days, suggesting that only 11 intercalations were introduced.

Table 1. Naranjo summary. Is = Initial Series, LS = Lunar Series, DD = Dedicatory Dates, PE = Period Ending. The letters E, D, C, A stay for Glyphs E, D, C, and A respectively. Combined with Glyph C are its lunar patrons denoting: s = skull (God A), m =Young Maize (Tonsured) God, and j= Jaguar God of the Underworld. Glyphs E, D, and A denote the number of days, numerical coefficients of Glyph C denote numbers of lunar months. Data (IS and DD dates, events and protagonists) according to Closs 1985, 1989; Grube 2004; Martin and Grube 2008. The IS Date on Stela 6 is tentative only and will not be taken into account.

No.	Monument	IS Date	LS	Event	Protagonist	DD	Uniformity
1	Stela 24	9.12.10.5.12	18D 1Cj A10	Arrival	Lady Six Sky	9.13.10.0.0	U
2	Stela 22	9.12.15.13.7	20ED 1Cf	Birth	K'ahk' Tiliw Chan Chaak	9.13.10.0.0	U
3	Stela 23	9.13.18.4.18	15D 5Cf A9	War against Yaxhá	K'ahk' Tiliw Chan Chaak	9.14.0.0.0	U + 1
4	Stela 2	9.14.1.3.19	13D 5?Cf A10	1st k'atun anniversary of accession	K'ahk' Tiliw Chan Chaak	9.14.1.3.19	U + 1
5	Stela 29	9.12.10.5.12	19D 6Cf A9	Arrival	Lady Six Sky	9.14.3.0.0	U + 1
6	Stela 30	9.14.3.0.0	4D 4Cj A9	PE, bloodletting	K'ahk' Tiliw Chan Chaak	9.14.3.0.0	U
7	Stela 28	9.14.4.7.1	4D 3Cj A10		Lady Six Sky	9.14.5.0.0/ 9.14.8.0.0.	U
8	Stela 31	9.14.4.12.7	21ED 6Cj A9	K'atun PE	Lady Six Sky	9.14.10.0.0	U
9	Stela 18	9.14.15.0.0	13D 6Cj	Lajuntun PE	Lady Six Sky	9.14.15.0.0	U
10	Stela 6	9.16.4.10.18?	23D 1Cf A10	Accession	K'ahk' Ukalaw Chan Chaak	9.17.10.0.0	?
11	Stela 13	9.17.10.0.0	27ED 4Cs A9	Lajuntun ending	K'ahk' Ukalaw Chan Chaak	9.17.10.0.0	U
12	Stela 14	9.17.13.4.3	7D 2Cf A9	Accession	Itzamnaaj K'awiil	9.18.0.0.0	U
13	Stela 8	9.18.10.0.0	21D 2Cj A9	Lajuntun ending	Itzamnaaj K'awiil	9.18.10.0.0	U

Table 2. Lunar computations between the IS dates from Stela 24 and 29.

Monument	IS date, Maya Day Number	LS	Difference, (days)	4784-day formula	Copán formula
Stela 24	9.12.10.5.12 1,386,112	18D 1Cj A10	11768	18D 1Cj 10: 29.5308642 = 398L + 14.72d, then 211x 30 + 187 x 29 + 15 = 3D 4Cj 9,	18D 1Cj 10 : 29.53020134 = 398L + 14.98d, then 210 x 30 + 180 x 29 + 16 = 4D 4Cj 9
Stela 30	9.14.3.0.0 1,397,880	4D 4Cj III. 4 9			
				3 D = 12 intercalations, 4 D = 11 intercalations	11 intercalations

Two monuments, Stelae 24 and 29, recorded the same date 9.12.10.5.12 (August 687 AD) which associated with two different Lunar Series (see Table 1). The uniform system is listed as 18D 1Cj 10, whereas the non-uniform as 19D 6Cm 9. The Uniform system increases the numerical coefficient of Glyph C by one, simultaneously decreasing the age of the Moon by one and shifting the length of the lunar month from 29 to 30 (see Thompson 1954: 48). Both lunar data are, in some sense, equivalent to each other. What is important here is that both series, the Uniform and the non-uniform one, utilized the same, 4784-day intercalary cycle (see Table 3). Since first uniform dates (Stelae 24 and 22) were retrospectively counted from the same base at 9.13.10.0.0 (January 702 CE), I assume this date seems to be appropriate to mark the introduction of the Lunar Uniformity to Naranjo. All non-uniform texts are counted retrospectively from the period between 9.14.0.0.0 and 9.14.3 0.0 (711 - 714 CE). Thus, the estimated period when Lunar Uniformity and 4784-day intercalary cycles were introduced to Naranjo (roughly 702 - 714 CE) coincides with K'ahk' Tiliw Chan Chaak efforts to recover the political control over Naranjo's neighbors (see Beliaev 2000: 76).

Not only lunar information from Naranjo follows the structure of the Lunar Table from Xultun, but also the Uniformity Period, as identified at Naranjo, is based on the same principle. It shows that the reckoning of the lunar months was the result of a gradual process which at Naranjo began at least a century earlier than the elaboration of the Lunar Table at Xultun.

As mentioned above, Stela 24 portrays Lady Six Sky Queen in the guise of the Lunar Goddess. Several scholars observed that impersonators of the Moon Goddess displayed the regalia similar to those of the Maize God costume. The diamond-like or jade net pattern is often seen on the images representing either by both deities or their impersonators. The overlapping or conflation of both divinities is often attributed to the ancestral-mythical character that considers them as the First Father and First Mother of the gods (Freidel et al. 1993: 279, 464; Looper 2002: 177-181; Schele and Miller 1986: 309; Tate 2000: 1060). Mutual relations observed between those two deities are attributed to the widespread use of the moon phases for planting and harvesting in pre-Hispanic Mesoamerica (Milbrath 1999: 133-135; Taube 1992: 68-69). Be as it may be, the overlapping costume elements reveal individual associations existing between this plant and the Moon. By extension, the costumes worn by Maya rulers and other persons which displayed Maize God forms sometimes merged with Moon Goddess representations causing a certain ambiguity among different female and male impersonators (Looper 2002: 177-184; Quenon and Le Fort 1997: 894-895; Zender 2014: 8). Recently Nikolai Grube (2016) has shown that "Six Sky" epithets may refer to the Maize God rather than to the Moon Goddess. Following Tokovine's (2013: 116) identifications of the place names containing number six with the Maize God, Grube observes that, on occasions, the name of the female ruler should be read as Ix(ik) Wak Jalam Chan, "Lady Six Weaver Sky". Therefore while the epithet "six sky" may refer to the Maize God, the epithet "Six Sky Weaver or "Six Weaver Sky" are exclusively associated with the Moon Goddess.

Her arrival is introduced by the same glyph HUL which is associated with the age of current moons (Glyphs E and D). Houston et al. (2006: 261) proposed the verb *hul-i*, meaning "arrival to" first described the arrival of the moon in the sky, but later compared the celestial movement to royal processions. The Maya imagined that like a human pilgrim on the earth, the Lunar Goddess traveled across the sky, and in Colonial times tens of pilgrims crossed Yucatan to get to the Ixchel (Lunar Goddess) shrine on the Island of Cozumel. Colonial dictionaries from Yucatan tell us these pilgrims were known as *hula* (Barrera Vásquez et al. 1980: 242-243).

Table 3. Lunar Series at Naranjo. Left – Lunar Uniformity, right non-uniformity.

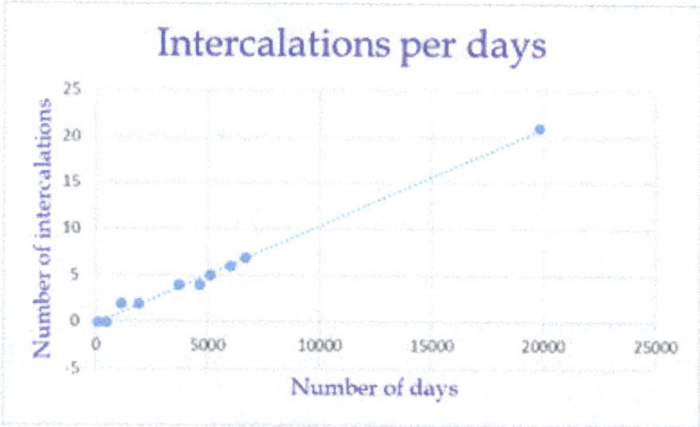

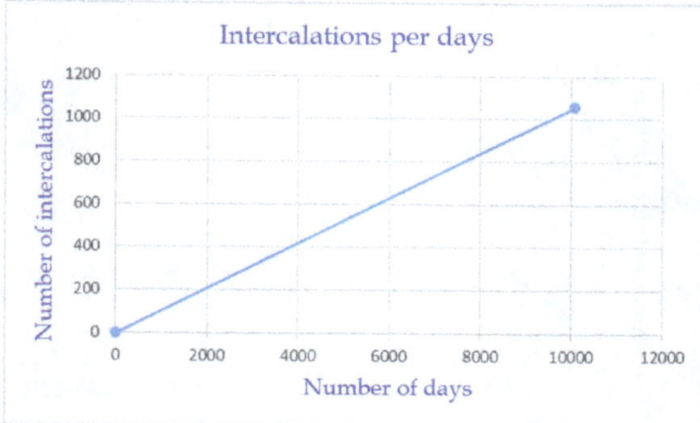

According to this interpretation, the introduction and eventual fixation of lunar uniformity and intercalations are closely related to the Lady Six Weaver Sky Queen "arrival" to the Naranjo-Sa'al kingdom. Grube (2016) identified a number of other Classic Maya queens bearing the name of Six Weaver Sky, each associated with a different polity and respective ruling family. The question remains whether Moon Goddess identities of those princesses can also be related to the fixing of intercalation and/or lunar uniformity.

4. OXPEMUL

The name of Ix(ik) Wak Jalam Chan, "Lady Six Weaver Sky", appears at Oxpemul, the Classic city located in the vicinity of Calakmul, one of the Maya superpowers. Stela 5 displays her name, but this monument bears no lunar data (Grube 2008: 201-202) and due to the erosion cannot be dated. According to Robichaux (2005: 31), the reference to the name of the ruler from Stela 5 can be found on Stelae 18 and 19, dedicated on 9.16.5.0.0 (756 CE). Whereas between 9.16.0.0.0 (751 CE, Stela 9) and 9.16.5.0.0 (756 CE) Oxpemul adopted a 4784-day cycle comprising five intercalations, by 9.17.0.0.0 (771 CE) this system was abandoned (consult Table 4). Unfortunately, the lack of Moon Ages on several inscriptions does prevent us from drawing graphical timelines.

5. LA CORONA

Still another queen with the same name is found in La Carona, where she became the wife of Chakaw Nahb Chan who reigned from 658 to 667 CE (Stuart et al. 2014: 438-439). It may be observed that between 677 (Panel 1) and 721 CE (Panel 6) ritual specialists at La Corona utilized both the Lunar Uniformity and the 4784-day intercalary cycle (see Tables 4 and 5).

Table 4. The Lunar Series at Oxpemul and La Corona.

Monument	IS Date	LS	Uniformity
		Oxpemul	
Stela 9	9.16.0.0.0	4Cj 29	U+1
Stelae 18/19	9.16.5.0.0	5D 5Cs 10	U+1
Stela 19	9.16.5.0.0	?D 5Cs 10	
Stelae 2	9.17.0.0.0	?D 3Cm 10	U
Stela 4	9.17.0.0.0	?D 3Cm 9	
Stela 7	10.0.0.0.0	13D 6Cj 9?	U
		La Corona	
Panel 193	9.9.2.0.8	5D 2Cm 9	U-1
Panel 1	9.12.5.7.4	22ED 1Cm 10	U
Panel 6	9.14.9.9.14	25ED 5Cs 10	U

6. YAXCHILAN

Lady Six Weaver Sky is also the name of one of the wives of Bird Jaguar IV, one of the greatest lords of Yaxchilan. Here she is described as ajaw, meaning "queen", that is, Lady Six Weaver Sky Queen (Grube 2016: 5; Martin and Grube 2008: 131). She originated from Motul de San José, in Petén, Guatemala. The first date recording Xultun formula appears on Lintel 56 from Structure 11 which was raised on 738 CE but from that time onwards all monuments commissioned by Bird Jaguar IV bear the lunar data computed by the scheme known from the Xultun Lunar Table (see Table 5). In this way, the institution of an intercalary cycle was associated with the authority with which the queen was endowed.

Table 5. Intercalation trends at Yaxchilan and La Corona during the Uniformity.

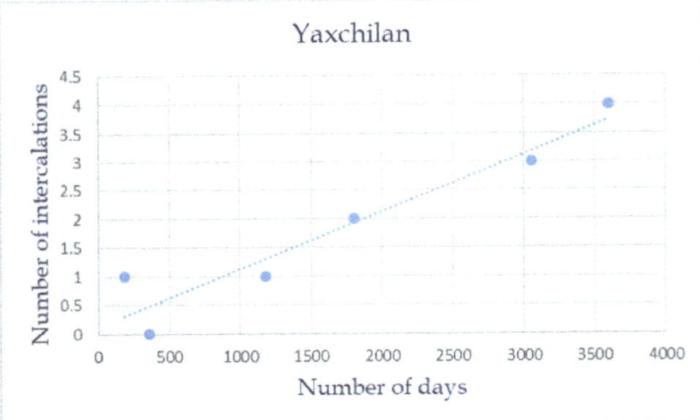

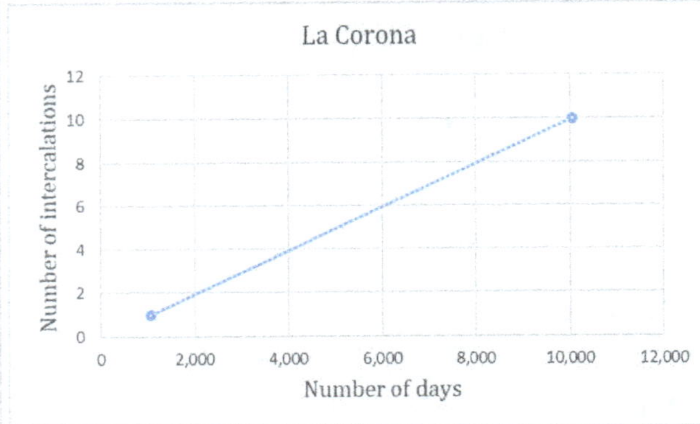

7. DISCUSSION

We see, therefore, that arrivals of various Maya queens bearing the name of Six Weaver Sky to new political seats to become wives of local rulers could have been associated with the dissemination of the knowledge of proper moon reckoning. Of course, my sample of records used in this paper is reduced to the cases listed by Grube (2016) who enumerates the queens from five Mayan centers: Naranjo, La Corona, Yaxchilan, Oxpemul, and Rio Azul. Since the name of a lady from Rio Azul lacks "Six Sky" element, it may be dismissed. So, interestingly, during the periods when the Maya queens were acting, Maya scribes utilized 4784-day intervals to follow the Moon.

8. CONCLUSIONS

Sources such as the Lunar Series are far more informative about astronomic computing rules of the Maya than it might be expected. The data in the Lunar Series can be treated as reliable, as they were meticulously computed and recorded; so there is no need to convert them to equivalent Julian dates. The knowledge of a 4784-day intercalary cycle was the result of a gradual process which already was in progress in the seventh and eighth centuries CE. Indeed, the Lunar Table from Xultun was not a sudden endowment of the early ninth century.

The use of the number "six" in the names of Mayan queens portrayed as embodying the Moon Goddess, may reflect the idea that the Maize God and the Moon Goddess were considered as parallel beings. It may be conjectured that the adoption of a rigid pattern of intercalations was somewhat instigated by the Mayan queens who migrated to new cities to re-establish local dynasties. Geographic distribution of these Maya polities may suggest, that the region where the Xultun Lunar Table was produced was related to the central Petén, Guatemala.

Finally, returning to lunar computations, it may be concluded the widespread use of the Lunar Table from Xultun makes futile any attempt to find a day-to-day correlation constant.

ACKNOWLEDGMENTS

The study of the Lunar Series at Naranjo, Oxpemul, Yaxchilán and La Corona was supported by the Instituto Nacional de Antropología e Historia through the Project 296 "La teoría lunar en astronomía maya".

REFERENCES

Aldana, G. V. (2006) Lunar Alliances: Shedding Light on Conflicting Classic Maya Theories of Hegemony. In *Viewing the Sky Through Past and Present Cultures*, T. W. Bostwick and B. Bates (eds.), Phoenix, City of Phoenix Parks and Recreation Department, pp. 237-258.

Aveni, A. F., Saturno, W., and Stuart, D. (2013) Astronomical Implications of Maya Hieroglyphic Notations at Xultun. *Journal for the History of Astronomy*, Vol. 44, No. 1, pp. 1-16.

Barrera Vásquez, A., Bastarrachea Manzano, J. R., and Brito Sansores, W. (1980) *Diccionario maya Cordemex*. Merida, Ediciones Cordemex.

Barthel, T. (1951) Maya Astronomie: Lunare Inschriften aus dem Südreich. *Zeitschrift für Ethnologie*, Vol. 76, pp. 216-238.

Beliaev, D. (2000) Wuk Tsuk and Oxlahun Tsuj: Naranjo and Tikal in the Late Classic. In *Sacred and the Profane: Architecture and Identity in the Maya Lowlands*, P.R. Colas (ed.), Markt Schwaben, A. Saurwein, pp. 63-81.

Bricker, V. R., Aveni, A. F., and Bricker, H.M. (2014) Deciphering the Handwritting on the Wall: Some Astronomical Interpretations of The Recent Discoveries at Xultun, Guatemala. *Latin American Antiquity,* Vol. 25, No. 2, pp. 152-169.

Bricker, H.M., and Bricker, V.R. (2011) *Astronomy in the Maya Codices.* Memoirs of the American Philosophical Society, Vol. 265, Philadephia, American Philosophical Society.

Closs, M. P. (1985) The Dynastic History of Naranjo: The Middle Period. In *Fifth Palenque Round Table, 1983,* V.M. Fields (ed.), The Pre-Columbian Art Research Institute, San Francisco, Monterey, CA, Herald Printers, pp. 65-77.

(1989) The Dynastic History of Naranjo: The Late Period. In *Word and Image in Maya Culture,* F. Hanks and D.S. Rice (eds.) Salt Lake City, University of Utah Press, pp. 244-254.

Freidel, D., Schele, L, and Parker, J. (1993) *Maya Cosmos. Three Thousand Years of the Shaman's Path.* New York, William Morrow.

Grube, N. (2004) La historia dinástica de Naranjo, Petén. *Beiträge zur allgemeinen und vergleichenden Archäologie,* Vol. 24, pp. 195-213.

(2008) Monumentos esculpidos: epigrafía e iconografía. In *Reconocimiento arqueológico en el sureste del estado de Campeche, México: 1996-2005,* I. Šprajc (ed.), BAR International Series, 1742, British Archaeological Association, Oxford, pp. 177-231.

(2016) The Logogram JALAM. *Research Note 3. Textdatenbank und Wörterbuch des Klassischen Maya.* Arbeitsstelle der Nordrhein-Westfälischen Akademie der Wissenschaften und der Künste an der Rheinischen Friedrich-Wilhelms-Universität Bonn.

Houston, S., Stuart, D. and Taube, K. (2006) *The Memory of Bones. Body, Being, and Experience among the Classic Maya.* Austin, University of Texas Press.

Iwaniszewski, S. (2011) El arribo de la Señora Seis Cielo a Naranjo y las Series Lunares. In *Los investigadores de la Cultura Maya, 19.* Campeche, Universidad Autónoma de Campeche, Vol. 19, tomo II, pp. 155-167.

(2012) Los ciclos lunares y el calendario maya. *Arqueología Mexicana,* Vol. 19, No. 118, pp. 38-42.

(2013) Las Series Lunares en Petén y en el sur de Campeche a la luz de la Tabla Lunar de Xultun. In *Los investigadores de la Cultura Maya, 21,* Campeche, Universidad Autónoma de Campeche, Vol. 21, Tomo 1, pp. 259-270.

(2015) La Tabla Lunar de Xultun y su comparación con las Series Lunares. In *Diferentes povos, diferentes saberes na América Latina,* L.C. Borges (ed.), Rio de Janeiro, Museu de Astronomia e Ciências Afins – MAST, pp. 16-31.

Looper, M. G. (2002) Women-men (and Men-Women): Classic Maya Rulers and the Third Gender. In *Ancient Maya Women,* T. Ardren (ed.), Walnut Creek, CA, Altamira Press, pp. 171-202.

Martin, S. and Grube, N. (2008) *Chronicle of the Maya Kings and Queens: deciphering the Dynasties of the Ancient Maya,* Second Edition. London, Thames and Hudson.

Milbrath, S. (1999) *Star Gods of the Maya: Astronomy in Art, Folklore, and Calendars,* Austin, University of Texas Press.

Quenon, M. and Le Fort, G. (1997) Rebirth and Resurrection in Maize God Iconography, In *The Maya Vase Book,* J. Kerr (ed.), New York, Kerr Associates, Vol. 5, pp. 884-902.

Robichaux, H. R. (2005) Las inscripciones de Oxpemul, In *Los investigadores de la cultura maya.* Campeche, Universidad Autónoma de Campeche, Vol. 13, No. 1, pp. 29-43.

Saturno, W., Aveni, A.F., Stuart, D., and Rossi, F. (2012a) Ancient Maya astronomical tables from Xultun, Guatemala, *Science,* Vol. 336, No. 6082, pp. 714-717.

Saturno, W., Stuart, D., Aveni, A.F., and Franco Rossi, F. (2012b) Supplementary Materials for Ancient Maya Astronomical Tables from Xultun, Guatemala, *Science,* Vol. 336, No. 6082, pp. S1-S8.

Saturno, W., Hurst, H., Rossi, F., and Stuart, D. (2015) To set before the king: residential mural painting at Xultun, Guatemala, *Antiquity,* Vol. 89, No. 343, pp. 122-136.

Schele, L. and Miller, M. (1986) *The Blood of Kings: Dynasty and Ritual in Maya Art.* Fort Worth, TX, Kimbell Art Museum.

Stuart, D., Mathews, P., Canuto, M., Barrientos Q., T., Guenter, S., and Baron, J. (2014) Un esquema de la historia y epigrafía de La Corona. In *XXVII Simposio de Investigaciones Arqueológicas en Guatemala, 2013,* B. Arroyo, L. Méndez Salinas, and A. Rojas (eds.), Guatemala City, Museo Nacional de Arqueología y Etnología, pp. 435-448. Available: www.asociaciontikal.com (accessed 20 February 2018).

Tate, C. (2000) Writing on the Face of the Moon: Women as Potters, Men as Painters in Classic Maya Civilization. In *The Maya Vase Book,* J. Kerr (ed.), New York, Kerr Associates, Vol. 6, pp. 1056-1071.

Taube, K. (1992) *The Major Gods of Ancient Yucatan,* [Studies in Pre-Columbian Art and Archaeology, 32]. Washington, D.C., Dumbarton Oaks.

Teeple, J. E. (1931) *Maya Astronomy. Contributions to American Archaeology,* No. 2:29-115. Publication 403. Washington, Carnegie Institution of Washington.

Thompson, J.E.S. (1954) Memoranda on Some Dates at Palenque, Chiapas. *Notes on Middle American Archaeology and Ethnology,* No. 120, Washington, D.C., Carnegie Institution of Washington, pp. 45-52.

Tokovine, A. (2013) *Place and Identity in Classic Maya Narratives,* [Studies in Pre-Columbian Art and Archaeology, Number 37], Washington, D.C., Dumbarton Oaks Research Library and Collection.

Zender, M. and Skidmore, J. (2012) Unearthing the Heavens: Classic Maya Murals and Astronomical Tables at Xultun, Guatemala. Available: *Mesoweb Reports,* www.mesoweb.com/reports/Xultun.pdf. (accessed 1 July 2012).

Zender, M. (2014) On the Reading of Three Classic Maya Portrait Glyphs. *The PARI Journal,* Vol. 15, No. 2, pp.1-14. Available: www.mesoweb.com/pari/journal.html (accessed 10 February 2017).

6

URBAN PLANNING AND RITUAL ACTION IN *COLONIA ULPIA TRAIANA* (XANTEN, GERMANY): UNDERSTANDING A NON-SOLAR ORIENTATION PATTERN

David Espinosa-Espinosa*[1], A. César González-García[2] and Marco V. García-Quintela[1]

[1]*University of Santiago de Compostela*
[2]*Institute of Heritage Sciences, Incipit-CSIC, Santiago de Compostela, Spain*

*Corresponding author: David Espinosa-Espinosa (davidespinosaespinosa@gmail.com)

ABSTRACT

There is increasing evidence to suggest that cosmological factors were applied in the planning and orientation of Roman towns, at least under Augustus. Among others, this is the case of *Colonia Augusta Praetoria Salassorum* (Aosta) in *Italia, Colonia Urbs Iulia Nova Carthago* (Cartagena) in *Hispania Citerior Tarraconensis, Colonia Copia Claudia Augusta Lugdunum* (formerly *Colonia Copia Felix Munatia,* Lyon) in *Gallia Lugdunensis, Colonia Augusta Treverorum* (Trier) in *Gallia Belgica,* and *Colonia Claudia Ara Agrippinensium* (formerly *Ara Ubiorum,* Cologne) in *Germania Inferior.* For the sake of strengthening the sample of cities studied, and identifying orientation patterns from a chronological and astronomical perspective, a number of public structures from *Colonia Ulpia Traiana* (Xanten) in *Germania Inferio*r were measured. This town was a Roman colony, founded in A.D. 98 by Trajan with a contingent of veteran soldiers and a group of Germanic people. The result was the establishment of a typical Roman settlement with an orthogonal urban grid, whose planning and orientation took cosmological factors into account. In this case, in contrast to the previous examples, we propose that the *decumanus maximus* was not oriented directly according to the solar arc, but that instead it was possibly linked with other celestial bodies. In addition, the Gallo-Roman temple supposedly dedicated in this town to the *Matronae* or the *Matres* was oriented according to the major lunar standstill ("lunistice"). Therefore, this study aims to present the first results regarding the urban orientation of *Colonia Ulpia Traiana* according to a non-solar pattern, and attemps to provide a preliminary explanation for it from a cultural perspective.

KEYWORDS: Roman Towns, Urban Orientation, Limes Germanicus, Trajan, Matronae, Major Lunar Standstill, Venus.

1. INTRODUCTION AND RELATED WORK

The application of cosmological criteria in the orientation and planning of Roman towns has recently been identified in a number of case studies, such as *Colonia Augusta Praetoria Salassorum* (Aosta) in Italia (Bertarione and Magli, 2015), *Colonia Urbs Iulia Nova Carthago* (Cartagena) in *Hispania Citerior Tarraconensis* (González-García et al., 2015), *Colonia Copia Claudia Augusta Lugdunum* (formerly *Colonia Copia Felix Munatia*, Lyon) in *Gallia Lugdunensis* (García-Quintela and González-García, 2014), *Colonia Augusta Treverorum* (Trier) in *Gallia Belgica* (Espinosa-Espinosa et al., 2016), and *Colonia Claudia Ara Agrippinensium* (formerly *Ara Ubiorum*, Cologne) in *Germania Inferior* (Espinosa-Espinosa and González-García, 2017). In all of them, the apparent movement of the sun across the sky throughout the year would have been chosen (usually at sunrise) to mark important public festivals associated with human and agricultural cycles, as well as with several deities from the Roman pantheon, including the Emperor. The intention behind this practice would have involved the commemoration of specific public events that were deeply significant for the population, but also favouring prosperity and maintaining the political and religious cohesion of the local communities.

Despite what has been traditionally argued by scholars such as Joëll Le Gall (1975), Roman towns, at least under Augustus, seem to have been planned and oriented according to solar criteria (González-García et al. 2018). In Ancient Greece, this practice is implicit in Callimachus' *Hymn* (2, 55-57) to Apollo: "And Phoebus it is that men follow when they map out towns. For Phoebus evermore delights in the founding of cities, and Phoebus himself doth weave their foundations." Regardless of whether the design of towns in the Roman world were subject to a series of empirical guidelines that were intended to provide the healthiest possible conditions for the people who lived in them –as we know from Vitruvius (*De arch.* 1, 4–6) and Cicero (*De Rep.* 2, 5–11)–, cosmology seems to have been a fundamental conditioning factor which determined the orientation of towns. This discipline, included by Vitruvius (*De arch.* 1, 1, 10) among the skills that an architect had to cultivate with the aim of knowing "the points of the heavens, the laws of the celestial bodies, the equinoxes, solstices, and courses of the stars," generates a series of ideas and beliefs that are expressed in different ways in the spheres of culture and religion. According to these, human beings view the world in a specific way, a situation that affects the manner in which space and life in society (i.e., the city) are understood and organized (Rykwert 1988; Carandini 2003; Humm, 2004; De Sanctis, 2012; Gargola, 2017). In the case of Rome, following Maurizio Bettini (2011: 91), "[the Romans] created a cosmogony in virtue of the *mundus*, which related their city to the celestial vault and to the realm of the dead, while the circular form of the furrow, together with the *pomoerium*, recalled the form of *orbis* that the city of Rome in some way reproduced."

2. *COLONIA ULPIA TRAIANA* (XANTEN, GERMANY)

Colonia Ulpia Traiana has been chosen as a new case study for the sake of strengthening the sample of towns studied, testing the application of cosmological criteria in Roman towns after Augustus, and identifying orientation patterns from a chronological and astronomical perspective. This Roman colony was founded in A.D. 98 by Trajan, nearly a century after the above mentioned towns, with a contingent of veteran soldiers and a group of *Cugerni/Baetasii* (Müller et al., 2008: 243-255; Eck, 2014; Schalles, 2014). *Colonia Ulpia Traiana* is located in modern Xanten (Germany), in the heart of present-day Europe (Figure 1). Due to its latitude (51 degrees north), it is one of the most northerly Roman towns in the area, together with *Ulpia Noviomagus Batavorum* (Nijmegen, in the Netherlands). *Colonia Ulpia Traiana* was established close to an ancient branch of the River Rhine, in the centre of a large river plain. It was connected with the mouth of the river and the capital of *Germania Inferior* (*Colonia Claudia Ara Agrippinensium*) by a Roman road that ran from north to south to the right of the *limes Germanicus*.

As regards its historical context, Trajan (A.D. 98-117) founded *Colonia Ulpia Traiana* in his own honour after being acclaimed *Princeps* in *Germania Inferior* following the death of his adoptive father, Nerva (Müller et al., 2008: 243-255; Eck, 2014: 106 and 109; Schalles, 2014: 126-131). This settlement, with an area of 73 ha, was the second Roman colony of *Germania Inferior* after the founding of *Colonia Claudia Ara Agrippinensium* by Claudius in A.D. 50. According to scholars such as Werner Eck (2014: 108), *Colonia Ulpia Traiana* would have been a demonstration of power by Trajan, due to its large dimensions and the fact that he was assured the loyalty of the war veterans and local auxiliaries who had settled there. As a result, as Werner Eck has suggested, Trajan would have created a permanent monument in honour of his imperial beginnings in this province (Eck, 2014: 108).

Like all new Roman towns, *Colonia Ulpia Traiana* had an orthogonal urban layout that was determined by the course of the River Rhine at that time. The

Roman authorities built whatever was needed to enjoy a comfortable and quiet life: a solid town wall, a forum, a river port, an amphitheatre, *domus* (houses), *tabernae* (shops), wide streets with a sewer system, public baths, and temples (Müller et al., 2008: 277-470). The countryside surrounding the colony produced cereals and legumes (among other crops) together with cattle, sheep and pigs, and so it can be assumed that apart from the veterans, most of the people from *Colonia Ulpia Traiana* were farmers and peasants, mainly of Gallo-Germanic origin (Müller et al., 2008: 21-48; Derks, 1998: 55-66).

Figure 1. Location of Colonia Ulpia Traiana. Map: GifeX.

3. ARCHAEOASTRONOMICAL MEASUREMENTS

The methodology employed involves measuring the archaeological remains that may help to identify the orientation of the urban layout. After determining the direction from which the orientation was based, the azimuth and the horizon altitude are taken for this direction, using tandems that include a precision compass and a clinometer. As the instruments used are magnetic, the magnetic declination readings are corrected. When this type of reading is not possible, the magnetic declination is estimated for the fieldwork dates from the models available from the National Oceanic and Atmospheric Administration website.[1] The data can then be compared with estimates for the celestial objects visible in this section of the horizon. To perform this comparison, our measurements are converted into astronomical declinations, resulting in an error estimate of approximately one day in the case of the sunrise/sunset. When the horizon is blocked due to the presence of modern buildings, a digital terrain model is used to reconstruct the horizon.[2]

To calibrate the orientation of the structures in *Colonia Ulpia Traiana*, measurements were taken in situ in different parts of the town. The areas that were measured included the north gate, the *cardo maximus*, the *decumanus maximus*, several *cardines* and *decumani minores*, the town baths, the *Hafentempel*, the forum, the *capitolium*, the *Matronentempel*, the amphitheatre, and the town wall. Based on an orthogonal urban layout, the purpose of gathering data in these areas is to examine a wide and representative sample of measurements, and to verify the existence of a correlation between them within a reasonable margin of error. Once these data were analysed, the urban orientation was established according to the direction of the *cardines* and *decumani* as shown below (Table 1).

Table 1. Astronomical orientation results of Colonia Ulpia Traiana.

Colonia Ulpia Traiana Lat. 51° 40′N	A (±¼°)	h (±½°)	δ (±0.75°)	Event
Cardines	134°	½°	−25.5°	Moon?
Cardines	314°	0°	25°	Moon?
Decumani	44°	0°	26°	Moon?
Decumani	224°	0°	−27°	Venus/Moon?

The columns show the structure measured, the azimuth (A), the horizon altitude (h), and the calculated astronomical declination (δ). The last column shows the possible astronomical event or date. In view of the results, there are two aspects that stand out. On the one hand, *Colonia Ulpia Traiana* has a rather peculiar orientation, where neither the *cardines* nor the *decumani* face the rising or setting sun. This is one of the few Roman towns studied that was oriented according to this type of pattern. Although we would expect that cardinal orientations would be the most common orientation for Roman towns, according to what the *agrimensores* prescribed (see e.g. Front., *De limit.* 8, 30-33; 10.27-28), the most common orientation of the Roman towns studied to date is actually towards sunrise at the winter solstice (WS) (see for example González-García and Magli, 2015; Rodríguez-Antón et al., in this volume). We would therefore argue that in this case it seems as if the Roman *agrimensores* would have planned an orthogonal layout according to true north, but rotated 45° towards the east (i.e., bisecting a 90 degrees angle)

[1] http://www.ngdc.noaa.gov/geomag-web/

[2] http://www.heywhatsthat.com/

for reasons as yet unknown. Perhaps the ancient course of the River Rhine (the end of the sewer system), or even a previous Julio-Claudian settlement, may have determined this choice (Müller et al., 2008: 171-209; Schalles, 2014: 114-117). Despite this, the orientation of the *decumani* towards the northwest and southwest could be compatible with events related to the moon (as they are facing a part of the horizon where the sun does not rise or set, but where the moon does) and Venus, in particular with the cycle of Venus and the movements of this planet in the sky (synodic period). Interestingly, this possibility can be seen more clearly in the south-western horizon, as the planet has a larger visible elongation towards the west than the east with respect to the sun on the horizon (see e.g. Sprajc, 2015).

In addition to the urban orientation, it is interesting to consider the orientation of a significant religious building that would have played an important role in the daily life of the *Traianenses*. This is the so-called *Matronentempel* (*Matronae* temple) whose stone foundations are still preserved today (Freigang, 1991; Zille 1995; Müller et al., 2008: 319-324). This building was a Gallo-Roman temple designed according to the Celtic tradition, which has a slight but clear deviation from the orthogonal layout orientation. In this respect, the orientation of this temple, based on its two axes of symmetry, is as follows (Table 2):

Table 2. Astronomical orientation results of the Matronentempel in Colonia Ulpia Traiana. The temple entrance is along the major axis, facing the Northern Major Lunistice.

Matronentempel Lat. 51° 40′N	A (±¼°)	h (±½°)	δ (±0.75°)	Event
major axis	38 ½°	0°	28.5°	NML
major axis	218 ½°	0°	−29.8°	SML
minor axis	128 ½°	0°	−23.2°	WS
minor axis	308 ½°	0°	22.2°	- - -
major *temenos* wall	38 ½°	0°	28.5°	NML
major *temenos* wall	218 ½°	0°	−29.8°	SML
minor *temenos* wall	131 ½°	½°	−24.2°	WS
minor *temenos* wall	311 ½°	0°	23.8°	SS

The major axis is oriented toward moonrise at the Northern Major Lunistice (NML) in winter, as well as towards the setting of the moon at the Southern Major Lunistice (SML) in summer (Figure 2). In contrast, the minor axis is aligned toward the sunrise at the Winter Solstice (WS). The latter is also the orientation of the minor *temenos* wall, as well as towards the Summer Solstice (SS). As can be seen, the orientation of the *Matronentempel* is extremely complex as it combines a Roman solar orientation and what is probable a Gallo-Germanic lunar orientation, something that has been documented in Gallo-Roman temples such as the *Lenus Mars* temple in *Colonia Augusta Treverorum* (Espinosa-Espinosa et al., 2016: 236-238). For the sake of brevity, we will focus our attention on the lunar orientation of the temple, in an attempt to understand its cultural significance.

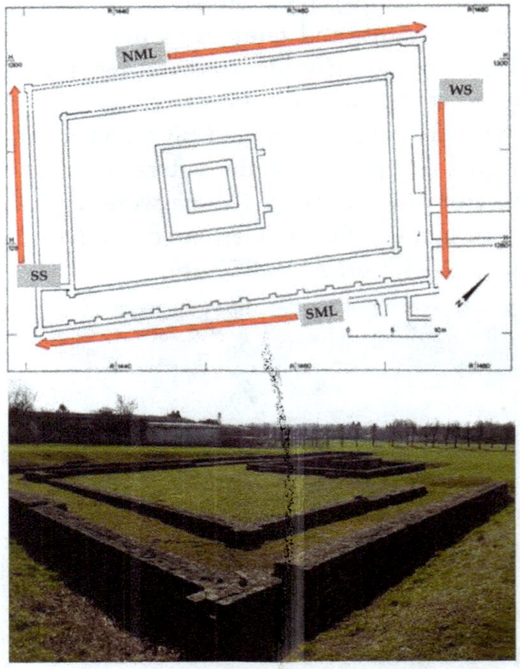

Figure 2. Matronentempel enclosure. Plan: Zille 1995: 107. Photo: David Espinosa-Espinosa.

Finally, as in the previous case, the so-called *Hafentempel* (harbour temple) also had a different orientation of the minor *temenos* wall in relation to the minor axis (Müller et al., 2008: 315; Schalles, 2014: 116-117). This building was a hexastyle Roman temple surrounded by columns (*peripteros*) with podium, adapted to the orthogonal orientation of the urban layout. The orientation of the *Hafentempel* enclosure, whose titular deity is unknown (presumably *Mars* or the deified Trajan) (Müller et al., 2008: 311-318), is as shown below (Table 3):

Table 3. Astronomical orientation results of the Hafentempel in Colonia Ulpia Traiana.

Hafentempel Lat. 51° 40′N	A (±¼°)	h (±½°)	δ (±0.75°)	Event
major axis	134°	½°	−25.5°	Moon?
major axis	314°	0°	25°	Moon?
minor axis	44°	0°	26°	Moon?
minor axis	224°	0°	−27°	Venus/Moon?
major *temenos* wall	134°	½°	−25.5°	Moon?
major *temenos* wall	314°	0°	25°	Moon?
SE minor *temenos* wall	41°	0°	27.5°	Moon?
SE minor *temenos* wall	221°	0°	−28.4°	SML
NW minor *temenos* wall	39 ½°	0°	28°	NML
NW minor *temenos* wall	219 ½°	0°	−29.1°	SML

The temple's major and minor axes, as well as the major *temenos* walls, have the same orientation as the urban layout. However, the minor *temenos* walls have a different orientation. According to the astronomical hypothesis, these orientations would coincide with moonrise on the Northern Major Lunistice (NML) during winter and towards moonset at the Southern Major Lunistice (SML) during summer (Figure 3), in the same way as the major axis and the major *temenos* walls of the *Matronentempel*. To date, no plausible explanation has been found for these differences in orientation. In any case, no indications of a previous sacral use of the area as a possible reason for this design have been found (Müller et al., 2008: 315). The answer to this problem is to try to understand this orientation pattern according to astronomical criteria based on religious reasons.

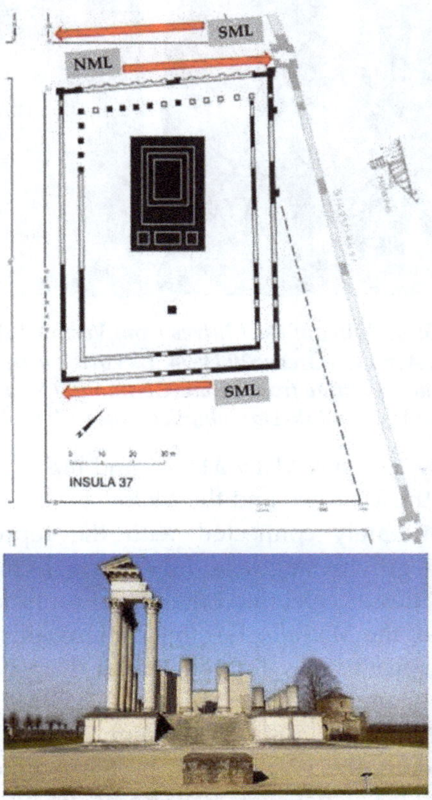

Figure 3. The Hafentempel enclosure. Plan: Müller et al., 2008: 314. Photo: David Espinosa-Espinosa.

4. DISCUSSION

In light of what we have seen in this paper, it is not easy to establish a reliable proposal to interpret these results. On the one hand, in the case of the urban orientation, the *decumanus maximus* of *Colonia Ulpia Traiana* was not oriented in direct compliance with the solar arc, but could be related to a cardinal orientation skewed by 45°. More speculatively, it could also be related with the apparent movement of the moon and/or Venus across the sky. This is both interesting and possibly intentional as the so-called *nova urbs* of *Italica* (Santiponce, Spain) in *Hispania Citerior Tarraconensis* has a quite similar orientation (González-García et al., 2014: 118). Located at a latitude of 37° 26′ N, the azimuth of the *decumanus* is 57 ¾° towards the eastern end and 237 ¾ towards the western parts, which translate into declinations of 25° and -24° respectively. Indeed, the orientations differ by some degrees, but it is remarkable that despite the difference in latitude and azimuth both towns seem to be facing similar parts of the sky, in any case slightly outside of the solar arc. This circumstance is highly significant and suggestive as *Italica* (Trajan's birthplace) became a Roman colony under Hadrian, his successor (Aul. Gell. *Noct. Attic.* 16, 13, 4). Unfortunately, due to the lack of references in ancient texts, we do not know what could have been the meaning of the moon and/or Venus as the evening star for the people of *Colonia Ulpia Traiana*, or with which political, religious, or seasonal event it was associated. However, based on the results from the *Matronentempel*, it is possible to suggest a likely working hypothesis, at least about the moon.

In this respect, it seems to be no coincidence that this temple was consecrated to the *Matronae*, probably the *Matronae Aufaniae* based on epigraphic remains dedicated to these goddesses (Figure 4) (Clauss, 1976: 30 nº 45; *AE* 1977, 566; Freigang, 1991: 119; Müller et al. 2008: 319-324). The *Matronae* (the "matrons"), identified with the *Matres* (the "mother-goddesses") (*contra* Garman, 2008: 34 and 85-86), were Celtic female deities venerated in Cisalpine Gaul, *Gallia Narbonensis*, *Tres Galliae*, the Rhineland, *Hispania* and Britain from the first to the fifth centuries A.D. (Thévenot, 1968: 165-176; Bauchhenss and Neumann, 1987: 1-30 and 242; Raepsaet-Charlier, 1993: 33-36 and 2015: 204-217; Derks, 1998: 119-130; Gómez Pantoja, 1999; Zaccaria, 2004; Garman, 2008: 17-19 and 29; Bermejo, 2014). They were depicted in many votive altars (asking for protection for the family, fertility or professional success, or giving thanks for being saved from danger or illness), reliefs and statuettes in groups of one, two or three in both sitting and standing position, dedicated by people of indigenous or Roman origin, including women and members of the civil and military Roman administration (Thévenot, 1968: 167-176; Garman, 2008: 44-45 and 68-69).

As regards the attributes and their accompanying symbols, the representation of the *Matronae* tends to emphasize general fertility and wealth, while the *Matres* are depicted with strongly reproductive imagery (Garman, 2008: 30 and 35). The *Matronae* appear holding baskets of fruit, loaves of bread, purses of coins, and weaving equipment, while the *Matres* hold children on their laps and napkins (as in the case of the *Matres* of Vertault, Côte-d'Or, France): the first one holds a child wrapped in swaddling clothes, the second one a baby's flannel blanket, and the third one a washbowl and a sponge (Figure 5: A) (Aldhouse-Green, 2004: 73-74). Another difference between them is that the *Matres* are shown as three women who are similar in age and dress, while depictions of the *Matronae* show them as being of different ages, garbed appropriately to their age and status (Aldhouse-Green, 2004: 74 and 203-204; Garman, 2008: 31). The latter is clear in the case of the *Matronae Aufaniae* from Bonn/Cologne: two middle-age women, wearing huge circular headdresses, flank a young girl with a footstool, bare head, and long flowing hair (Figure 5: B). They are seated on a high-backed bench covered by cushions and blankets. The two older women are wearing neck ribbons or necklaces with pendants shaped like half-moons (Aldhouse-Green, 2004: 203; Garman, 2008: 35). All three hold baskets of fruit or bread. However, regardless of these differences, the *Matres* and the *Matronae* share some common elements which are usual in the representations of the maternal figures of the Roman/Italic pantheon such as *Fortuna/Tyche*, *Iuno* and *Mater Matuta*: the *cornucopia* or horn of plenty, the *patera*, the globe and the rudder. In fact, *Fortuna* was depicted as a "nursing mother" (*kourotrophos*) in the statue of the baby Jupiter sitting together with his sister Juno on the lap of *Fortuna*, reaching for her breast (Cic. *De Div.* 2, 85), and her worship was related to that of *Mater Matuta* in the area of Sant'Omobono (Gagé, 1963: 25-26, 57 and 228-235; Champeaux, 1982a: 257-265 and 314-317). All of these elements appear to indicate that the *Matres* and the *Matronae* were related to fertility, prosperity, nourishment, the passage of time, fate, growth, and protection (Thévenot, 1968: 167-178; Champeaux, 1982a: 46; Aldhouse-Green, 2004: 74-76 and 203-205; Bermejo 2014).

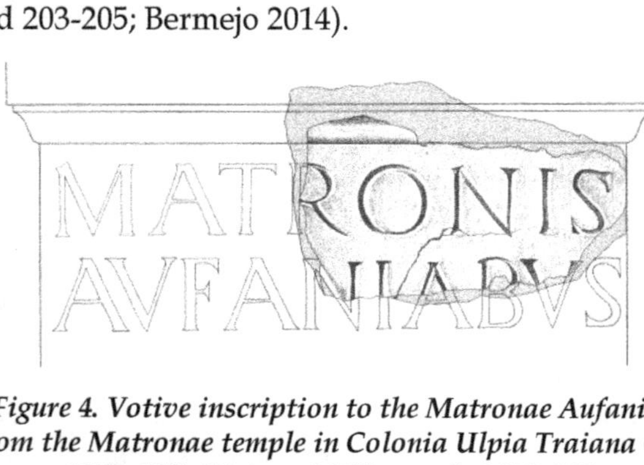

Figure 4. Votive inscription to the Matronae Aufaniae from the Matronae temple in Colonia Ulpia Traiana (AE 1977, 566). Picture: Müller et al., 2008: 323.

Having established the link between the *Matronae* and *Fortuna*, it is easier to understand the role that was probably played by the moon in the orientation of the *Matronentempel* in *Colonia Ulpia Traiana* and possibly, by extension, of the urban layout. In this respect, according to Leonardo Magini (2015: 27), *Fortuna* had lunar characteristics. In fact, this goddess, identified with the moon, facilitated the encounter between *Mars* and *Rhea Silvia* eclipsing the sun (Plut. *De Fort. Rom.* 8; *Rom.* 12, 5). Also, according to this scholar, the rapid succession of the moon's changing appearance is embodied by three divine figures in the Roman religion: *Fortuna* and her two handmaidens, *Necessitas*, who precedes her and *Spes*, the last of the goddesses (Hor. *Od.* 1, 35). They would be represented in the heavens respectively by the new moon, the waning moon, and the waxing moon, equated with *Fortuna caeca*, *Fortuna Respiciens* and *Fortuna Obsequens* (Magini, 2015: 27-31). These goddesses would have precisely played an important role as tutelary and benefactor deities (Champeaux, 1982b: 97-105 and 115).

Figure 5: A. Relief of the Matres from Vertault (France). Photo: Alhouse-Green, 2004: 74. B. Votive altar to the Matronae Aufaniae from Bonn (Germany) (AE 1930, 19). Photo: Epigraphik-Datenbank Clauss-Slaby EDCS.

Finally, *Fortuna*, *Mater Matuta* and *Iuno*, closely related to the *Matronae* and the *Matres* as shown above, were intimately connected with the reproductive cycle, the growth stages, and the moon. In this sense, both goddesses celebrated their feast on June 11, i.e., the day of the *Matralia*. For the Romans, this was the most auspicious day for conception (Gagé, 1963: 21 and 25; Dumézil, 1974: 66-72 and 343-344; Champeaux, 1982a: 250, 281-333; Magini, 2015: 47). Based on this date, the ideal day to be born coincides on March 9 (273.25 days later). In Rome, no public feast was held on that date, since the birth of a common citizen remained a private event. However, two feasts were held one *nundina* before and one *nundina* after March 9 (Macr. *Sat.* 1, 16, 36). In the *nundina* before birth, on March 1, pregnancy completes the 9th and enters the 10th lunar month. As Macrobius (*Sat.* 1, 8, 5) explains, this is the moment when "in the uterus the fertilized seed ripens into life in the 10th month, retained by delicate natural bonds until it issues into the light." This is the moment when, approaching the start of labour and all the associated perils for mothers and newborn infants, the feast of *Matronalia* consecrated to *Iuno Lucina* was celebrated, marking the start of the lunar year at the Calends of the first month of the year (Gagé, 1963: 5, 13 and 66-80; Dumézil, 1974: 302-310; Cid, 2007; Magini, 2015: 54). As noted by Georges Dumézil (1974: 302-303), "la déesse de l'enfantement preside aussi au début du mois, à la 'renaissance' de la lune." On the other hand, the *nundina* after birth, on March 17, was the moment when males born on March 9 underwent the ritual of purification and naming. This *dies lustricus* coincides with the feast of *Liberalia,* when the birthday of those born at the same time and in the same age group was celebrated, and fertility rites were made to *Ceres, Liber* and *Libera* (Dumézil, 1974: 382-385; Magini, 2015: 55).

Based on the above, the orientation of the *Matronentempel* in *Colonia Ulpia Traiana,* and also possibly the urban layout according to the apparent movement of the moon across the sky, may have become astronomical markers related to natural and human reproductive cycles associated with procreation and birth, among other aspects. As Ton Derks (1998: 55) has proposed for the northern part of Roman Gaul, religious concepts and practices could have been interwoven with specific characteristics of the landscape and the daily life of the majority of the population in the rural parts of Northern Gaul. This seems reasonable as most of the people that lived in this town would have been farmers and peasants. There is therefore a clear correspondence between all the data. Dispensers of fruits and nurses of newborn children, the beneficial action of the *Matronae* and the *Matres* would have been essential to human, animal and plant life. As *Fortuna, Mater Matuta* and *Iuno Lucina,* they would have had an immediate and specific impact on reproductive forces. The moon, included on the pendants that adorn the depictions of the *Matronae,* must have played crucial role in the myth and ritual action of their worship, as well as in the continuation, cohesion, control, and prosperity of the political and human community, to the point that its celestial movements may have served to define the orientation of the *Matronentempel* and the urban layout of *Colonia Ulpia Traiana.*

ACKNOWLEDGEMENTS

David Espinosa-Espinosa is a Postdoctoral Fellow of the Galician Regional Government (Spain), and A. César González-García is a Ramón y Cajal Fellow of the Spanish MINECO (Spain).

REFERENCES

Aldhouse-Green, M. (2004) *An Archaeology of Images. Iconology and Cosmology in Iron Age and Roman Europe,* London/New York, Routledge.

Bauchhenss, G. and Neumann, G. (1987) *Matronen und verwandte Gottheiten. Ergebnisse eines Kolloquiums veranstaltet von der Göttinger Akademiekommission für die Altertumskunde Mittel- und Nordeuropas,* Köln/Bonn, Rheinland-Verlag.

Bermejo, J. (2014) Un santuario a las *Matres* en el foro de *Arucci*: la constatación de las *Rixamae* en la *Baeturia Celtica. Revista Onoba* 2014/2, pp. 107-125.

Bertarione, S. V. and Magli, G. (2015) Augustus' Power from the Stars and the Foundation of *Augusta Praetoria Salassorum, Cambridge Archaeological Journal* 25/1, pp. 1-15.

Bettini, M. (2011) Missing Cosmogonies: the Roman Case?, *Archiv für Religionsgeschichte* 13/1, pp. 69-92.

Carandini, A. (2003) *La nascita di Roma. Dèi, lari, eroi e uomini all'alba di una civiltà,* Turin, Einaudi.

Champeaux, J. (1982a) *Fortuna. Le culte de la Fortune à Rome et dans le monde romain. I. Fortuna dans la religion archaïque,* Rome, École Française de Rome.

Champeaux, J. (1982b) *Fortuna. Le culte de la Fortune à Rome et dans le monde romain. II. Les transformations de Fortuna sous la République,* Rome, École Française de Rome.

Cid, R. M. (2007) Imágenes y prácticas religiosas de la sumisión femenina en la antigua Roma. El culto de *Juno Lucina* y la fiesta de *Matronalia, Studia Historica. Historia Antigua* 25, pp. 357-372.

Clauss, M. (1976) Neue Inschriften im Rheinischen Landesmuseum Bonn, *Epigraphische Studien* 11, pp. 1-39.

De Sanctis, G. (2012) 'Urbigonia'. Sulle tracce di Romolo e del suo aratro, *I Quaderni del Ramo d'Oro on-line,* Numero Speciale (Per un atlante antropologico della mitologia greca e romana), pp. 105-135.

Derks, T. (1998) *Gods, Temples and Ritual Practices. The Transformation of Religious Ideas and Values in Roman Gaul,* Amsterdam, Amsterdam University Press.

Dumézil, G. (1974) *La religion romaine archaïque avec un appendice sur la religion des Étrusques,* Paris, Payot.

Eck, W. (2014) Traians Herrschaftsbeginn in Germania inferior und seine Städtepolitik in dieser Provinz. In *Trajan und seine Städte. Colloquium Cluj-Napoca, 29. September – 2. Oktober 2013,* I. Piso, and R. Varga (ed.), Cluj-Napoca, Mega, pp. 101-109.

Espinosa-Espinosa, D. and González-García, A. C. (2017) *a. d. VIIII Kalendas Octobres, dies natalis Augusti.* Some Considerations on the Astronomical Orientation of Roman Cologne and the Imperial Cult, *Numen. International Review for the History of Religions* 64/5-6, pp. 545-567.

Espinosa-Espinosa, D., González-García, A. C. and García-Quintela, M. V. (2016) On the Orientation of Two Roman Towns in the Rhine Area, *Mediterranean Archaeology and Archaeometry* 16/4, pp. 233-240.

Freigang, Y. (1991) Das Heiligtum der Insula 20 in der Colonia Ulpia Traiana (Xanten), *Archäologische Informationen* 14, pp. 118-120.

Gagé, J. (1963) *Matronalia. Essai sur les devotions et les organisations cultuelles des femmes dans l'ancienne Rome*, Brussels, Latomus.

García-Quintela, M. V. and González-García, A. C. (2014) Le 1er août à *Lugdunum* sous l'empire romain: bilans et nouvelles perspectives, *Revue archéologique de l'Est* 63, pp. 157–177.

Gargola, D. J. (2017) *The shape of the Roman order: the republic and its spaces*, Chapel Hill, The University of North Carolina Press.

Garman, A. G. (2008) *The Cult of the Matronae in the Roman Rhineland: An Historical Evaluation of the Archaeological Evidence*, Lewiston, Edwin Mellen Press.

Gómez-Pantoja, J. (1999) Las madres de Clunia. In *Pueblos, lenguas y escrituras en la Hispania prerromana. Actas del VII Coloquio sobre lenguas y culturas paleohispánicas (Zaragoza, 12 a 15 de Marzo de 1997)*, F. Villar and F. Beltrán (ed.), Salamanca, Ediciones Universidad de Salamanca, pp. 421-432.

González-García, A. C., Noguera Celdrán, J. M., Belmonte J. A., Rodríguez Antón, A., Ruiz Valderas, E., Madrid Balanza, M. J., Zamora, E. and Bonnet Casciaro, J. (2015) *Orientatio ad sidera*: astronomía y paisaje urbano en *Qart Hadašt/ Carthago Nova, Zephyrus. Revista de Prehistoria y Arqueología* 75, pp. 141-162.

González-García, A. C., Rodríguez Antón, A. and Belmonte, J. A. (2014) The Orientation of Roman Towns in Hispania: Preliminary Results, *Mediterranean Archaeology and Archaeometry* 14/3, pp. 107-119.

González-García, A. C., Rodríguez-Antón, A., Espinosa-Espinosa, D., García Quintela, M. V., Belmonte, J. A. (2018) Establishing a New Order: the orientation of Roman Towns Built at the Age of Augustus. In *The Archaeoastronomy of the Roman World*, G. Magli, A.C. González-García, J.A. Belmonte and E. Antonello, (ed.), Springer, in press.

Humm, M. (2004) Le *mundus* et le *Comitium*: représentations symboliques de l'espace de la cité, *Histoire urbaine* 2004/2 (n° 10), pp. 43-61.

Le Gall, J. (1975) Les Romains et l'orientation solaire, *Mélanges de l'École Française de Rome. Antiquité* 87/1, pp. 187-320.

Magini, L. (2015) *Stars, Myths and Rituals in Etruscan Rome*, Cham, Springer.

Müller, M., Schalles, H.-J., Zieling, N. (ed.) (2008) *Colonia Ulpia Traiana. Xanten und sein Umland in römischer Zeit*, Mainz, Phillip von Zabern.

Raepsaet-Charlier, M.-Th. (1993) *Diis Deabusque Sacrum. Formulaire votif et datation dans les Trois Gaules et les deux Germanies*, Paris, De Boccard.

Raepsaet-Charlier, M.-Th. (2015) Cultes et territoire, Mères et Matrones, dieux 'celtiques': quelques aspects de la religion dans les provinces romaines de Gaule et de Germanie à la lumière de travaux récents, *Antiquité classique* 84, pp. 173-226.

Rodríguez-Anton, A., González-García, A. C., and Belmonte, J. A. (2018) An Archaeoastronomical Approach to Roman Urbanism: Orientation of Roman Settlements Across the Empire. *Mediterranean Archaeometry and Archaeology*, in press.

Rykwert, J. (1988) *The Idea of a Town. The Anthropology of Urban Form in Rome, Italy and the Ancient World*, Cambridge (Massachusetts), MIT Press.

Schalles, H.-J. (2014) Urbanisierung an der Nordgrenze des Imperium Romanum. Die Ausstattung der Colonia Ulpia Traiana mit öffentlichen Bauten, ihr Gründungsdatum und fie Präsenz des Kaisers. In *Trajan und seine Städte. Col-loquium Cluj-Napoca, 29. September – 2. Oktober 2013*, I. Piso, and R. Varga (ed.), Cluj-Napoca, Mega, pp. 111-137.

Šprajc, I. (2015) Alignments upon Venus (and other planets) - identification and analysis. In *Handbook of Archaeoastronomy and Ethnoastronomy*, C. L. N. Ruggles (ed.), New York, Springer, pp. 507-516.

Thevenot, E. (1968) *Divinités et sanctuaires de la Gaule*, Paris, Fayard.

Zaccaria, Cl. (2004) Alla ricerca di divinità 'celtiche' nell'Italia settentrionale in età romana. Revisione della documentazione per le Regiones IX, X, XI. In *Die Kelten und ihre Religion im Spiegel der epigraphischen Quellen. Akten des 3. F.E.R.C.A.N.-Workshops (Vitoria-Gasteiz, September 2000)*, J. Gorrochategui and P. de Bernardo Stempel (ed.), Vitoria, Servicio Editorial. Universidad del País Vasco, pp. 129-164.

Zille, M. (1995) Das Matronenheiligtum auf der Insula 20. In *Tatort CUT. Die Spur führt nach Xanten*, M. Zelle (ed.), Cologne, Rheinland-Verlag, pp. 106-118.

7

THE PREHISTORIC VILLAGES OF THE AEOLIAN ARCHIPELAGO AND MILAZZO: ASTRONOMY AND LANDSCAPE, PRELIMINARY RESULT

Andrea Orlando*[1], Sebastiano Tusa[2] and Davide Gori[3]

[1] *Institute of Sicilian Archaeoastronomy, Italy*
Fondazione Floresta Longo, Italy
[2]*Regional Department of Cultural Heritage and Sicilian Identity, Italy*
[3]*Wood plc, Italy*
Institute of Sicilian Archaeoastronomy, Italy

**Corresponding author: Andrea Orlando (orlando@archeoastronomia.com)*

ABSTRACT

The archipelago of the Aeolian Islands is located in the Tyrrhenian Sea, to North of the Sicilian coast. Since 2000 the Aeolian Islands are an UNESCO World Heritage Site. The human presence in the archipelago is attested since very ancient times (Bernabò Brea, 1958). The prehistoric people were in fact attracted by the presence of large quantities of obsidian, glassy substance of volcanic origin through which the Aeolian Islands were the center of flourishing trade routes (Tusa, 1999).
In this paper we analyze the positioning of the prehistoric villages of the Eolian Islands taking into account archaeoastronomy and the aspects of landscape archeology. In particular, the study is aimed at understanding if in the choice of the site used for the construction of the various villages, the profile of eastern or western horizons may have played a determining role or not. Here we present preliminary results of the study made with the satellite data (Google Earth) and data measured by the first measurement campaigns in the field (april 2017). The in situ measurement campaign will be completed in autumn 2018.

KEYWORDS Aeolian Islands, prehistoric villages, Bernabò Brea, Capo Graziano, Punta Milazzese, topographic orientation, astronomical orientation.

1. INTRODUCTION

The Aeolian Islands are a volcanic archipelago in the Tyrrhenian Sea (North of Sicily). The Aeolian Islands consisting of seven real islands, to which are added islets and rocks emerging from the sea. The seven islands are arranged in Y-shaped form, with the rod pointing westward (Figure 1). There are two active volcanoes, Stromboli and Vulcano; Milazzo instead, with its beautiful promontory, is considered the natural 'access door' to the Aeolian Islands.

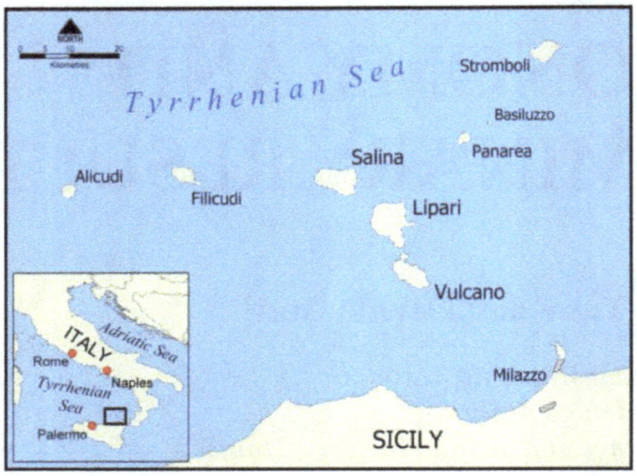

Figure 1. Map of the Aeolian archipelago with its position.

The seven islands are Lipari, Salina, Vulcano, Stromboli, Panarea, Alicudi and Filicudi. First of all, it is worth remembering that the chronology of Sicilian prehistory is different from the Italian chronology (Gullì, 2009), so we show a summary of the various periods in table 1. The prehistoric villages involved in the study are the following:

- to Lipari - *Castellaro* (Middle Neolithic) (Bernabò Brea and Cavalier, 1957) and *Acropoli* (Late Bronze Age) (Bernabò Brea and Cavalier, 1979);
- to Panarea - *Punta Milazzese-Cala Junco* (Middle Bronze Age) (Bernabò Brea and Cavalier, 1968);
- to Stromboli - *San Vincenzo* (Ancient-Middle Bronze Age) (Cavalier, 1981);
- to Salina - *Portella* (Middle Bronze Age) (Bernabò Brea and Cavalier, 1968);
- to Filicudi - *Filo Braccio* (Ancient Bronze Age) (Martinelli et al., 2010) and *Capo Graziano* (Ancient Bronze Age) (Bernabò Brea and Cavalier, 1991);
- to Milazzo - *Viale dei Cipressi* (Ancient Bronze Age) (Tigano, 2009).

A similar study about the orientation of a prehistoric village in Sicily was realized in Ustica island. As regards the *Villaggio dei Faraglioni* (Middle Bronze Age) was proposed that the site might have been selected because it allowed a quite privileged position enabling to follow the rising and setting of the Sun in the months close to the winter solstice (Martin and Magli, 2016).

Table 1. Chronological table of Sicilian prehistory.

Period	Years (BC)	Facies
Lower Paleolithic	2.5 millions	
Medium Paleolithic	120.000-40.000	
Upper Paleolithic	30.000-10.000	Epigravettian
Mesolithic	10.000-7.000	
Ancient Neolithic	6.500-5.500	
Middle Neolithic	5.500-4.500	Stentinello
Late Neolithic	4.500-3.500	Diana
Early Copper Age	3.500-3.000	San Cono-Piano Notaro-Grotta Zubbia-Malpasso-Serraferlicchio-
Late Copper Age	3.000-2.000	Sant'Ippolito
Ancient Bronze Age	2.000-1.450	Castelluccio
Middle Bronze Age	1.450-1.250	Thapsos
Late Bronze Age	1.250-1.150	Pantalica I
	1.150-1050	Pantalica II
Final Bronze Age	1.050-900	Pantalica III-Cassibile
Iron Age	900-734	Pantalica Sud-Finocchito
Beginning of the Greek colonization	734	Licodia Eubia

2. ARCHAEOASTRONOMICAL ANALYSIS: FIRST DATA

A first analysis using satellite data (Google Earth) was carried out in December 2017, while a first survey on Salina, Filicudi and Panarea islands was carried out in April 2017. For the prehistoric villages visited we show the panoramic view of the principal horizons with azimuth data obtained with a professional compass. For the others villages that are on Stromboli and Lipari islands and Milazzo promontory we will only do a few preliminary considerations.

2.1. *Salina island*

Salina is formed by six ancient volcanoes, it has the first and third highest reliefs of the archipelago. The *Portella village* is located in the North-East part of the island (Figure 2a). The prehistoric village is of the Middle Bronze Age. In the map of the village (Figure 2b) it is possible to see the huts and the chosen point (red hot) for the observations of the horizons. This point has an altitude of 197 meters about sea level.

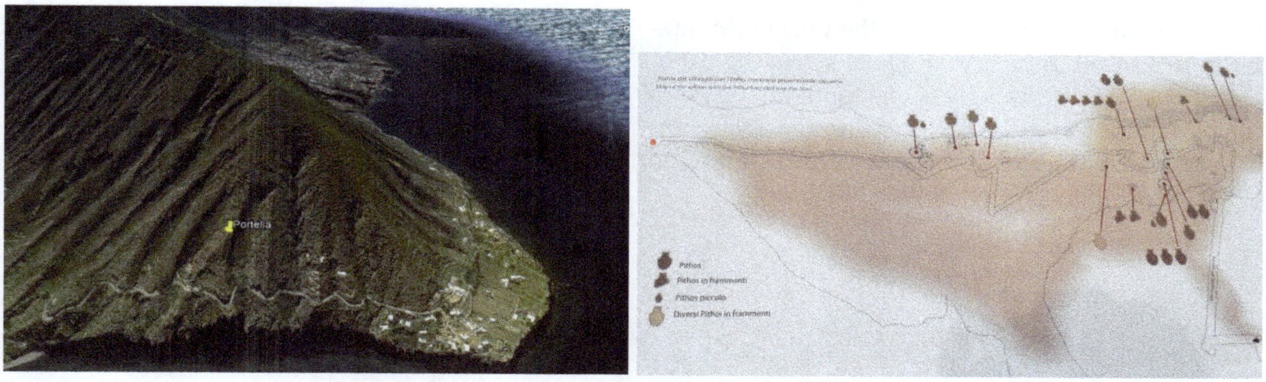

Figure 2. a (up) The position of the Portella village from Google Earth; b (down) map of the village with the position of the pithoi founded into the huts (from Martinelli, 2005).

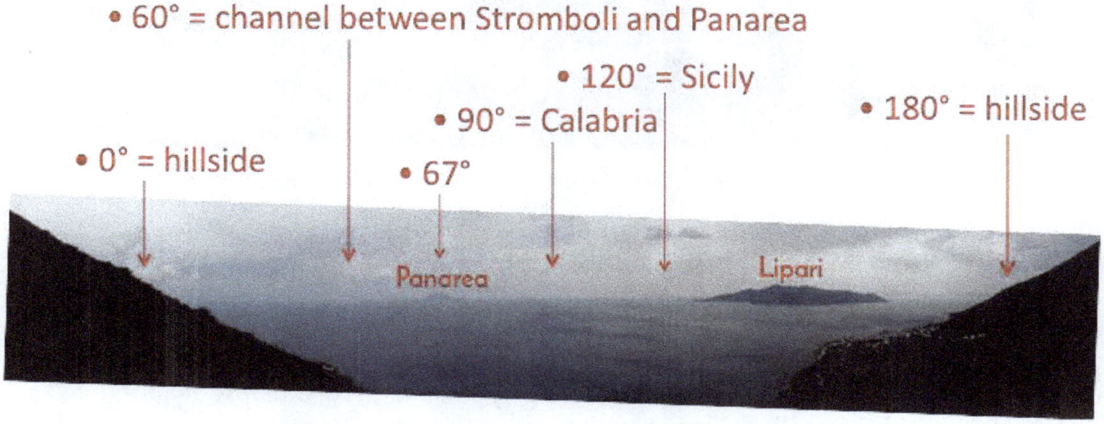

Figure 3. Panoramic view with azimuth data from Portella village (photo and elaboration by authors).

The azimuth data are shown in Figure 3. The observation point has been chosen considering the highest point of the village, at the height of the staircase that joins the various huts today visible.

From the Portella village the Eastern horizon is completely visible while the Western horizon is covered by the mountain behind the village (Fossa delle Felci Mount, 962 masl). As it is possible to see in Figure 3 there is one topographic point (the Panarea island) that is situated inside the solar sunrise range. From the horizon's observation, the azimuth of about 60° (linked to Stromboli-Panarea channel) is interesting, while the azimuth linked to the East and to the sunrise of the winter solstice are in correspondence of unspecified (covered by clouds) points on the Calabria and Sicily mountains.

2.2. *Filicudi island*

Filicudi is dominated by the Fossa Felci Mount, a volcano off high 773 meters. In the island there are two important prehistoric villages: *Capo Graziano* and *Filo Braccio*, both of the Ancient Bronze Age. Capo Graziano is located on the Montagnola promontory (South-East of the island) (Figure 4a), where there is a village with 27 huts (Figure 4).

In the map of the prehistoric village (Figure 4b) it is possible to see the huts and the chosen point (red hot) for the observations of the horizons. This point has an altitude of 102 meters about sea level.

Figure 4. a (up) The position of Capo Graziano village from Google Earth; b (down) map of the village (from Martinelli, 2015).

From the Capo Graziano village the Western horizon is partially visible while the Eastern horizon is covered by the mountain behind the village (Montagnola - 174 masl). As it is possible to see in the panoramic photo (Figure 5) there is one topographic point (Alicudi island) that is located inside the solar sunrise range and in particular it seems to be linked to the sunset at the equinoxes (Alicudi - Punta Stimpagnato channel). There are also a U-shaped depression and a V-shaped depression that could be interesting at the summer solstice sunset.

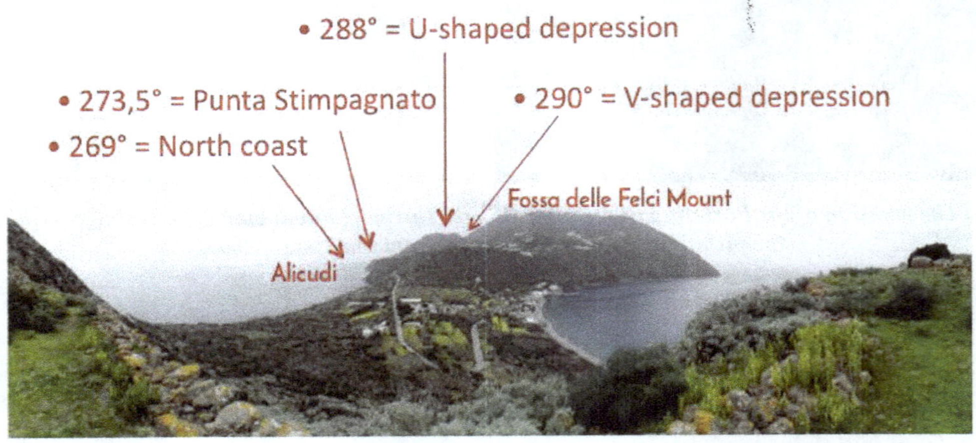

Figure 5. Panoramic view with azimuth data from Capo Graziano (photo and elaboration by authors).

The village of *Filo Braccio* extends on the coast of the peninsula of the Piano del Porto, that is on the Filo Braccio, one of the crags on the lava flows forming the island of Filicudi (Figure 6). It was excavated by two great archaeologists of the 20th century: Luigi Bernabò Brea and Madeline Cavalier, who dug up most of the Aeolian prehistoric villages between the 50s and 60s of the last century. In the map of the village (Figure 7) it is possible to see the huts and the chosen point (red hot) for the observations of the horizons. This point has an altitude of 12 meters about sea level.

Figure 6. The position of Filo Braccio village from Google Earth.

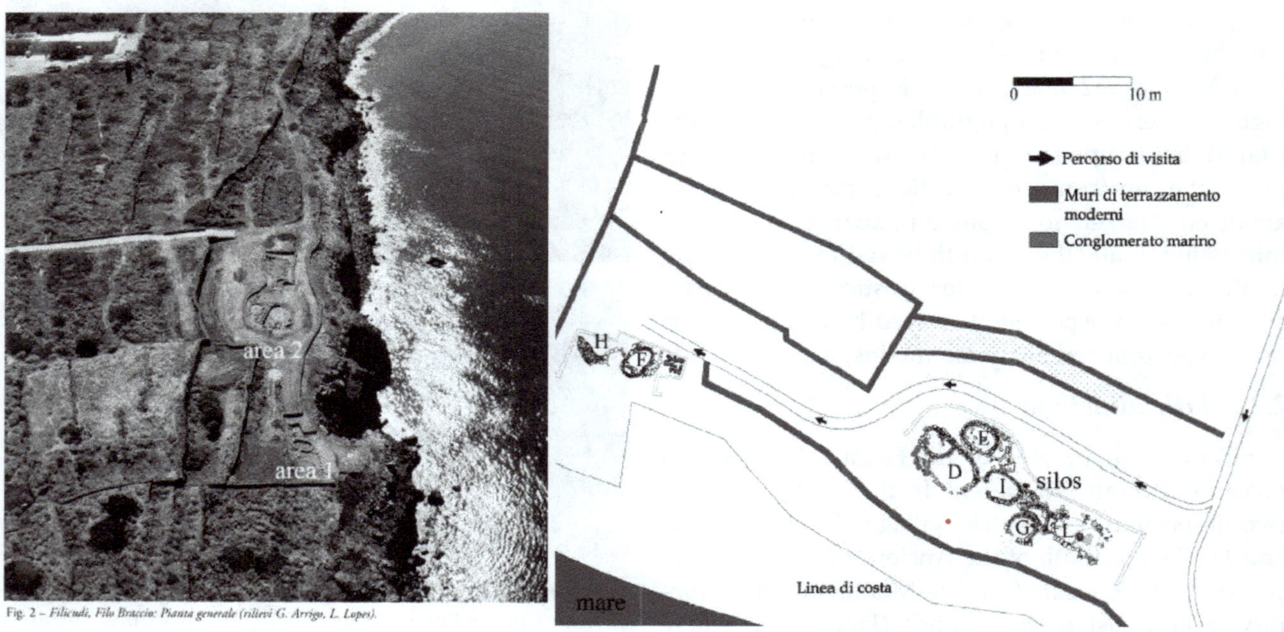

Figure 7. A (up) aerial view of Filo Braccio village (from Martinelli et al, 2010); b (down) map of the village (from Martinelli et al, 2010).

From the Filo Braccio village the Western horizon is partially visible while the Eastern horizon is covered by the mountain behind the village (Montagnola of Capo Graziano). As it is possible to see in Figure 8 there is one topographic point (Alicudi island) that seems to be a sunset marker for the equinoxes. There are also Punta Stimpagnato and a U-shaped depression that could be interesting (azimuth connected to the sunset of the equinoxes and the summer solstice respectively).

Figure 8. Panoramic view with azimuth data from Filo Braccio (photo and elaboration by authors).

2.3. *Panarea island*

Panarea is the smallest and the lowest island of the Aeolian archipelago. On the promontory of *Punta Milazzese*, at South-West of the island, there is the village of the Middle Bronze Age of Cala Junco (Figure 9a). The particular position of the little plateau, stretched out to the sea and protected by high walls facing the sea made it an ideal place for settlement.

In the village, where the remains of about twenty huts can be seen and visited, have been found materials of Mycenaean origins, bearing in mind the role-played, even in antiquity, by the Aeolian archipelago, at the heart of the main commercial routes of the Mediterranean Sea. In the map of the village (Figure 9b) it is possible to see the huts and the chosen point (red hot) for the observations of the horizons. This point has an altitude of 10 meters about sea level.

From the Punta Milazzese village the eastern and western horizons are visible. As it is possible to see in the panoramic photo (Figure 10) there are two topographic points (Dattilo islet and Punta di Lingua on Salina island), moreover there is a rocky peak near 270° that could be interesting, as linked to the sunset of the equinoxes. This site, with the characteristics of the horizons just exposed, may have been chosen precisely for its peculiarity to create a true calendar of the horizon.

Figure 9. a (up) aerial view of Punta Milazzese with the prehistoric village; b (down) map of the village (from Martinelli and Mastelloni, 2015).

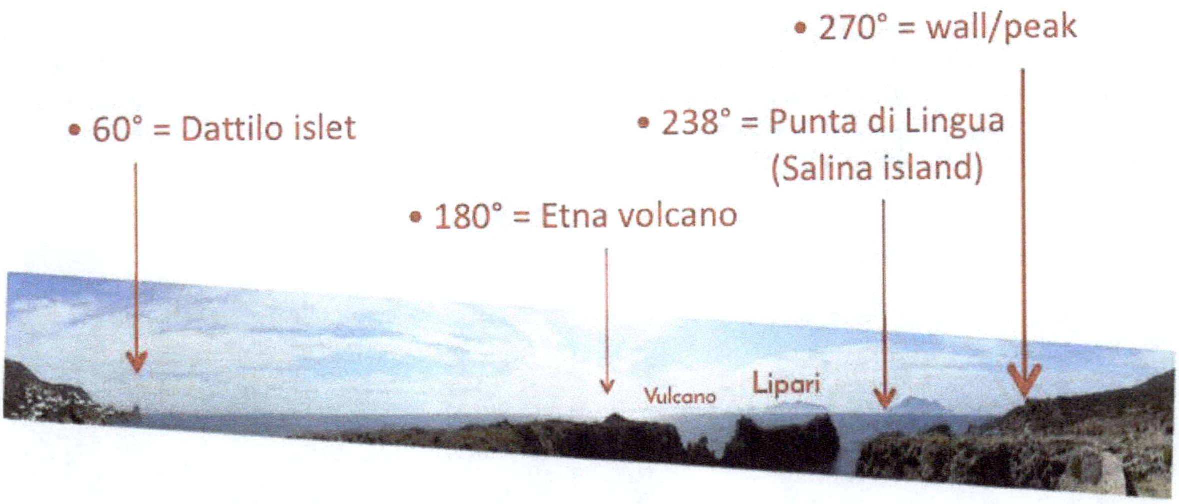

Figure 10. Panoramic view with azimuth data from Punta Milazzese (photo and elaboration by authors).

2.4. *Stromboli island*

In Stromboli we find the San Vincenzo village (Ancient/Middle Bronze Age) (Figure 11), which has been excavated in recent years and was then covered (Cannavò et al., 2017). The eastern horizon is visible, and it is possible to see the Calabria coast, while the western horizon is mainly covered by the volcano cone (the highest point is 926 masl). The field surveys will be carried out in autumn 2018.

Figure 11. The position of Filo Braccio village from Google Earth.

2.5. *Lipari island*

Lipari is the main island of the archipelago. It has two prehistoric sites, one situated in the western part of the island, the *Castellaro site* (of the Middle Neolithic Age), where excavations of the last years have returned lithic and ceramic materials, while the huts have never been traced; the other site is the *Acropoli*, situated in the eastern part of the island, on the hill that dominate the city center of Lipari (Figure 12).

Figure 12. The positions of Filo Castellaro and Acropoli villages from Google Earth.

Today on the *Acropoli* (Figure 13) it is possible to see the prehistoric village of the Late Bronze Age, but the site is surrounded by modern buildings, including the archaeological museum, the most important of Sicily, along with that of Syracuse. Using satellite data we will study only the eastern horizon, because the western is covered by hills.

Figure 13. Aerial view of Acropoli village from Google Earth.

2.6. *Milazzo*

And finally we will consider the Ancient Bronze Age village which is located at Milazzo in Viale dei Cipressi, in eastern part of the promontory (Figure 14). For about 5 years a tensile structure has been installed to protect the village, which has a particular hut (Figure 15), in fact some studies have emphasized its ritual character. The main's hut (called 1) entrance bisector has an azimuth of 202°, a value out of the solar range. In addition to the orientation of the village, the orientation of the entrance of this hut will be studied.

Figure 14. The position of Viale dei Cipressi village from Google Earth.

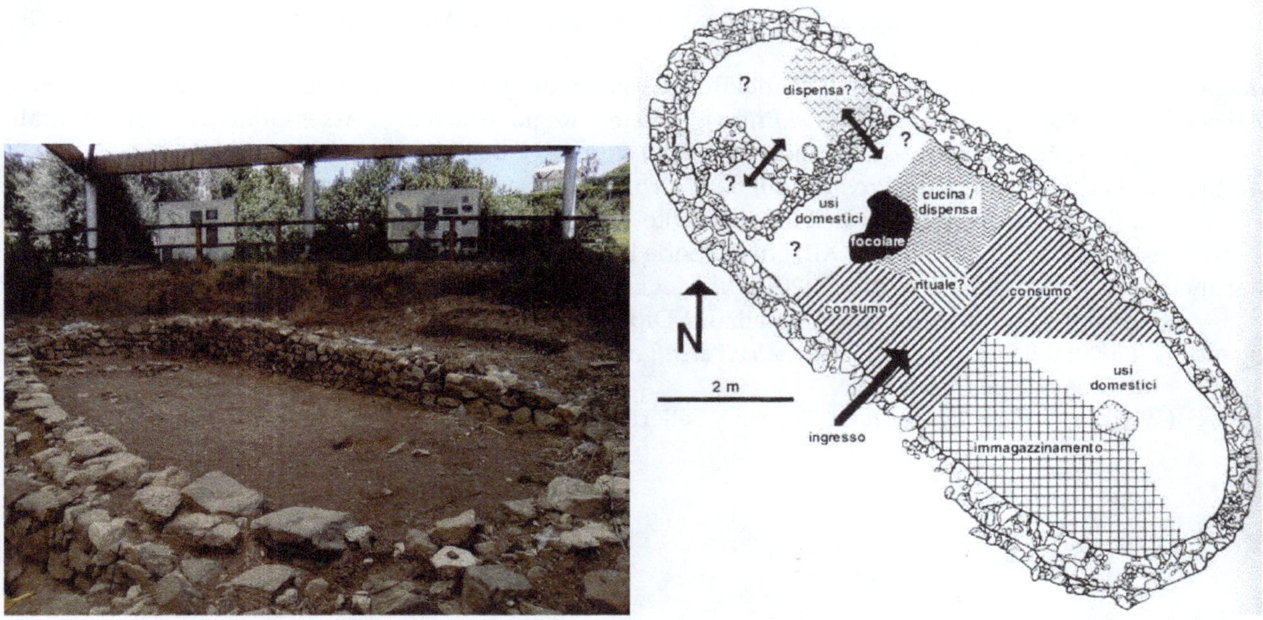

Figure 15. a (up) Milazzo, Viale dei Cipressi village: the main hut; b (down) relief of the main hut (from Tigano, 2009).

3. CONCLUSIONS

In conclusion we can do some very preliminary observations:

a) first of all the horizon profiles are in most cases characterized by the topography of the islands. Observe markers linked to the landscape and to the apparent movement of the Sun during the year is in many cases evident, in fact the rising and setting of the Sun at the equinoxes and solstices is often linked to topographic points;

b) Panarea's village seems to have a complete "horizon calendar" (from East to West), which could lead to thinking that the choice of the site has been conditioned by this aspect as well;

c) the villages of Filicudi have interesting western profile, where there seems to be a rather evident marker (Alicudi island) for sunset observation to equinoxes.

It should also be noted that until recently, only an archaeoastronomical study on a Sicilian prehistoric village was carried out, ie the Bronze Age village of Ustica (Martin and Magli, 2016). The preliminary study proposed on prehistoric villages of the Aeolian Islands can open new scenarios of scientific research, for example to understand if the choice of sites for the construction of the villages was also dictated by astronomical knowledge. The in situ measurement campaign that we will carry out in the autumn 2018 will allow us to complete the study and conclusions, also considering a possible astronomical orientation of the urban layout of each prehistoric villages present in the Aeolian archipelago and in Milazzo.

ACKNOWLEDGEMENTS

We want to thank Dr. Maria Amalia Mastelloni, ex Director of the Archaeological Eolian Park, for her availability and cultural sensitivity. We also thank the Milazzo Refinery (RAM) for the patronage of the scientific and cultural initiative.

REFERENCES

Bernabò Brea, L. (1958) *La Sicilia prima dei greci*. Milano, Il Saggiatore.

Bernabò Brea, L. and Cavalier, M. (1957) Stazioni preistoriche delle Isole Eolie, *Bullettino di Paletnologia Italiana*, N. S. XI - Vol. LXVI, Roma.

Bernabò Brea, L. and Cavalier, M. (1968) *Meligunìs Lipàra III. Stazioni preistoriche delle isole Eolie. Panarea, Salina, Stromboli*. Palermo.

Bernabò Brea, L. and Cavalier, M. (1979) *Meligunìs Lipàra IV. L'Acropoli di Lipari nella preistoria*. Palermo.

Bernabò Brea, L. and Cavalier, M. (1991) *Meligunis Lipara VI. Filicudi: insediamenti dell'età del bronzo*. Palermo.

Cannavò V., Bettelli M, Di Renzoni A., Ferranti F., Levi Sara T., Ollà A. e Tigano G. (2016) – *San Vincenzo - Stromboli (Lipari, ME). Campagne 2015-2017*. Notiziario di Preistoria e Protostoria - 2017, 4.III, pp. 91-99.

Cavalier, M. (1981) Villaggio preistorico di San Vincenzo. Stromboli, *Sicilia Archeologica*, N.46-47, pp. 27-54.

Foresta Martin, F. and Magli, G. (2016) - Astronomy and landscape at the prehistoric settlementVillaggio dei Faraglioni, Ustica, Sicily. Mediterranean Archeology and Archaeometry, Vol. 16, No. 2, pp. 167-172.

Martinelli, M.C. (2005) Il villaggio dell'età del Bronzo medio di Portella a Salina nelle Isole Eolie, *Origines*.

Martinelli, M.C. (2015) *Filicudi nell'età del Bronzo*. Palermo: Regione siciliana, Assessorato dei beni culturali e dell'identità siciliana, Dipartimento dei beni culturali e dell'identità siciliana.

Martinelli, M.C., Fiorentino, G., Prosdocimi, B., D'Oronzo, C., Levi, S.T., Mangano, G., Stellati, A. and Wolff, N. (2010) Nuove ricerche nell'insediamento sull'istmo di Filo Braccio a Filicudi. Nota preliminare sugli scavi 2009, *Origini* XXXII, Nuova Serie IV, pp. 285-314.

Martinelli, M.C. and Mastelloni M.A. (2015) *Il Museo Archeologico*. Palermo: Regione siciliana, Assessorato dei beni culturali e dell'identità siciliana, Dipartimento dei beni culturali e dell'identità siciliana.

Tigano, G. (2009) *Mylai II - Scavi e ricerche nell'area urbana*. Soprintendenza Beni culturali ed Ambientali di Messina, Sicania, Messina.

Tusa, S. (1999) *La Sicilia nella preistoria*. Palermo, Sellerio Editore.

8

SURVEY, ARCHAEOASTRONOMY AND COMMUNICATION: THE MAUSOLEUM OF GALLA PLACIDIA IN RAVENNA (ITALY)

M. Incerti[1], G. Lavoratti[2], S. D'Amico[1] and S. Giannetti[1]

[1]*Architecture Department of Ferrara (Italy)*
[2]*Architecture Department of Florence (Italy)*

Corresponding author: manuela.incerti@unife.it

ABSTRACT

The Mausoleum of Galla Placidia (built pre-450), is one of Ravenna's UNESCO protected monuments, renowned worldwide for the extraordinary mosaic decorations that cover its internal surfaces. The famous starry vault profoundly engages and inspires the observer. Its accurate representation of the real sky has been analysed and described, as has its mystical and symbolic meaning in relation to the iconographic tradition of the time.

Archaeoastronomical research was carried out on the building by Giuliano Romano, who measured its orientation. The azimuth value of 180.2° highlights a north-south bearing, unique among the other Byzantine buildings in Ravenna. This paper also examines other architectural elements beyond the orientation, with particular attention paid to the building's small slit windows, investigating their possible archaeoastronomical significance.

A functional 3D model was developed from the archaeoastronomical analysis data to display the original morphology of the building (the floor was about 140 cm lower due to subsidence), and astronomical phenomena, and to provide a multimedia way of communicating the scientific content produced. This final part of the contribution forms part of the trials conducted by the research group into interactive and non-interactive multimedia communication based on virtual models as an edutainment tool for the enjoyment of cultural sites and artefacts.

KEYWORDS: Galla Placidia, architectural survey, archaeoastronomy, edutainment, digital museum, dome

1. INTRODUCTION (M.I.)

The small, beautiful monument in Ravenna is attributed to the Roman empress Galla Placidia (c. 390 - Rome 450), the daughter of Theodosius I, and was very probably one of two sacella located at the extreme north and south of the narthex of the Basilica di Santa Croce (c. 417-425). The religious complex, built in several stages (David, 2013), bears witness to Augusta's ardent devotion to the relic of the True Cross, discovered by Helena, the mother of the emperor Constantine I, who commissioned the Basilica di Santa Croce in Gerusalemme in Rome.

Today the mausoleum is a small, isolated building in a Latin Cross shape, with wings covered with barrel vaults and a dome at their intersection. This study reflects on various aspects of the building, based on survey data collected especially for the research.

2. THE ARCHITECTURAL SURVEY (G.L.)

The survey involved the use of a FARO Focus 3D scanner and a digital SLR camera on a tripod. In a few days of work, this gathered all the data required to create a digital reconstruction of the monument. Recording the 23 point clouds and aligning the 459 photographs produced a morphometrically and chromatically reliable model, from which horizontal and vertical sections passing through particularly important points were then extracted. These were used to analyse the measurements and geometry of the building and its decoration. Integrating the digital photogrammetry with the laser scan, now a widely used surveying technique, turned out to be essential in this case, as documenting the starry skies located on the intrados of the roofing system was one of the main aims of the whole process.

3. THE TWO STARRY SKIES (M.I.)

The mausoleum contains two starry skies, which differ greatly both in their appearance and their meaning: the dome at the intersection of the transept is decorated with a sky filled with 567 golden stars (Figures. 1, 2.1, 2.2, 2.3), while the barrel vaults of the main nave and the apse have a pattern that alludes to a more geometric and abstract starry sky (Figures. 1, 2.1). The dome, in particular, has been analysed in some detail in the past, described both as representing the real sky and a mystical, symbolic sky, in relation to the iconographic tradition of the time. Remaining in this area of research, it is interesting to compare certain aspects of the representation of stars in the mausoleum with other major archaeological sites and, in particular, to investigate the geometric rules that underpin the morphology of the stars and their distribution in space.

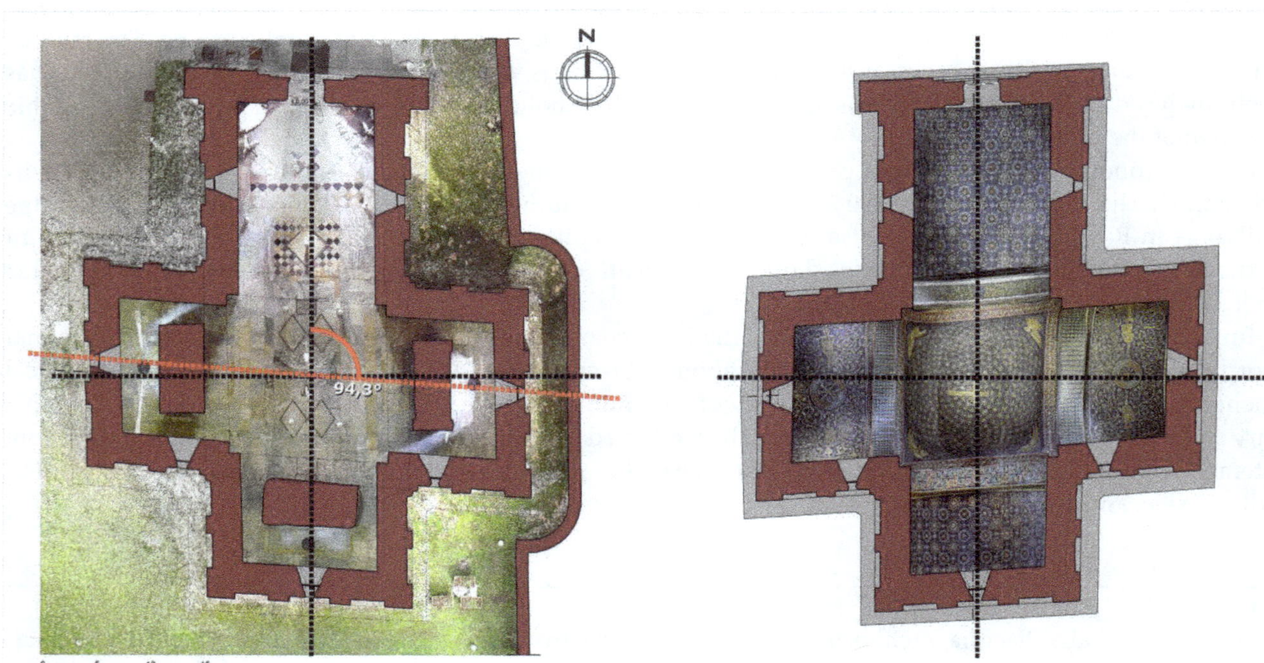

Figure 1. Plan of the ground floor, +1.20 meters (material restitution from the pointcloud) scale 1:50; plan of the dome (material restitution by photomodelling software, M. Incerti) scale 1:50 (drawings by G. Lavoratti).

Figure 2. The stars of Galla Placidia (1-2-3, ph. M. Incerti); Baptistery of Dura Europus (4, ph. Marsyas); Mithraeum of Dura Europus (5, by Swift, and Alwis); Baptistery of San Giovanni in Fonte (6, ph. M. Cyron); Santa Maria di Casaranello (7, Ph. M. Santi Amantini); Diagram of the stars A: 45° pattern; Diagram B: 90° pattern.

3.1. Barrel vault

The pattern here is based on the alternating repetition of two figures with an octagonal form (and which therefore can be traced back to the 'rotated square') which allude to two types of multicoloured star/flower: the first has a compact corolla and a central circle in warm colours, while the second has eight long, slender, golden, tripartite leaves and cold colours at the centre (Fiures. 1, 2.1). The stars follow the direction of the arch of the vault, and are staggered so as to expand the theme of the eight directions in the space: a row of eight stars of the first type is followed by a row of nine of the second type. The large stars alternate with small white flowers. The dominant lines of the square mesh pattern obtained thanks to the alternation between the two main figures are those at 45° to the walls (Figure 2.A), and evoke the precious nature of oriental silks and fabrics (Ranaldi, 2011, p. 23)[1]. The design of the alternating rows of stars, intentionally drawn perpendicularly to the axis of the nave, is interrupted asymmetrically and abruptly at the arches that connect with the dome due to the morphology of the space (a parallelogram), while in the apse the direction of the alternating rows of stars is in line with the back wall.

3.2. Dome

The surface is covered with 567 eight-pointed golden stars (Figures 2.1, 2.2, 2.3), the dimensions of which get smaller as one moves from the impost to the keystone: this is undoubtedly a figurative device used to expand the perspective effect enjoyed by the viewer, in other words making the space seem higher. Each individual star comprises a circular golden nucleus from which eight curved rays extend. The bright figures stand out against the deep blue background, which is made from tesserae laid out in well-defined circles that, although they are not visible to the naked eye, nevertheless evoke the expansive motion of the light.

Comparing the Galla Placidia golden stars with iconography of the same period immediately reveals the unique and innovative features of the work.

Although they are based on the geometry of a rotated square (eight rays), like almost all of the stars listed in Table 1, the Galla Placidia stars have an unbroken structure, with no gaps between the various elements (the centre, the rays and outer points), which are generally 'separate' in the other examples, probably to suggest the brightness of the stars and the expansive movement of the light. Here, meanwhile, this visual effect is achieved by using infinite, exquisite gold tesserae, laid out to form a compact and clean image.

From the orthophoto of the vaulted surfaces, obtained using digital photogrammetry, it can be deduced that the geometric structure of the starry sky comprises five concentric circumferences, which are used as the starting points for spherical curves. This is undoubtedly the most complex geometric layout of all the examples listed in the table.

---

[1] On the same topic, see the patterns of the mosaics in the church at Kartmin monastery, which also contains a jewelled cross set against a starry sky (with stars based on a square form) (Hawkins, Mundell, and Mango, 1973).

Tab. 1. The starry skies (M. Incerti).

Building	Medium	Architectural element	Number of rays	Rays/central part	Small circles outside the ray	Colours	Layout	Central element	Other elements
Galla Placidia Ravenna, before 450	Mosaic	Barrel vault		Multicoloured flowers	-	Multicoloured	Quadrangular mesh pattern at 45°, obtained using staggered rows: "n" type 1/"n-1" type 2 *diagram fig. 2.A*	-	-
Galla Placidia Ravenna, before 450	Mosaic	Dome on pendentives	8	Circle with rays extending from it	-	Gold	See description in text	Latin Cross (transept axis)	Corners: 4 beings of the Apocalypse
Dura-Europos, Baptistery, first half of 3rd century	Painting	Barrel vault/flat ceiling	8	Linear rays detached from the central circle (type 1, large) Pointed rays detached from the central circle (type 2, small)	Between rays (type 1)	White/ clear	Quadrangular mesh pattern at 45°, obtained using staggered rows: "n" type 1/"n-1" type 2 *diagram fig. 2.A (reconstruction by Yale University)*	Moon?	-
Dura-Europos, Mithraeum	Painting	Barrel vault	8	Linear rays detached from the central circle (type 1, large) Linear rays detached from the central circle (type 2, small)	Between rays (type 1 and 2)	White/ clear	Quadrangular mesh pattern, type 1/type 2 (same row), rows aligned with the directions of the vault *diagram fig. 2.B (reconstruction by Yale University)*	-	-
Dura-Europos, Synagogue	Painting	Wall	8	Linear rays detached from the central circle	At the end of the ray	White/ clear	7 stars near the crescent moon, on the side opposite the sun	-	Moon, Sun
Battistero di San Giovanni in Fonte, Naples, early 5th century (before Galla Placidia)	Mosaic	Dome	8	Petal-style rays and different/same colour central circle	At the end of the ray	White/ blue yellow red (centre)	Circular, just for the 2 outer rows Blue sky	Alpha and omega cross monogram	Corners: 4 beings of the Apocalypse (Bishop Sotere, after Galla Placidia)
Sepolcro Canosa Arcosolium al (first half of 5th century)	Painting	Small barrel vault	8	Petal-style rays separated from a central circle of the same colour	At the end of the ray	Alternating yellow/white	Quadrangular mesh pattern at 45°, obtained using staggered rows: "n" type 1/"n-1" type 2, with alternating colours *diagram fig. 2.A* Blue sky	Fragment of a clipeus	
Santa Maria di Casaranello, Lecce, 5th-6th century	Mosaic	Dome on pendentives	6	Petal-style rays detached from a central circle of the same colour	At the end of the ray	Alternating white/ clear and gold	4 concentric rows on 3 strips of sky, with fading colours	Latin Cross (church axis)	Corners: 4 beings of the Apocalypse
Battistero di Albenga, Savona, c. 450	Mosaic	Barrel vault	8	Central circle with linear rays extending from it	At the end of the ray?	White	Quadrangular mesh pattern: 5 rows on each side, starting from the barrel vault impost	Christogram + alpha and omega (x 3) + 12 doves	12 doves, symbolising the 12 apostles
Sant'Apollinare in Classe, Ravenna, 549	Mosaic	Apse semi-dome	6	Central circle with linear rays extending from it	At the end of the ray	White and gold	Irregular circular layout	Jewelled cross + alpha and omega	
Santa Maria Mater Domini in Vicenza (5th century)	Painting	Dome on pendentives	6 and 8	Rosettes surrounded by a circle + 8-pointed stars	-	Red/yellow (8-pointed star)	Currently illegible	Red circle?	Sun(?), moon Corners: 4 beings of the Apocalypse

The complicated 45° quadrangular mesh pattern of the barrel vault can actually be traced directly onto the surface by dividing the generator arch into equal parts. Drawing horizontal lines (Albenga and Dura-Europos) or a series of concentric circumferences (Naples and Casaranello) is equally simple, usually confined to the upper part of the dome, and never found in the pendentives or the joins between the semicircle dome and the quadrangular volume beneath (tab. 1). The creators of the small mausoleum's renowned eye-catching and profoundly moving starry ceiling set themselves the challenging aim of creating a geometrically continuous design, perceptively similar to a ribbed vault, decorating the entire surface all the way up to the joining surfaces (the pendentives) with orderly stars. This must have required an excellent knowledge of geometry and projection, as will be illustrated below.

At the top, inside the first circle, there are seven stars laid out around a cross, of which three are at the bottom right, three are at the bottom left, and one is at the top left[2]. The glorious, large Latin Cross in the centre is, unusually, at 90° to the axis of the mausoleum: this golden symbol can undoubtedly be traced to the worship of the True Cross (to which the

[2] The same number of stars found in the Dura-Europos synagogue (Table 1) which has been posited as referring to the seven planets, but which could instead relate to various constellations.

church is dedicated)[3], but also to the symbol of *Christus Victor*: indeed, the long arm faces eat-west, thus forcing the observer to turn their back to the west and look east to see it in its correct orientation. The layout of this cross supports and gives significant weight to the theory that it represents the sun, something seen in the depictions of 'Christ the Sun' in the Cappella di Sant'Aquilino in the Basilica di San Lorenzo (Milan), attributed to Galla Placidia[4], and in the Mausoleo dei Giulii (Rome)[5]. Indeed, both representations can be traced to the worship of the *Sol Invictus*, celebrated at the winter solstice (*Dies Natalis Invicti*) and promoted by Emperor Aurelian, who in 274 dedicated a temple in Rome to this divinity (Salles, 1997).

The numerous symbolic images in the mausoleum include the depictions of the four beings of the Apocalypse (on the four pendentives) and Christograms surrounded by wreaths in the transept's two barrel vaults (on the east and west sides). The latter are the Christian chi-rho symbol, formed from the letters X and P, the spread of which is tied to 'Constantine's dream' (Eusebius, 2009), where this precise symbol appeared to the emperor above the midday sun (*In hoc signo vinces*). The Christogram is also found on various Galla Placidia coins (inside a laurel wreath, the symbol of victory). Another symbol that recurs in the coinage that can be traced to her is a large Latin Cross, sometimes with a star on top (Gerke, 1966). This makes an astronomical event linked to the death of the Empress, described in Andreas Agnellus' *Liber Pontificalis Ecclesiae Ravennatis* (9th century), all the more interesting: according to the text, an extraordinarily bright star shone in the sky after Galla Placidia died. Yet another astral symbol associated with the woman who commissioned the small star-covered treasure (*Liber Pontificalis Ecclesiae Ravennatis, Vita Johannis*, 42).

4. FORM AND ARCHAEOASTRONOMY (M.I.)

The irregular forms of this piece of architecture, which are particularly clear in the plan, cannot be ascribed to a lack of care in their creation, but instead must be attributed to a precise compositional aim in the tracing of the axes, which are clearly not perpendicular to one another.

Giuliano Romano (Romano, 1995) carried out archaeoastronomical research into the building, including measuring its orientation: the value of the principal azimuth, equal to 180.2°, reveals a north-south bearing that is highly unusual compared to other sacred Byzantine buildings in Ravenna. The transept is not perpendicular to the nave; it has a bearing of 94.3° (Fig. 1). The azimuth of the transept was aligned with the sunset on 26-27 March and 13-14 September at the time of its construction (before 450 AD). The former date is very close to the 25 March, the Annunciation, a feast day documented in the fourth- or fifth-century *Martyrologium Hieronymianum*. 14 September, meanwhile, is the date of the Feast of the Cross, to which the main church in the complex is dedicated.

A further aspect of the research involves the intentional creation of light effects on significant dates of the astronomical or liturgical year. The building's narrow windows were reopened as part of the restoration work carried out during the last century (Iannucci, 1995) and are divided up as follows:

- 7 windows in the lower level (2 east, 3 south, 2 west);
- 3 windows in the lunettes (1 east, 1 south, 1 west);
- 4 windows in the tholobate (1 east, 1 south, 1 west, 1 north).

The light that comes from these openings is today dimmed by the wonderful alabaster slabs donated to the building by King Victor Emmanuel II in 1911.

As the material they are made of is only partially transparent, the light conditions, while undoubtedly captivating, are completely different from how they were originally, and with them the way the colours are seen: the sun's rays are highly filtered, and it is no longer possible to observe their progress over the surfaces as time passes. Using 2D and 3D simulations limited to the south-facing windows, it is possible to establish that at midday on the winter solstice a beam of light illuminates the door of the ancient sacellum at a height of -1.4 m (lunette and vault windows, Figure 3)[6], while the sun starts to enter at midday on the equinoxes.

The choice of this date to create a hierophanic light effect is undoubtedly linked to the iconography

[3] For information on the sources and iconography behind the legend of the True Cross in Europe from the 4th to the 15th century, see Baert Barbara, 2004.

[4] Ranaldi, 2011, p. 24. Various archaeoastronomical inferences and important examples of astral iconography have been uncovered in the Cappella di Sant'Aquilino: the Roman portal, from left to right, depicts the Sun, Jupiter, Mars, Victory or Nemesis, Venus and the Moon.

[5] See also Christ as the sun god Helios/Sol Invictus driving a carriage (a third-century mosaic in the Vatican Grottoes beneath St. Peter's Basilica, on the floor of the tomb of Pope Julius I).

[6] The restoration work (Iannucci, 1995; Ricci, 1914) allowed the original level of the flooring to be identified, 1.4 m below the current level. Due to major subsidence (which caused the raising of the groundwater level) it was no longer possible to return to the initial heights and, for this reason, the proportions of the current space are profoundly different from the original design.

on display, supporting the hypothesis that the effect could be attributed to the worship of the *Sol Invictus* in its relationship with the theme of the 'solstice gates of heaven': in Roman tradition the guardian of the gates (*ianuae*) and passageways (*iani*) was the double-fronted god Ianus (*Janus*).

Figure 3. Rendering of the interior of the original space (the floor has been lowered by 1.4 meters compared to today's level); on the left: view from the apse to the door (from south to north); on the right: perspective view from the top to the floor (the door is in the upper part of the image). At midday of the Winter Solstice, a ray of light illuminates the door (south facing windows of the second and third level, north-south apse, render by S. Giannetti).

5. THE GEOMETRY OF THE STARRY DOME (S.G.)

The geometry of the dome of the Mausoleum of Galla Placidia is difficult to interpret due to the lack of structural elements and its characteristic rounded corners, which hide the generatrices and directrices of its surface.

It was possible to get a better idea of its geometry from further analysis of the laser scanner survey, which involved building a mesh model obtained by triangulating the point cloud. The following issues were encountered:

1. The dome comprises a slightly raised spherical cap;
2. The dome sits on a parallelogram at the impost;
3. Analysing the parallel sections of the dome, perpendicular to the axis, reveals an unclear course, that seems to find an average between the parallelogram of the impost and the circumference at the top.

Figure 4. The orthogonal projection extracted from the textured mesh model produced during the photo modelling process. The stars do not lie on parallel planes, but instead on courses that can be defined as "festooned". These curves are obtained from the intersection of the cones and the overall surface of the dome (S. Giannetti).

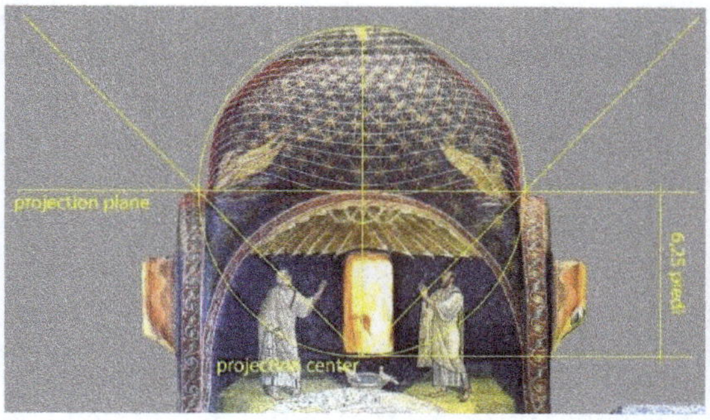

Figure 5. The diameter of the theoretical hemisphere is 12.5 Roman feet. The centre of projection (the antipodal point of the polar stereographic projection) is located at 6.25 feet from the projection plane (S. Giannetti).

Figure 6. The viewpoint placed in the centre of projection. At this point all the rings of stars, which on the surface are irregular, appear perfectly circular (S. Giannetti).

From these premises it can be established that the structure is a sort of raised ribbed vault, geometrically very different from a portion of a sphere. To analyse it, however, one must simply start from the diagonal arches. These are lancet arches, with a rounded off keystone. Dissecting the vault with a sheaf of planes perpendicular at every point to the tangents of the diagonal arch, and therefore passing through its centre, one can see that the sections of the roof are also made up of circular arcs. However, it is clear to see that the surfaces cannot comprise matching circular arcs throughout the entire quarter ceiling, or rather that the surfaces are not portions of a torus. In fact, it is more likely that these surfaces are obtained by running circular generatrices over directrices that are also curved, thereby generating surfaces with a value inside the segment, which makes up an eighth of the ceiling. This building process is very similar to that used for the raised groin vaults typical of the later years of the Italian gothic style, when the geometry of the generatrices changed from straight to curved.

The mosaic that depicts the starry sky sits on top of this complex ceiling structure. In the 1970s the mosaic was detached from the surface and then reattached. The photos of the restoration reveal various tesserae connected by metal wire (Ranaldi, 2011), probably positioned to indicate the layout of the stars before they were detached. In its current state, seen from below, the mosaic looks like it is laid on a perfectly hemispherical surface. This effect is generated by the apparent distribution of the stars in concentric circles and aligned on loxodromic spirals on the sphere.

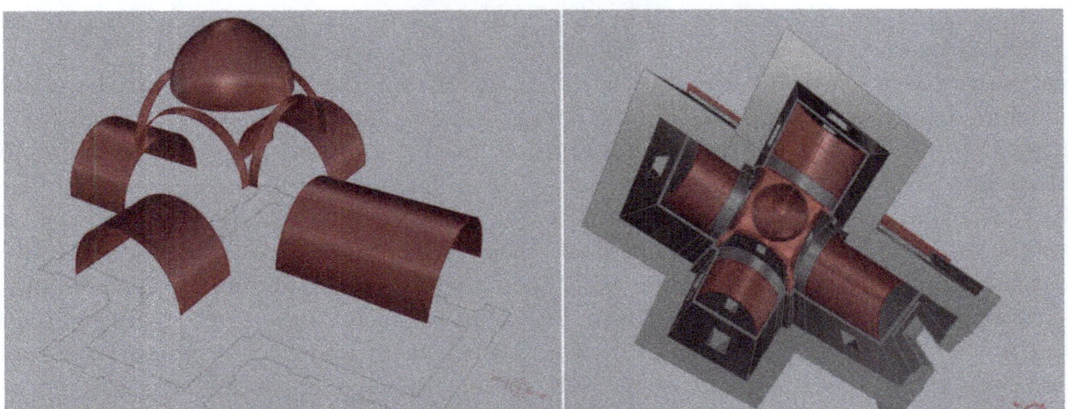

Figure 7 .The model in Rhinoceros: to the left, the inner surfaces of the vaults; to the right prospective view from the bottom (S. D'Amico).

However, detailed analysis, based on the textured mesh model, of any section along a plane that passes through the axis of the dome, reveals that the stars do not lie on parallel planes, but instead on courses that can be defined as 'festooned'. From this observation it can be deduced that a central projection was used for the layout of the stars on the surface, and the particular curves along which the stars are located are the intersection of the cones (with vertices at the centre of projection and directrices corresponding to the circles lying on the impost plane) and the surface of the dome.

Various studies, in particular Bartoli's research on the dome of the Pantheon (Bartoli, 1994), have looked at the coffers in Roman domes, noting that they were created using a polar stereographic projection. As the surface of Ravenna's starry ceiling is not really hemispherical, as stated above, the diameter of the theoretical hemisphere had to be hypothesised.

In this case this was the same as the diameter of the real dome at its impost, 12.5 Roman piedi (Figure 5). As a result, the centre of projection can be located at approximately 6.25 piedi from the 'equator' of the sphere, or from the impost plane described above. When the viewpoint is placed in the centre, all the rings of stars, which in reality are irregular, appear perfectly circular (Figure 6). In addition, the effect is heightened by the points of the stars tilted at 45° aligned with the loxodromic curves of the sphere, which in the stereographic projection look like logarithmic spirals on the plane. This occurs with a high level of reliability until the fifth ring of stars from the cross. From here on it seems that the stars, although laid out on concentric circles, are no longer aligned along spirals.

6. A DIGITAL ANALYSIS AND VERIFICATION MODEL (S.D.)

The research carried out on the Mausoleum of Galla Placidia combined various IT tools, both when managing the large quantity of data collected and in the subsequent geometric and dimensional reconstruction; following an archaeoastronomical survey, the representation of the building was dealt with digitally and also through the three-dimensional re-elaboration of its main formal features and measurements. The choice of modelling applications had to guarantee the production of a discretised digital model, which would be used for subsequent investigations and therefore had to be sufficiently descriptive, immediate and easy to consult. The studies and surveys illustrated here confirmed the benefits of systems that allow the virtual reconstruction of architectural artefacts, particularly during the analytical phases of the research. Some brief considerations regarding the geometrical characteristics of the mausoleum provide good examples of this: the building's distribution and volumetric system is effectively structured around a clear displacement, along the longitudinal axis, between the two eastern and western portions (Figure 7).

This apparent irregularity generates adaptations in the surfaces of the internal ceilings, in particular between the barrel vaults of the four wings and the central dome. The latter, set on a rhombic base, also has variable vertical profiles, so use of the model maker and a systematic comparison between two-dimensional representations and the corresponding three-dimensional transposition allowed the researchers to hypothesise the most plausible generative geometric patterns. With this in mind, it is worth pointing out that the choice of 3D software was made based on the need to construct relatively complex curved surfaces -the internal ceilings- and particularly their formal geometric connections.

It is worth reiterating that digital three-dimensional representation has both an instrumental and investigative value: the model, in fact, makes organic and simultaneous control of the whole complex and the diversified data set resulting from the surveys possible.

7. COMMUNICATION OF CULTURAL HERITAGE (G.L)

Documenting, safeguarding and managing Italy's vast tangible cultural heritage remains a very pressing topic, of great public interest. Although the cultural heritage system extends to virtually all areas of the country, major logistical, economic, and stability (linked to sites' safety) issues mean that in reality many monuments are difficult to access, with little opportunity for them to be enjoyed by the general public.

In recent years, digital architectural models, as well as describing buildings much more accurately in terms of their morphometry and colour, have also become useful tools for sharing the results of research carried out on monuments. They create virtual environments which can be navigated and consulted, able to interact with a widespread and varied user base who, even at great distances, can freely enjoy a piece of cultural heritage that would otherwise be inaccessible. At the same time, the opportunity to continually input data makes them extremely important tools for checking and monitoring the monuments themselves.

In the specific case of the Mausoleum of Galla Placidia, the digital model allowed the actual geometry of the architectural structures to be verified, with every single deformity compared to an 'ideal' model of the building. It also allowed the extent and precision of various specific light effects that appear within the room at certain times of the year and at certain times of day to be examined, thanks to specific simulations calculated and recreated on the model. The reconstructions, in turn, have become a valuable way of publicising the work, and helped to identify further areas of research that are yet to be explored.

In the case in question, moreover, the visualisation, or indeed, the communication of the results obtained from the research itself has been beneficial: here, therefore, the virtual reconstruction has not remained exclusively a material reproduction of the architectural object but has, above all, become an illustration of different meanings, reflecting technical, cultural and symbolic parts of our historical and monumental heritage (on these issues: Kiourt, 2016).

REFERENCES

Baert, B. (2004) *A Heritage of Holy Wood: the Legend of the True Cross in Text and Image.* Leiden, Brill.

Bartoli, M.T. (1994). Scaneographia vitruviana: il disegno delle volte a lacunari tra rappresentazione e costruzione, *Disegnare, Idee, Immagini*, 9–10, pp. 51-62.

Eusebius, C. (2009). *Vita di Costantino.* Ed. L. Franco. Milan, BUR.

Hawkins, E. J. W., Mundell, M. C., & Mango, C. (1973) The Mosaics of the Monastery of Mar Samuel, Mar Simeon, and Mar Gabriel near Kartmin with A Note on the Greek Inscription. *Dumbarton Oaks Papers*, Vol. 27, 279.

Kiourt, C., Koutsoudis, A., Markantonatou, S., Pavlidis, G. (2016) The "synthesis" virtual museum, Mediterranean Archaeology and Archaeometry, 16 - 5, pp. 1-9, DOI: 10.5281/zenodo.204961.

Ranaldi, A. (2011) Dalla realtà sensibile all'astrazione. La volta stellata del mausoleo di Galla Placidia. In L. Kniffitz (Ed.), *Architettura e mosaico. Atti della Giornata di studi, Ravenna 9 ottobre 2010*. Ravenna, pp. 20–42

Rizzardi, C. (2005) Il cielo stellato del mausoleo di Galla Placidia. In S. Pasi & A. Mandolesi (Eds.), *Studi in memoria di Patrizia Angiolini Martinelli. Alma Mater Studiorum, Università di Bologna, Dipartimento di Archeologia.* Bologna, pp. 277–288.

Salles, C. (1997) Le culte de sol invictus "soleil invaincu." In Y. Le Bohec (Ed.), *L'Empire romain: de la mort de Commode au Concile de Nicée,* Paris, Éd. du Temps, pp. 281–294.

Swift, E., & Alwis, A. (2010) The role of late antique art in early Christian worship: a reconsideration of the iconography of the "starry sky" in the "Mausoleum" of Galla Placidia. *Papers of the British School at Rome,* Vol. 78, pp. 193–217.

9

CASE STUDIES WITH ARCHAEOASTRONOMIC APPROACH IN THE STATE OF TABASCO, MEXICO

Hans Martz de la Vega

National School of Anthropology and History, Mexico

Corresponding author: pequenosolin@hotmail.com

ABSTRACT

In the year of 2016 started the Research Project Archeology, Archaeoastronomy, Calendar and Landscape Olmec and Maya (PIAACPOM). It focuses on archaeoastronomy, and in this field season, research was steered in the state of Tabasco, Mexico.

The work is centred on the methodology of the case studies, which allow to know in greater detail each of the archaeological zones. The places researched in this paper are the Olmec site La Venta and the Zoque like archaeological zone of Malpasito, and also San Claudio and Comalcalco, these representatives of the Mayan culture. All of them have an architecture in which a series of measurements have been carried out. The objective is to know in more detail the intention of their orientation and, to a certain extent, the relations with the surrounding landscape, in these cultural regions.

The work will cover cases of orientations ranging from the relationship between settlements, such as La Venta, to the general profile of a site and the relationship it has with the landscape, in the case of Malpasito and San Claudio. The architectural component is also studied, such as a burial chamber, a temple structure at Comalcalco or the *orientation axis* of some structures at San Claudio. The different levels of information complement each other and help to better understand the intention that existed about orientation in Mesoamerica.

In summary, it is possible to see how the orientation was considered in the planning of spaces, constructions, building traces, location, among others, from the smallest to the largest structures. In this work it will be shown with different cases how the result is in some way the same. The orientation was an aspect of great relevance in Mesoamerica, and it reveals that it can be associated with iconography, architecture, settlement and landscape.

KEYWORDS: Solstice, equinox, starting step, orientation axis, landscape.

1. INTRODUCTION

The sky and the landscape were inseparable parts of ancient Mesoamerican civilizations, as has already been demonstrated by the most enthusiastic academics, for example, Franz Tichy, Anthony Aveni, and Johanna Broda of the old guard, as well as Stanisław Iwaniszewski, Ivan Šprajc, Pedro Francisco Sánchez Nava and Jesús Galindo Trejo, of the new generation. Also, similar conclusions have been reached by those who have dealt with the topic of the general orientations of ancient cities such as Ignacio Marquina, James Dow and René Millon, and Horst Hartung.

With this in mind, I proposed to launch an archaeoastronomical and landscape research project based on the study of alignments of the architecture placed in archaeological zones, opened to the public and remaining under the custody of the Mexican National Institute of Anthropology and History (INAH). The project focuses on Mayan and Olmec cultures and some possible cultural continuity found in the present Mixe-Zoque communities at Tabasco. Four sites are Mayan: Pomona, Moral-Reforma, San Claudio, and Comalcalco, one is Olmec: La Venta and one is of a possibly Zoquean origin: Malpasito (Figure 1).

Till now, only the first season, in 2016-2017, has been carried out. In the present paper, I will limit myself to the short description of simple examples from those four sites. The objective of this work is to present some progress on the results. I will focus on three main topics: families of orientations, prominent landforms used to mark calendrical intervals, solstices and equinoxes, and aspects of symmetrical relationships among diverse architectural structures. In some cases, these examples extend beyond solar alignments and include possible alignments to the Moon and Venus, like in sites of Comalcalco and San Claudio.

2. METHODOLOGY

The research project is based mainly on a methodology established over decades by scholars such as those mentioned above. The measurements in the field were made with a theodolite brand YOM3, model 4T3OP-10, the year 2002, assembled in Mexico, whose error of the readings is 30 seconds of arc in each. The theodolite was placed at a certain distance from one end of the wall. Each horizon point was measured, both the altitude and azimuth angles were recorded. Observations of the Sun were made to convert the measured sightlines into the azimuths relative to true north. The error of the measurement was always compared with the data provided by NASA and other specialized astronomy software. The procedure used to calculate the results is that of Positional Astronomy that has been used from the beginning by specialists such as Aveni (1991) or Šprajc (2001). The results were realised on the date of the measurement (2017), and then they were projected backward in time to the epoch when the structures were built. Finally, the results were also compared with the data administered by NASA and astronomical software (*Sky Charts*).

The geographical coordinates are displayed in the WGS84 datum. To obtain the geographic coordinates was handled a GPS Garmin brand, model etrex. The abbreviations used in this work are δ= declination (apparent), A= azimuth (astronomical or true), and h= height of the local horizon (without considering the correction by refraction).

The project studies architectural structures of the past that have been exposed for hundreds of years to various weather destructive causes, and alterations made by wild and domesticated animals and people. That is why the alignments are not 100% reliable.

Before entering the subject, some fundamental concepts should be defined. A common problem in Mesoamerican archaeoastronomy is that of measuring wall azimuths for association with calendrical and astronomical events. The proper definition of the *orientation axis* of architectural features (buildings, plazas) is one of the main issues that can affect the results since this depends on how it is conceived of and on the decisions regarding components that should be taken into consideration. In contrast to the symmetry axis, the *orientation axis* does not necessarily rely on the alignments of the structure's lateral sides. Aveni and Hartung (1989) determined that the symmetry axis would be the sightline perpendicular to the facade of the structure. In my opinion, the *orientation axis* is determined by the architectural segment that creates the link between the structure and the level on which was built. For example, elevated shrine erected on the top of the pyramid would undoubtedly possess conceptually different symmetry axis than the steps located on the level of the plaza. This is because any person ascending the structure and crossing different spaces would differently perceive the links with his/her actual position and the surroundings. The *starting step* is the architectural element that marks the position of a stairway, located on the ground level. This architectural element is usually in good condition, and therefore seems to be a reliable source of data. Another concept is that of astronomical-calendrical family. Each such family is based on three aspects: the interval must be fixed (e.g. 52), the dates associated with that interval and the initial counting point must be specified (e.g. December 21) and there must be a universe of cases. Each family is defined through the concept of intervals of days. Intervals of days are counted from the nearest solstice and refer to the numbers of days between those solstices and particular dates.

To date, more than 150 measurements were recorded in six archaeological sites of Tabasco. However, in this paper, for each site, only one or two measurements were selected to show their potential. Orientation patterns can be identified because:

1) An archaeological context that has been talked about for decades but its importance has been deciphered gradually and despite having suffered looting or destruction, the information it has provided is very rich. No doubt the true orientation is an aspect that cannot be left aside. For example, the Tomb of the Nine Lords of Temple IX has been a question since 1926, by Frans Blom and Oliver La Farge, or the long mound or elongated platform D8 and mound D1 of La Venta since 1952 by Phillip Drucker.

2) The presence of intervals of significant days for calendrical such as the interval of 73 days, such as the case of La Venta.

3) The presence of the mountain as a marker of time. Indicate solar events such as equinoxes and solstices. In this regard, the cases of San Claudio and Malpasito are presented.

4) The presence of the mountain playing a role of first importance in the general layout of the settlement as it happens in Malpasito.

3. COMALCALCO

In the year 649 A.D., the city of Comalcalco was conquered by Tortuguero, an ally of the kingdom of *B'aakel*, whose capital was the city of Palenque. In fact, the Palace of Comalcalco is very similar to that of Palenque, especially in its proportions and appearance. Thus, the architecture, the sculpture, the stucco ornamentation are similar on both sites, corresponding to the Late Classic Period (600-900 A.D.). For example, in the fine and well-finished stucco representation of the nine lords over the three walls of the burial chamber of Temple IX resembles much of the Palencan wall painting.

The funerary chambers or funerary crypts of Comalcalco are spaces of rectangular shape with corbel arch or vaulted ceilings (Gallegos Gómora and Armijo Torres, 2014:17-18), located inside the architectural structure. Besides, they are usually boarded up (covered) by a staircase that ascends to the sanctuary. In Comalcalco, two of the three known chambers were oriented to the West.

3.1 *The Tomb of the Nine Lords of the Temple IX*

The first reference of the burial chamber of Temple IX was made by Frans Blom and Oliver La Farge in 1926 (Andrews, 1989:95-102; Blom and La Farge, 1926:106). Scholars found it desecrated and ransacked (pillaged), even with some remains of bones and oysters on the floor. In this, the surface was painted on red color and they reports four bricks that were thought to be the basis for the sarcophagus made of perishable materials. The three walls were painted in red too. Today, most of the structure no longer exists and it can be observed as a reconstruction of the 1960s in the archaeological site.

Temple IX is located at the Great Acropolis. The two structures, the Palace and the Temple IX, are in front of the plaza. In addition the two structures are oriented to the plaza. Therefore, it is possible that Temple IX is related to the Palace.

The stucco-patterned decoration of the camera features three scenes with three human characters (celebrities) each. There are three of them on the north wall, three on the south wall and three on the east wall (Andrews, 1989). In the past they had been called "The Nine Lords of the Night" because the characters were associated with in Central Mexico, an issue that has been overcome. The specialists have already indicated that it is the tomb of a *k'uhul ajaw*, "sacred lord", of Comalcalco and the other characters are their priests and the *ajk'uhuun*, "conservatives", who are part of the court (Armijo Torres, 2003:36). According to the epigraphic reading of Marc Zender (Armijo Torres, 2016:249), five characters carry the titles of *ajk'uhuun* and *yajawk'ahk'*. The researchers associated the characters with the Nine Lords of the Night because the number nine was the sacred number of the underworld, and so far it is only possible to associate the number nine with the Mayan culture and not for the deities of the night, because they have not been identified. In other words, the characters do not represent deities because they are human beings and have rather been identified as priests by their attire, but the number nine if it is associated with the underworld and therefore the funeral contexts. Therefore, the characters are priests and belong to the Comalcalco elite; they are the companions and the portraits of the royal court of this ancient city.

The relationship between the funerary chamber and the underworld seems to be supported in the mythical state. The Mayan rulers, especially those from the lowlands, became the god of corn to represent their resurrection from the underworld (Juárez Cossío et al., 2016:48), content aspect in the *Popol Vuh*. This implicitly makes manifest the incarnation of the cosmic tree in his person (Schele and Freidel, 2011:98). So, the ruler recovers its essence from the underworld as an intermediary between the forces over natural and the terrestrial world (Florescano Mayet, 2014:49). In other words, he legitimizes himself both in its rulership role and before the eyes of other cities. The speech could represent the investiture or legitimate enthronement, of ruler of Comalcalco, in the underworld or emerging from there. If we consider the previous thing, the scene of the personages is one of the best representations of the entire site and for that reason it is so important to know all the aspects including, which is the orientation.

First, it highlights the intention of its builders to orientate it to the west. That is, it is a funerary temple with access from the west, in which nothing obstructs the visual of the local horizon. Within it, there are nine characters modelled in stucco distributed in three of its walls.

At present the burial chamber is closed by a grid that does not allow measurements with great accuracy but if you obtain values close to the true, approximate. Once making the measurements of the south and north walls and obtaining averages, it is concluded that their orientation is between 24° and 27° of declination.

The values of the North Wall are δ = 24°.3, A = 295°.9 and h = 0° and the coordinates are 18°.277638 N, 93°.200444 W and 26 m asl. The result, for sunset, is circum-solstitial because it means the azimuth is circum-solstitial, because the correct value of the azimuth of the summer solstice sunset at Comalcalco in the year 600 A.C. is 294°.9 at the height of 0°. The error is around 1°. The other data is δ = 23°.6.

The values of the South Wall are δ = 27°, A = 298°.8 and h = 0° and the coordinates are 18°.277638 N, 93°.200444 W and 26 m asl. The south wall could be associated with Venus because it is maximum western elongation since their values ranged from 24°.1667 to 27° and this astronomical phenomenon was known and favoured by the orientations in Mesoamerica with greater emphasis in the seventh century and beyond, and appeared as a star in the afternoon before the solstices, between April and June in the north extreme (Šprajc, 1998:25). The error with respect to the extreme north is negligible. However, in this temple there are no elements so far to justify temples connexions to this planet, but in one of the associated temples there is indeed two large star representations (*Ek'*) that on occasion can be identified with Venus, when it is accompanied by *chaak* and means 'big/red star'. It´s the decoration of the facade of Temple VII, a structure that in its configuration corresponds to the same plaza together with Temple IX, at the Plaza of the Great Acropolis. The sunken space is limited to the north by the temples VI, VII and VIII, to the east by the Great Acropolis and to the west by a series of structures of few dimensions and little known. What is the relationship between Venus and the underworld? Venus can be a *nahual*, an animic entity that generates problems, sometimes it is a light of bad omen.

4. MALPASITO

4.1. Steam bath

The region of archaeological zone of Malpasito had a pre-Hispanic occupation between the years 650-900 A.D. and is partially attributed to proto-Zoque architecture (Cuevas Reyes, 2004). His steam bath is in the best characterized structures of the site and is one of the most representative examples of Mesoamerica. It is located at Structure 18 of the ballcourt.

The structure is rectangular with access to the west, just in the direction where a remarkable local horizon is found, since it has at least four prominent hills and whose forms and association, side by side, seem significant, among which the Hill of the Pava a height around 860 m asl. According to the *orientation axis* of the access stairway, the orientation to the west exceeds the 24° of declination. The data is δ = 24°.4, A = 293°.3 and h = 7°.4±20'. The measurement station has the coordinates 17°.338138 N, 93°.598888 W and 234 m asl. It could well be the summer solstice, where the Sun ends its path, because the correct value of the azimuth of the summer solstice sunset at Malpasito in the year 600 A.D. is 292°.5, so the error is of 0°.8. The other data is δ = 23°.6.

The steam bath is usually located near the ballcourt, and both form a ritual architectural unit and that is why it is not surprising that they can be associated with an astronomical event of such importance as the solstice.

4.2 The guiding axis of the site

The Malpasito settlement is located on a semi-vertical axis, which begins north of the site, in Structure 43; crosses the Main Plaza, in the centre of the site, passes over the ballcourt and the first part ends in the Acropolis, which represents the south and the highest part of the site with the presence of structures. Between the ballcourt and the Acropolis, the guiding axis continues through a series of levels through large staircases, or in other words, stepped overlays on the natural slope, which was enabled for these purposes. It is easily appreciated that the Acropolis is crowned by the highest part of the slope. The axis crosses said crown or top of the slope and continues its imaginary trajectory without distorting itself to reach its end, at the top of Hill Mono Pelado or Mono Pelón, the hill with one of the highest altitudes in the state of Tabasco. The axis has approximately A = 21° or 201°. Between the top of the hill and the Acropolis there are 2.77 km.

4.3 *The Acropolis and the local west horizon*

Located at the top of the stairway; that is to say, on the access of the Acropolis, it is possible to observe that the Sun sets on the top of Hill el Chato during the days of the equinoxes. The data are: δ = 0°.2, A = 264°.4, h = 18°.15. The measurement station has the coordinates 17°.337444 N, 93°.599388 W and 274 m asl. In the year 600 A.D., the sunset on the top of the hill occurred in March very close to the equinox. It happens on March 17 and the data are: A = 264°.1 at the same height, h = 18°.15, and δ = 0°, therefore the error is 0°.3.

It is one of the hills discussed above, distinguishable on the horizon and astronomical milestone. The other hills, la Copa, la Pava, las Flores as well as the Mono Pelado, are non-solar in nature, because they are outside the solar arc; they have a different meaning, but complementary, in the Mesoamerican worldview.

5. LA VENTA

5.1 *The hidden side of the landscape*

The following site is one of the earliest in Mesoamerica with a planned layout, 1200–400 B.C., and perhaps also one of the most enigmatic, for all that has already been said (González Lauck, 2014:31). It seems that La Venta privileged the path of the Sun towards the summer solstice at dusk, but in an unusual way. Of course, the above is valid for the observation point in the upper part of Structure C-1, and its coordinates are 18°.103250 N, 94°.040833 W and 44 m asl.

The Olmecs occupied Los Tuxtlas nuclear core, a set of volcanoes of great dimensions, among which San Martín Tuxtla (1680 msnm) and Santa Marta Tuxtla (1640 msnm), and bodies of water such as the great Catemaco Lagoon. In general, the Olmec sites had these volcanoes as part of their landscape (Šprajc and Sánchez Nava, 2015), and one of the most distant sites among them all, and larger, La Venta, maintained its importance on the local horizon.

What is interesting is that the Sun was behind the massif of the volcanoes during several months of the year. Everything seems to indicate that they could not be observed with the naked eye and it was only possible when the Sun was behind.

Day after day, during the sunsets, the Sun entered the mountain massif of Los Tuxtlas since mid-April. Then they could see how it illuminated the highest parts, approximately between May 3 and 7. Finally it was moving away from the mountainous massif towards the Gulf of Mexico at the beginning of June. But as the dimension that the Sun occupies from the equinox on the horizon, day by day it was shrinking to stop at the summer solstice, about ten days out, never really separated from Los Tuxtlas. Finally, the Sun was still there other days when it began its return, to leave the region of Los Tuxtlas by the end of August. The total count of days approaches that of one third of the year.

Los Tuxtlas would begin to see when the rainfall increased with respect to the driest months, March and April, as it has been in recent years according to the record of the *Meteorological Station 00027026*, of the National Meteorological Service, between 1951 and 2010. A few days later, the area of the three great volcanoes illuminates, towards the beginning of May, with the first increase in rainfall and shortly afterwards the first zenith step happened. A few days later, the heavy rains would have arrived to stay up to eight months. In fact, in our investigation we had to wait for the Sun to settle behind Los Tuxtlas to take the photographs because the days when it was necessary to make measurement were cloudy.

For the above works and some other, it is known that in La Venta the astronomical-calendrical principles were taken into account. Specifically, the presence of the interval of 73 days (Galindo Trejo, 2011).

5.1 *The E-Group*

Structures D1 and D8 have been identified as an E-Group by some researchers (Aimers and Price, 2006). What is an E-Group and how does it work? It is defined, in its simplest form (Blom, 1924), as a platform oriented north-south and east of it, is a mound from which you can make the observations of the sunsets behind the platform. It is the foundational set of the settlement and can be related to creation myths and deities. In the case of La Venta one of the sculptures, the Altar 4, could represent the birth of the gods.

In general, it has been thought that these types of architectural assemblages are of Mayan origin. However, they may well be present in other cultural areas such as the State of Guerrero (Martz de la Vega et al., 2017) or in the Puebla-Tlaxcala Valley (Flores Esquivel, 2010).

So far, the two structures, D1 and D8, have not been studied as a Group E. Only a simulation is presented based on its location and existing cartographic materials (González Lauck, 2014), and not from field measurements because the structures are covered by vegetation. Two important issues can be added to the data we already know. The first is that the platform would cover the entire solar arc for the sunrises and a little more, as far as the major lunar standstill to the north, as originally have been thought for this architecture. And the second is that the mound reached at least seven meters in height, growing more in height than the platform itself, as was the case of Lost World of Tikal (Fialko, 1988).

The coordinates of Structure D1 are 18°.098638 N, 94°.040722 W and 34 m asl. After making an approximation of the values that could be observed in the set, it is only necessary to say that they are significant with respect to the few data that are already supported for the site and that, in addition, they are of great importance, subject to doing a work with more detail in the field.

The center of the platform marks, in the sunrise, the dates close to March 4 and October 9, approximately the dates that correspond to the same interval of days that they found, each one by its side, Galindo Trejo (2011) in Structure C-1 and Šprajc and Sánchez Nava (2015) in Complex A, with date of April 9. The presence in the site of the two pairs of dates of the family of the interval of 73 days to the east is interesting. The approximate data are A = 96°.7 ±1° and h = 0°, remembering that it is only a simulation and that it is necessary to perform measurements in the field. On the other hand, in Martz de la Vega et al. (2017) we proposed to make, at least, one measurement from the East Platform in the direction of the Frontal Mound. If we rotate the azimuth 180°, we get 276°.7 ±1°. The altitude is also 0°. The result was surprising because the dates are near April 4 and September 7. The interval approaches that of the family of 78 days, one of the families of the *trecenas* as it was developed in Martz de la Vega et al. (2016). In that work the interaction between the two families was discussed, that of 73 and that of the *trecenas* 78 to be five days apart, because the five is a calendrical interval to consider. The *trecena* is based on the number thirteen, a sacred number in Mesoamerica related to the heavens.

6. SAN CLAUDIO

6.1 The Structure 12B and the starting step

Structure 12 contains two separate basements so we divide it into Structure 12A or south and Structure 12B or north.

The orientation of Structure 12 was studied by Sánchez Nava and Šprajc (2011:25). They concluded that the declination is greater than ±26° and therefore is not within the solar arc. Its values are to the east: δ = −26°.1, A = 117°.5 and h = 0°.6 and to the west: δ = 26°.1, A = 297°.5 and h = 0°.1, for the coordinates 91°.161111 N, 17°.337500 W and 70 m asl.

The Basement of Structure 12B can instead be associated with the winter solstice because its most remarkable element, a staircase with three steps of 10 cm of height that lead to the upper part to an enclosure with stone walls, in its *starting step* presents the declination of −24°.4 east over the part of the local horizon where the mountain descends to the level of the "zero horizon". The other data are A = 115°.6 and h = 0°.3. The coordinates of the station are 17°.337333 N, 91°.160333 W and 81 m asl. The declination of the winter solstice in San Claudio, in the year 500 A.D., was -23°.6, and the azimuth at a height of 0°.3 was 114°.9, then the difference is 0°.7. The exploration of this stairway few years ago, provided the following information:

"In the first of the steps of this stairway three spear points of the so-called eccentrics were found for showing whimsical and very elaborate shapes, two of them with jagged edges and a third whose silhouette outlines the body of a snake." (Romero Rivera, 2002:66).

That is to say, an offering was found consisting of three eccentrics below the *starting step*, and this is a further antecedent that the *starting step* has a particular importance. In fact, I have proposed that the *orientation axis* of the structures can be studied from the *starting step* or, in case of their absence, from the first steps. By courtesy of José Luis Romero Rivera it is possible to talk about the shape of one of the eccentrics, the SCLEX003, which reminds the scepters of Tláloc of the Center of Mexico, and which are associated archaeologically to the high parts of the mountains and deposits of water, as ichnographically to the thunder and water. This remember, with the presence of the lagoon and the hill, the complex hills-water-corn.

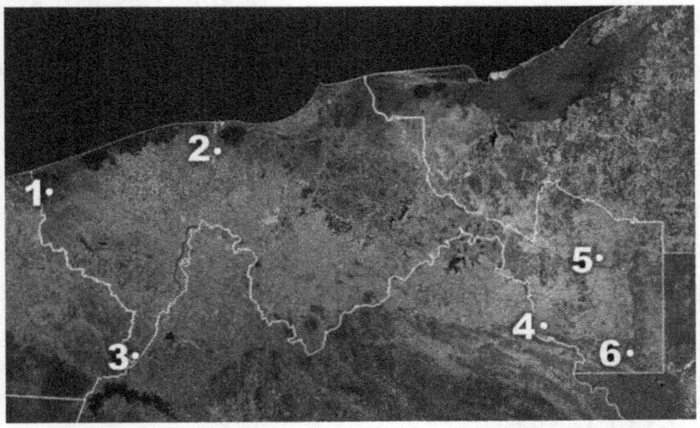

Figure 1. A map showing the locations of different archaeological zones in study. 1. La Venta, 2. Comalcalco, 3. Malpasito, 4. Pomoná, 5. Moral-Reforma, and 6. San Claudio. Adapted from http://www.cartocritica.org.mx/

6.2 Structures 1 and 69

Structure 69 was not measured because it is outside the perimeter zone of the National Institute of Anthropology and History (INAH); however, a simulation based on cartography is presented here. This structure has a singular interest because its dimensions are among the largest of the site along with Structure 1 (in Group II), considering both are of the temple type. In addition, it turned out to have a remarkable position with respect to Structure 1 and the local southwest horizon. In fact, there is a hill (SCL001), with no name detected so far, which observed from Structure 1 was the approximate location of the major lunar standstill of the Moon, δ = -28°.5 (the value of the declination was corrected by lunar parallax, because δ = -28°.9), A = 236°.9 and h = 6°.8. The declination of the standstill of the Moon in the year 500 A.D. on average it was 28°.8. The error seen as the difference of the declinations is 0°.3. The coordinates of Structure 1 are 17°.335944 N, 91°.158805 W and 78 m asl.

If the observation is made from Structure 69, the Sun would stand on the same hill, SCL001, very close to the winter solstice. The data for the year 500 A.D. are: A = 242°.6 ±1°, h = 6°.8 and δ = 23°.6. The solstice day, March 19. The Structure 69 has the coordinates 17°.334861 N, 91°.153066 W and 64 m asl, and is in Group I.

From Structure 4, the local horizon is very similar to Structure 1. Structure 4 is located in the same group as Structure 1, and they are facing each other. The difference is that the Structure 4 is the one closest to the San Claudio Lagoon, on the edge, a few steps from the water level. Its coordinates are 17°.335111 N, 91°.159 W and 72 m asl. In its local horizon, the value of the standstill of the Moon also approaches the top of Hill SCL001, since the declination is -27°.2. For the year 500 A.D. the average declination was -28°.8, with error 1°.6. However, what stands out most about Structure 4 is that from it was possible to observe, at the lowest point of Hill SCL001, the winter solstice at the sunset, since its declination is -23°.9 at a height of 5°.8 and an azimuth of 242°.7. The azimuth that is obtained when calculating the winter solstice at the same height, 5°.8, for the year 600 A.D. it is 243°. Therefore the error is 0°.29. In fact, the knowledge of the movement of the Sun and the Moon could work, among other things, for agriculture (Iwaniszewski, 2006).

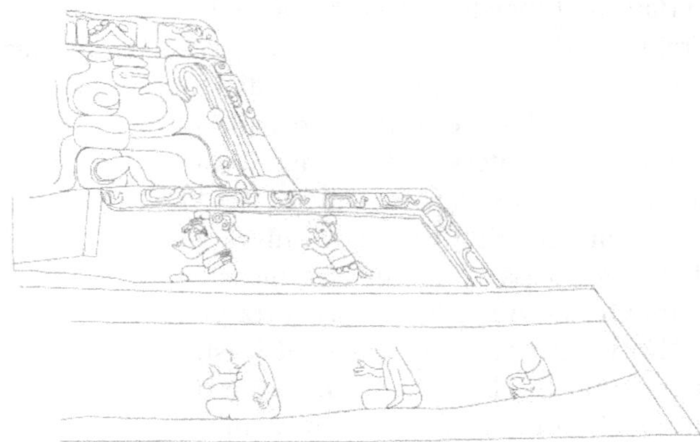

Figure 2. Reconstruction of the stucco decoration of Temple IX. The motif of Ek' (star) can be seen at the top left.

ACKNOWLEDGEMENTS

I want to thank Rocío de la Vega Folgarolas in a very special way. Also to Cecilia González Morales for her contributions in the field and in the cabinet work. To the Archaeologist Miguel Pérez Negrete.

A sincere recognition to the authorities of the National Institute of Anthropology and History (INAH), to Dr. Pedro Francisco Sánchez Nava, to the companions of the administration, of investigation and of the custody, since thanks to them it has been possible to realize the field work, in special to the archaeologists José Luis Romero Rivera, Francisco Cuevas Reyes, Ito Javier Kato Vidal and Janeth Lagunes Celis. To the authority of the National Institute of Anthropology and History of the State of Tabasco, Juan Antonio Ferrer Aguilar. To Dr. Rebecca González Lauck and Dr. Ricardo Armijo Torres as well as the authorities of the Postgraduate Program in Archaeology of the National School of Anthropology and History (ENAH), especially Dr. Stanisław Iwaniszewski.

To Dr. Juan Antonio Belmonte Avilés and Dr. César González García for all the attentions.

REFERENCES

Aimers, J. J. and Rice, P. (2006) Astronomy, ritual, and the interpretation of maya "E-group" architectural assemblages. *Ancient Mesoamerica*, No. 17, pp. 79–96.

Andrews, G. F. (1989) *Comalcalco, Tabasco, Mexico: Maya art and architecture*. California, Labyrinthos.

Armijo Torres, R. (2003) Comalcalco. La antigua ciudad maya de ladrillos. *Arqueología Mexicana*, Vol. XI, No. 61, pp. 30-37.

Armijo Torres, R. (2016) *Un katún de investigaciones en Joy'Chan (Comalcalco)*. Unpublished Ph.D. Dissertation, Escuela Nacional de Antropología e Historia, Mexico.

Aveni, A. F. (1991) *Observadores del cielo en el México Antiguo*. Fondo de Cultura Económica, Mexico.

Aveni, A. F. and Hartung, H. (1989) Uaxactun, Guatemala, Group E and Similar Assemblages: an Archaeoastronomical Reconsideration. In *World Archaeoastronomy*, A. F. Aveni (ed.), Cambridge University Press, pp. 441-461.

Blom, F. (1924) Report on the Preliminary Work at Uaxactun, Guatemala. *Carnegie Institution of Washington Yearbook*, No. 23, pp. 217–219.

Blom, F. and La Farge, O. (1926) Tribes and Temples. In *Middle America Research Institute*, No. 1, New Orleans, Tulane University.

Cuevas Reyes, F. (1994) *Informe del Proyecto Arqueológico Malpasito, Huimanguillo, Tabasco. Temporada 1993.* Unpublished report, Archivo Técnico of the Consejo de Arqueología, Mexico.

Cuevas Reyes, F. (2004) El juego de pelota de Malpasito, Huimanguillo, Tabasco. Arqueología, No. 33, pp. 47-59.

Drucker, P. (1952) La Venta, Tabasco. A Study of Olmec ceramics and art. *Bureau of American Ethnology*, No. 153, Smithsonian Institution, Washington.

Fialko, V. (1988) Mundo Perdido, Tikal: Un ejemplo de Complejos de Conmemoración Astronómica. *Mayab*, No. 4, pp. 13-21.

Flores Esquivel, A. (2010) Los complejos del Tipo "E" y su asociación con acrópolis o arreglos de tipo Triádico: Esbozos de un posible patrón urbano y sus posibles significados. In *XXIII Simposio de Investigaciones Arqueológicas en Guatemala*, B. Arroyo, A. Linares Palma and L. Paiz Aragón (eds.), Guatemala, Museo Nacional de Arqueología y Etnología, pp. 111-122.

Florescano Mayet, E. (2014) *Memoria mexicana.* Fondo de Cultura Económica, Mexico.

Galindo Trejo, J. (2011) Orientación calendárico-astronómica en el Preclásico: el caso de La Venta. In *Legado astronómico*, J. D. Flores Gutiérrez, M. Rosado Solís and J. Franco López (coords.), Universidad Nacional Autónoma de México, pp. 233-241.

Gallegos Gómora, M. J. and Armijo Torres, R. (2014) Prácticas funerarias en Tabasco: de las culturas prehispánicas a los indígenas *yokot´an* del siglo XXI. In *Practicas funerarias y arquitectura en espacio y tiempo*, A. Benavides Castillo and R. Armijo Torres (eds.), Universidad Autónoma de Campeche, pp. 10-25.

González Lauck, R. B. (2014) Zona Arqueológica La Venta, Tabasco: retrospectiva y perspectivas. In *Tabasco: una visión antropológica e histórica*, M. Á. Rubio Jiménez, R. Perales Vela and B. Pérez González (coords.), R. Vargas Fregoso and H. de Paz (support for the ed.), Mexico, CONACULTA, Instituto Estatal de Cultura de Tabasco, Universidad Nacional Autónoma de México, pp. 31-83.

Juárez Cossío, D., Velázquez Castro, A. and Valentín Maldonado, N. (2016) El joven dios del maíz. Tecnología y simbolismo de un pendiente de concha del Museo Nacional de Antropología. *Estudios de Cultura Maya*, No. XLVII, pp. 31-53.

Iwaniszewski, S. (2006) Lunar agriculture in Mesoamerica, *Mediterranean Archaeology and Archaeometry* Vol. 6, No. 3, pp. 67-75.

Martz de la Vega, H., Pérez Negrete, M. and Núñez Mejía, C. A. (2017) Un Conjunto Tipo E en el Estado de Guerrero. *Los Investigadores de la Cultura Maya. El comercio y otros temas*, pp. 371-383.

Martz de la Vega, H., Wood Cano, D. and Pérez Negrete, M. (2016) La familia del intervalo de 78 días, familia calendárico-astronómica de 260/105 días en su relación con la etnografía y con las fuentes. In *Perspectivas etnográficas e históricas sobre as astronomías*, P. Faulhaber and L. C. Borges (orgs.), Rio de Janeiro, Museu de Astronomia e Ciências Afins, pp. 77-94.

Romero Rivera, J. L. (2002) Investigaciones en el sitio arqueológico de San Claudio, Tenosique, Tabasco. In *Primer Encuentro Cultural del Usumacinta Memorias. Tenosique, Tabasco.* Ediciones del Programa de Desarrollo Cultural del Usumacinta-CONACULTA, pp. 53-68.

Sánchez Nava, P. F. and Šprajc, I. (2011) *Propiedades astronómicas de la arquitectura y el urbanismo en Mesoamérica: Informe de la temporada 2011.* Archivo Técnico of the Consejo de Arqueología, No. 33-215. Mexico, Instituto Nacional de Antropología e Historia.

Schele, L. and Freidel, D. (2011) *Una selva de reyes. La asombrosa historia de los antiguos mayas.* Fondo de Cultura Económica, Mexico.

Šprajc, I. (1998) *Venus, lluvia y maíz: simbolismo y astronomía en la cosmovisión mesoamericana.* Colección Científica No. 318, Instituto Nacional de Antropología e Historia, Mexico.

Šprajc, I. (2001) *Orientaciones astronómicas en la arquitectura prehispánica del centro de México.* Colección Científica No. 427, Instituto Nacional de Antropología e Historia, Mexico.

Šprajc, I. and Sánchez Nava, P. F. (2015) *Orientaciones astronómicas en la arquitectura de Mesoamérica: Oaxaca y el Golfo de México.* Ljubljana, Slovenia.

10

ΒΩΜΟΣ ΔΩΔΕΚΑ ΘΕΩΝ/ALTAR OF THE TWELVE GODS: AN ASTRO-ARCHAEOLOGICAL ANALYSIS

Vance R. Tiede

Astro-Archaeology Surveys

**Corresponding author: vance.tiede@aya.yale.edu*

ABSTRACT

The Altar of the Twelve Gods (ΒΩΜΟΣ ΔΩΔΕΚΑ ΘΕΩΝ) was constructed in the Athenian Agora, NW of the Acropolis, at the direction of Pisistratos the Younger, Archon of Athens, 522/521 BC (Thucydides). In 1934, archaeologists excavated a marble statue base *in situ* next to a rectangular limestone *hypaethral* edifice on the Panathenaic Way. Its inscription reads "Leagros, the son of Glaukon, made the dedication / To the Twelve Gods". Although the Altar is now buried under the Athens-Piraeus railway tracks, the author surveyed the exposed SW corner (12 September 2017) with a Gurley 20" theodolite (for solar observations) to Ground Truth published (magnetic compass) azimuths. This paper presents counter-evidence rejecting the interpretation that the Altar takes its orientation from the Panathenaic Way (Camp, 1986). Rather, the axis of the Altar's *peribolus* wall openings is oriented to horizon points consistent with Graeco-Roman *hypaethral* astro-architectural design principles (Vitruvius), viz.: Full Moon rise nearest the Midwinter Minor Standstill (+18.66° decl.) and set nearest the Equinox Minor Standstill (-5.1° decl.). The significance of these lunar standstill alignments is that they connote systematic *local* and *long-term* luni-solar observations in the century preceding the 19-year luni-solar cycle proposed by Meton of Athens (ca. 432 BC): i.e., the NE *peribolus* opening framed the last of a series of Midwinter Lunar Eclipses observable in Athens every 19-years, viz.: Partial Eclipse 29 December 560 BC; Total Eclipse 29 December 541 BC; and Partial Eclipse 30 December 522 BC— this last coinciding with the Altar's construction under Pisistratos the Younger.

KEYWORDS: Altar of the Twelve Gods, Athenian Calendar, Archaeoastronomy, eclipses, Greek architecture, Greek religion, Meton, Moon

1. INTRODUCTION

1.1 Background

Pisistratos the Younger, Archon of Athens 522/521 BC directed the Altar of the Twelve Gods (ΒΩΜΟΣ ΔΩΔΕΚΑ ΘΕΩΝ) be constructed in the Agora, NW of the Acropolis (Thucydides, 6.54). The Altar, also known as the Sanctuary of Pity, was the locus from which geographic distances from Athens were measured. In 1934, American archaeologists confirmed its location upon finding a marble statue base *in situ* inscribed "Leagros, the son of Glaukon, made the dedication /To the Twelve Gods" (Camp, 2010) (Figure 1).

Scholars have attributed the Altar's distinctive altar's architecture as the prototype for Imperial Rome's *Ara Pacis Augustae* (Müller, 1943; Thompson, 1952; Kleiner, 1992; Mauzy, 2006). Their shared architectural elements include: an altar enclosed by a rectangular wall (*peribolos*) with opposing twin openings; an axis of the openings running SW–NE; an enclosure "open to the sky" (*hypaethral, ὕπαιθρος*); and sited adjacent to a principal ceremonial avenue. The major difference between the two structures is that while only the limestone *peribolos* sill or foundation of Athen's Altar of the Twelve Gods remains, the *peribolus* of Rome's *Ara Pacis Augustae* retains its magnificent marble friezes. Recently, the "Tellus" frieze of the *Ara Pacis Augustae* has been reinterpreted as a sky map depicting the heliacal rise of the Morning Star phase of Venus (patron goddess of the Julian clan) seated with Aries and Taurus on the *peribolus* opening directly facing sunrise on the *Dies Natalis Romae* (21 April 9 BC), i.e. the 744th anniversary (or 93rd iteration of the 8-year Venus cycle) of the founding of Rome (753 BC) (Tiede, 2016). Therefore, if the Athenian altar were the architectural prototype for its subsequent Roman astro-architectural analogue, then it follows that the former may have been astronomically oriented as well; especially given the evidence to be found in Graeco-Roman literature, e.g.,

> *The early Greeks attached significance to the risings and settings of fixed stars and star-groups (Pleiades, Hyades, Orion, Sirius, Arcturus, etc.). Plato, Aristotle, Theophrastos all use risings of* stars *and solstices to specify times of the year….. Theophastos (Sign. I.4) tells us that Meton's teacher, Phaeinos, observed the solstices (τας τροπας) from Mount Lykabettos, and two other astronomoi from other mountains.* (Prichett, 2001).

> *…[O]ne who professes himself as an architect should be…acquainted with astronomy and the theory of the heavens…. From prescription (θεματισμῷ), in the case of hypaethral (ὕπαιθρος) edifices open to the sky, in honor of Jupiter Lightning, the Heaven, the Sun,* **or the Moon:** *for these are the gods whose semblances and manifestations we behold before our eyes in the sky when it is cloudless and bright.* (Vitruvius, I:I,3; II:ii,5)

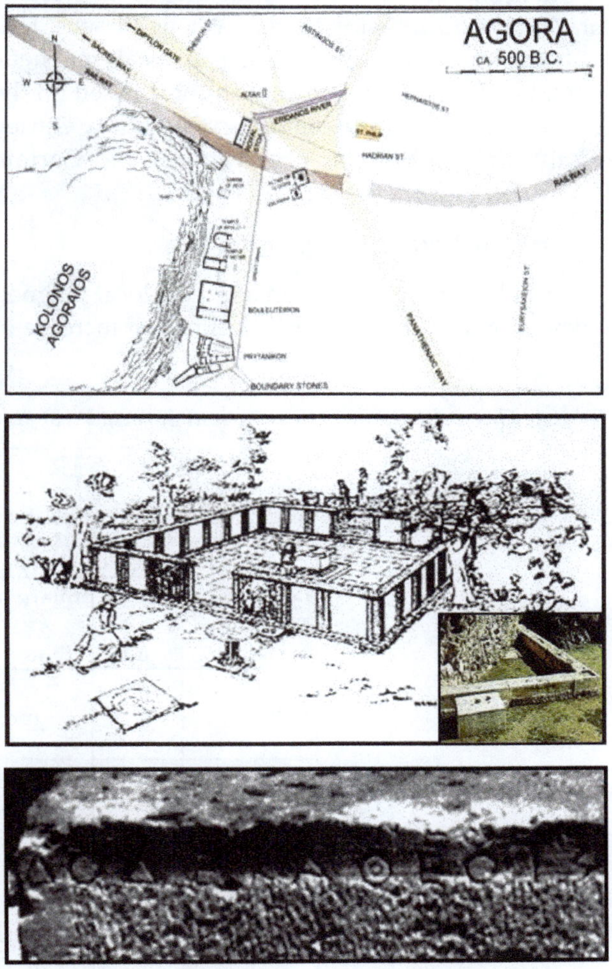

Figure 1. Agora plan; Altar XII gods reconstruction and SW corner photo; ΔΩΔΕΚΑΘΕΩΝ inscription. (Camp, 2010; McCallister, 1906; Meritt, 1966: http://www.greece-athens.com/page.php?page_id=258)

2. METHODOLOGY

2.1 Research Objectives

This paper investigates two research questions regarding the Altar of the Twelve Gods: (1) Does the Altar take "its orientation from the street" (Camp, 1986); and (2) Was the axis of the Altar's *peribolos* wall openings oriented to a coherent pattern of astro-calendric targets on the local horizon (Hawkins, 1968) when Pisistratos the Younger was Archon of Athens (522/521 BC)?

2.2 Research Design

Preliminary astro-archaeological data was compiled from the published 1935 site plans (Gadberry

1992) of the American School of Classical Study at Athens (ASCSA), NOAA Magnetic Field Calculator (IGRF12), NASA Index to Five Millennium Catalog of Lunar Eclipses, aerial/satellite imagery, and Google Earth's Digital Elevation Model (for horizon altitude). The (magnetic compass) azimuths plotted on published site plans are problematic due to the Altar's close proximity to the Athens-Piraeus railroad tracks potentially creating a local magnetic anomaly. Therefore the present study is based on the author's Ground Truth field survey with a Gurley 20″ theodolite (for solar observations) and Garmin eTrex GPS, 12 September 2017 (Tiede 2017).

2.3 *Ground Truth Survey Data*

The field survey confirmed a 1° East local magnetic deviation resulting in a +4° declination increase of the NE horizon (south slope of Mt. Likavitos). Final astro-archaeological data are presented in Table 1 and the following: Latitude = North 37° 58′ 32.89″; Observer Elevation = 54 meters ASL; SW-NE Axis True Azimuth = 71.7°; Date = 522/521 BC; NE Altitude ≈ +5° 16.8′; and SW Altitude ≈ +13° 34.2′. Program *Stonehenge* (Hawkins, 1983) calculated the two respective horizon declination vectors for the axis centered on the Altar's SW and NE *peribolus* openings. The declinations of three historical astro-targets (i.e.: Jupiter, Sun and Moon) prescribed for *hypaethral* edifices were compared with the Altar's SW and NE axis horizon intercept declinations for both years of the archonship of Pisistratos the Younger (522/521 BC), and modeled in Starry Night Pro Plus-6 digital planetarium.

Table 1. Theodolite (Solar Observation) Survey: West side of SW corner, Altar the Twelve Gods, Agora, Athens, Greece

Date: 12 September 2017
Crew: VRT
http://aa.usno.navy.mil/data/docs/AltAz.php
https://www.accuweather.com/en/gr/athens/
Time: UT+3h (CEE Time Zone B +2h +1h Summer Time)

Local Time h m	USNO AZ dd	Interpol. AZ dd	Obs. AZ dd	Δ AZ dd	Obs. AZ °	Obs. AZ '
14:00	196.9					
14:00:07		196.9	215.30	-18.4	215	18
14:01	197.4					
14:25	207.1					
14:25:09		207.1	225.42	-18.3	225	25
14:26	207.5					
14:29	208.6					
14:29:50		209.0	227.17	-18.2	227	10
14:30	209.0					
14:34	210.5					
14:34:34		210.7	228.90	-18.2	228	54
14:35	210.9					
14:51	216.6					
14:51:21		216.8	235.03	-18.3	235	2
14:52	216.9					
14:58	219.0					
14:58:47		219.2	237.50	-18.3	237	30
14:59	219.3					
15:04	220.9					
15:04:36		221.1	239.40	-18.3	239	24
15:05	221.3					
			Mean Average	-18.3	-18	-17
			+	360.0		
			SE-NW Axis True Azimuth	**341.7**	341	43
			-	270.0		
			SW-NE Axis True Azimuth	**71.7**	71	43

	Temperature (°C)			Long	Lat
Date	High	Mean	Low	East	North
12-Sep-17	32	27	22	23° 43'	37° 58'
Historical	28	24	19		

3. ANALYSIS

3.1 Street Orientation Hypothesis

If, as Camp (1986) states, the Altar of the Twelve Gods "...takes its orientation from the street", then the NE and SW sides of the Altar should be parallel to the extant curbing of the Panathenaic Way immediately adjacent. However, Figure 2 and three other published archaeological site plans of the Agora show that the Altar is not parallel to the Panathenaic Way, but is instead rotated approximately 31° from the adjacent curbing (see Crosby, 1949; Gadbery, 1992; Mauzy, 2006). Such a large deviation is sufficient to reject the hypothesis that the altar takes its orientation from the street.

3.2 Astro-Orientation Hypothesis

Table 2 compares the NE and SW vectors of the *hypaethral* Altar's *peribolus* openings at the time of its construction (522/511 BC) with the respective contemporary declinations of the Sun and Moon disc centers (i.e., disk radius = 0°). The horizon-target declinations differences for Jupiter and the Pleiades were out of range (Δ > 0.5° decl.) and are not included in this paper.

The Goodness-of-Fit of the luni-solar targets is less than the angular diameter (Δ <0.5° decl.) of the moon or sun with the exception of the rising Full Moon, 12 December 534 BC (where Δ = -0.93° decl.).

The NE opening of the Altar's *peribolos* wall is oriented to the south slope of Mount Likavitos. Possible astronomical targets include: (a) the sunrises 50 days after the Vernal and 53 days prior to Autumnal Equinoxes; and/or (b) the Full Moon rise nearest the Midwinter Minor Lunar Standstill (+18.66° decl., see Hawkins 1983) every 235 lunar Synodic months or 19 solar Tropical years, 534–496 BC. (Figure 3, Tables 2 & 3).

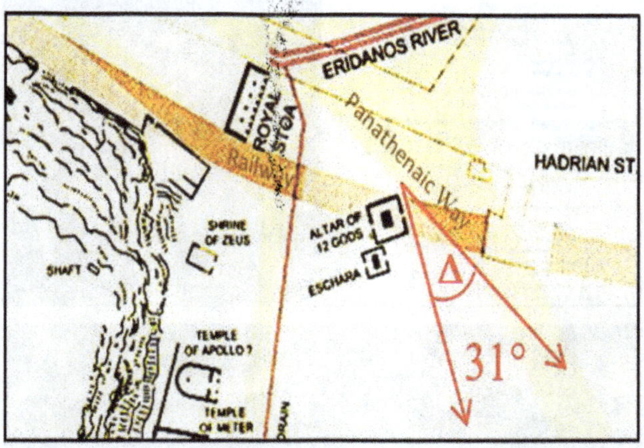

Figure 2. Orientation of the Altar of the Twelve Gods and Panathenaic Way, Athens (after Camp, 2010).

Table 2. Altar of the Twelve Gods Peribolus Wall Openings & Timing Luni-Solar Alignments

Target	Azimuth °	Δ °	Altitude °	Δ °	Declination °	Δ °	Local Time	Day-Month	Year BC	Year Δ	Interpretation
NE Horizon											
Rising	71.71		5.25		18.15						NE Peribolus Wall Opening Vector
Sun	71.71	0.00	5.64	0.39	17.86	-0.29	0649	18-May	522		Sunrise Spring/Summer Cross Quarter Day?
Sun	71.71	0.00	5.36	0.11	17.68	-0.47	0659	11-Aug	522		Sunrise Summer/Fall Cross Quarter Day?
Full Moon	71.71	0.00	13.33	8.08	22.49	4.34	1932	30-Dec	522		Midwinter **Partial Lunar Eclipse** (Saros 55)
Full Moon	71.71	0.00	4.62	-0.63	17.22	-0.93	1840	12-Dec	534		Midwinter Minor Lunar Standstill
Full Moon	71.71	0.00	5.24	-0.01	17.71	-0.44	1844	12-Dec	515	19	Midwinter Minor Lunar Standstill
Full Moon	71.71	0.00	5.35	0.10	17.68	-0.47	1836	12-Dec	496	19	Midwinter Minor Lunar Standstill
SW Horizon											
Setting	251.71		13.57		-4.88						SW Peribolus Wall Opening Vector
Full Moon	251.71	0.00	57.56	43.99	22.74	27.62	0347	10-Jan	522		Midwinter **Total Lunar Eclipse** (Saros 45)
Sun	251.71	0.00	14.37	0.80	-4.99	-0.11	1807	14-Mar	522		Vernal Equinox (27 Mar) - 14 days (inclusive)
Sun	251.71	0.00	14.78	1.21	-4.84	0.04	1743	11-Oct	522		Autumnal Equinox (29 Sep) +14 days (incl.)
Full Moon	251.71	0.00	14.75	1.18	-4.74	0.14	0625	18-Mar	539		Vernal Equinox Minor Lunar Standstill
Full Moon	251.71	0.00	14.97	1.40	-4.59	0.29	0624	18-Mar	520	19	Vernal Equinox Minor Lunar Standstill
Full Moon	251.71	0.00	15.24	1.67	-4.41	0.47	0624	18-Mar	501	19	Vernal Equinox Minor Lunar Standstill

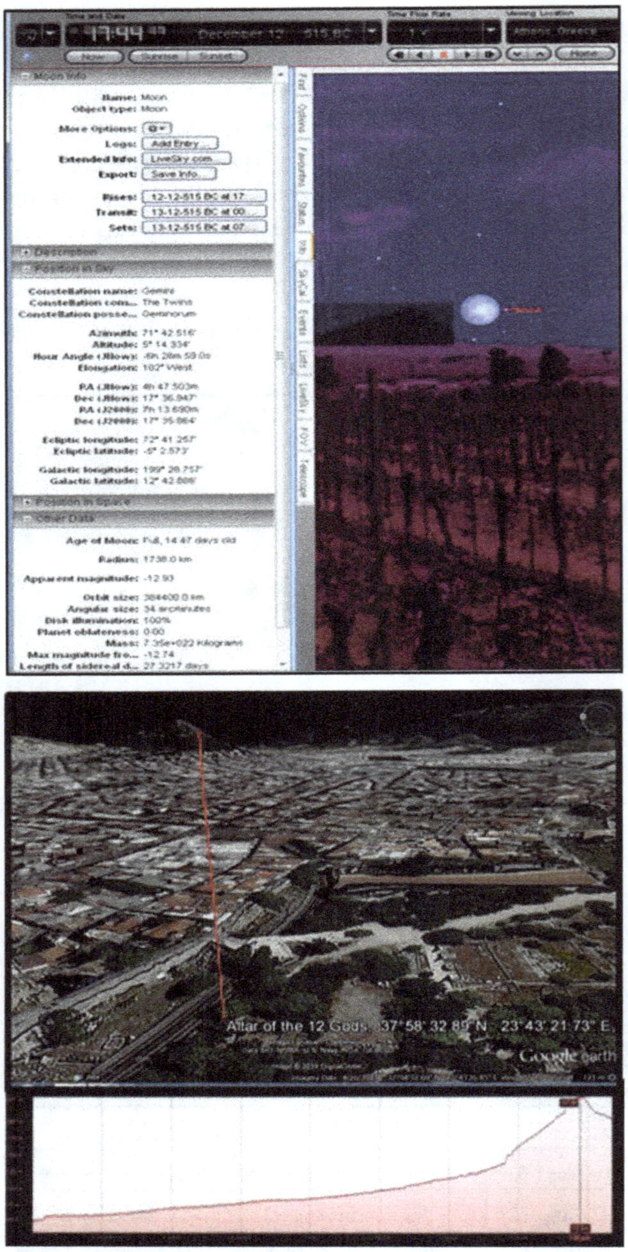

Figure 3. Full Moon nearest the Midwinter Minor Lunar Standstill rises from slope of Mount Likavitos, 12 December 515 BC. (Starry Night Pro Plus-6; Google Earth Pro)

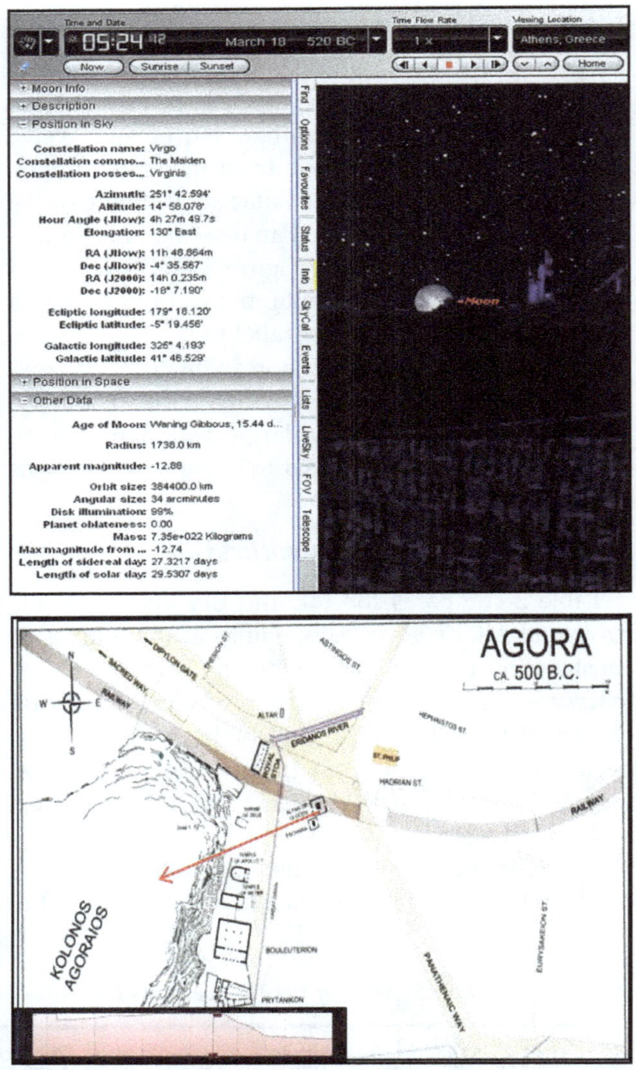

Figure 4. Full Moon nearest Equinox Minor Lunar Standstill sets over Kolonos Agoraios, 18 March 520 BC. (Starry Night Pro Plus-6; after Camp, 2010)

The SW *peribolus* opening axis (True Azimuth = 251.7°) is oriented to the nearby Kolonos Agoraios ridge (Altitude = +13.57°) between the Shrine of Zeus and Temple of Apollo (ca. 500 BC) where the Full Moon nearest the Equinox set at Minor Lunar Standstill (-5.15° decl., see Hawkins 1983) every 235 lunar Synodic months or 19 solar Tropical years, 539 – 501 BC (Figures 2 & 4; Tables 2 & 3).

Table 3. Luni-Solar Periodicities (after Hawkins, 1973; Sarton, 1952; https://eclipse.gsfc.nasa.gov/SEsaros/SEsaros.html; https://eclipse.gsfc.nasa.gov/SEsaros/SEperiodicity.html#section104

Unit	M Lunar Synodic Month	T Solar Tropical Year	E Luni-Solar Eclipse Year	Regression Lunar Nodes	Meton's Cycle	Halley's Saros Eclipse Cycle
Days	**29.53**	365.24	346.62	6793.85	6939.69	**6585.78**
M	1	**12.37**	**11.74**	230.06	**235**	**223.02**
T	12.37	1	0.95	**18.60**	**19**	**18.03**
E	11.74	**0.95**	1	17.65	**18.03**	**19**

If the Altar's *peribolus* NE and SW openings were oriented to the Midwinter moonrise and Equinox moonset at Minor Lunar Standstills, then might the purpose of the Altar have been to honor the Moon goddess Selene (*Σελήνη*), or perhaps even predict the periodic cosmic combat between Zeus and Typhon, i.e., lunar eclipses (Plutarch, ca. AD 120?) Curiously enough, a Total Lunar Eclipse (Saros 45) passed over Athens as the Midwinter Full Moon, 10 January 522 BC, six months prior to Pisistratos' one-year archonship (Figure 5; Table 2). Twelve months later (ancient Greeks counted inclusively) a second Midwinter Full Moon emerged as a Partial Lunar Eclipse (Saros 55) from behind Mount Likavitos and passed over (Altitude = +13° 20′) the marble altar and NE opening of the *peribolus* wall at sunset on 30 December 522 BC, i.e., in the middle of his archonship (Figure 6; Table 2).

The *hypaethral* Altar of the Twelve Gods (***dodeca**theon*, ΔΩΔΕΚΑΘΕΩΝ) may have been named to honor Heaven (vs. the 12 Olympian gods); inspired perhaps by the remarkable twin midwinter lunar eclipses of 522 BC bracketing the ***12-month*** (counting inclusively) cycle of Heaven through all ***12 signs*** of the Zodiac (cf. Camp, 1986). In Greek, "*Zodion* could mean either a sign of the zodiac or a constellation of the zodiac; when the Greeks wanted to be perfectly clear they used the word ***dodeca**temorion* **[twelfth** part] for a sign of the zodiac" (Thurston, 1994).

4. CONCLUSION

Map inspection confirms that the *peribolus* of the Altar of the Twelve Gods is some 31° azimuth more than would be required to "take its orientation from the street" curb of the Panathenaic Way. Therefore, alternate rationales for the Altar's orientation deserve consideration, e.g., the highest peak on the local horizon (i.e., Mt. Likavitos summit, or Parthenon), or calendric/horizon astro-alignment. The evidence for topo-geomantic orientation is limited to the NE vector intersecting the south *slope* (as distinct from the *summit*) of Mt. Likavitos, while the SE axis intersects the top of the relatively flat Kolonos Agoraios ridge. By contrast, an astronomical rationale is supported by the Altar's twin reciprocal Minor Lunar Standstill alignments with horizon altitudes >+5° (to minimize the effects of atmospheric refraction), as well as by primary historical evidence. According to Graeco-Roman architectural design prescription, *hypaethral* edifices were constructed explicitly to honor Jupiter Lightning (*Ζεύς*), the Heaven (*οὐρανός*), the Sun (*Ἥλιος*) ***or the Moon*** (*Σελήνη*). Moreover, by the Vth century BC, "Scientific astronomy, we might say, was Greek or perhaps late Babylonian, Chaldean" (Sarton, 1952). For example, Anaxagoras of Clazomenae ("Lord of the Agora," ca. 510–428 BC, the pre-Socratic natural philosopher who moved to Athens from Ionia) suggested that "the eclipses of the moon are caused by the shadow of the earth and other bodies; and that eclipses of the sun are caused by the moon." Meton of Athens (ca. 432 BC) proposed that 235 synodic lunar months closely approximate 19 solar tropical years, and thereby correlated luni-solar calendric cycles (Table 3; Thurston, 1994).

Given its historical context, the construction of a VIth century BC Athenian *hypaethral* edifice with reciprocal NE and SW vectors aligned to the Full Moon at the Minor Lunar Standstills nearest Midwinter (+18. 66° decl.) and Equinoxes (-5.1° decl.) appears more likely the product of deliberate astro-architectural prescription, than of either coincidence (cf. Schaeffer 2007) or topo-geomancy. One might even imagine that Athenian observations of minor lunar standstills in the VIth century were used by Meton to supplement Babylonian data for deriving or confirming his theory of a 19-year luni-solar period (cf. Greswell 1862; Sarton, 1952; Thurston, 1994).

In conclusion, the Altar of the Twelve Gods deserves recognition as monumental architecture to honor the Cosmos (see Prichett, 1947, 2001; Prichett and Neugebauer, 1947); if not as astro-architecture to time Midwinter celestial clashes between Zeus and Typhon (Plutarch, ca. AD 120) observed as lunar eclipses by Athenians in the preceding century at 19-year intervals:

Partial Lunar Eclipse 29 December 560 BC (Saros 35);
Total Lunar Eclipse 29 December 541 BC (Saros 45);
Partial Lunar Eclipse 30 December 522 BC (Saros 55)

– this last coinciding with the Altar's construction under Pisistratos the Younger, Archon of Athens, 522/521 BC (Figure 6).

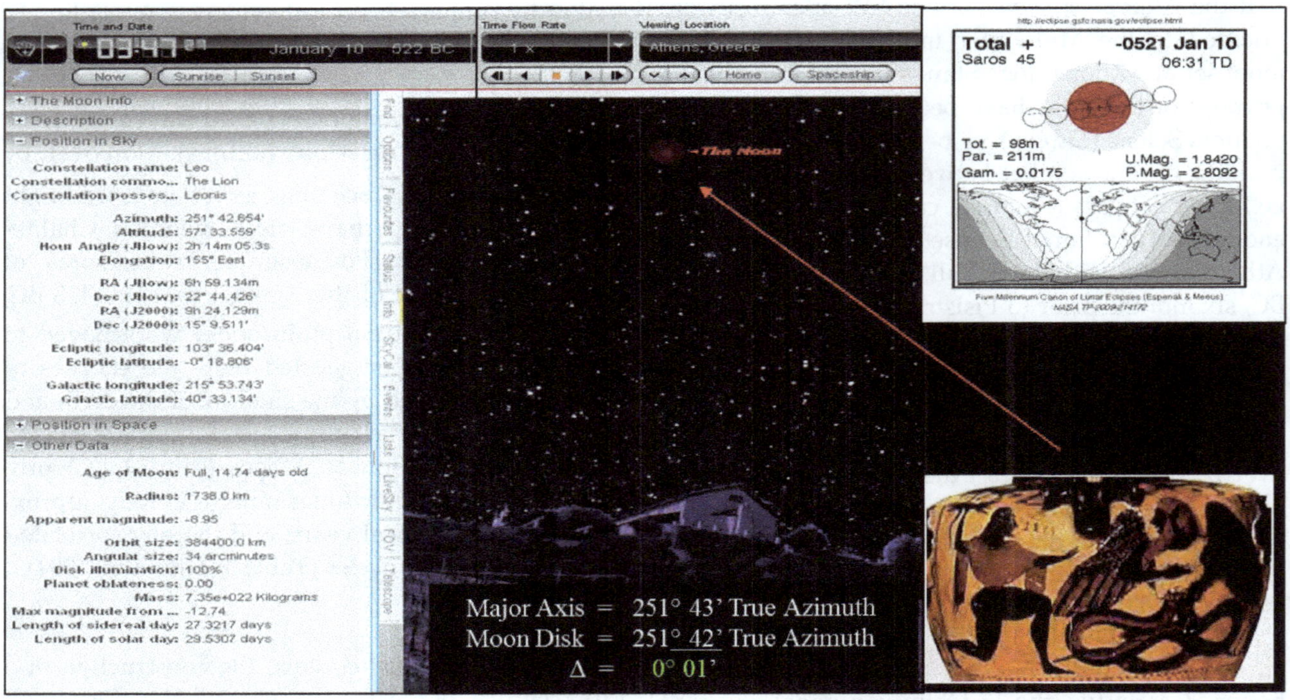

Figure 5. Typhon as Total Lunar Eclipse at Midwinter Full Moon over Athens, 10 January 522 BC. (Starry Night Pro Plus-6; https://eclipse.gsfc.nasa.gov/LEcat5/LE-0599--0500.html; Greek vase from Etruria, Italy, ca. 525 B.C., Staatliche Antikensammlungen Nr. 596, Munich)

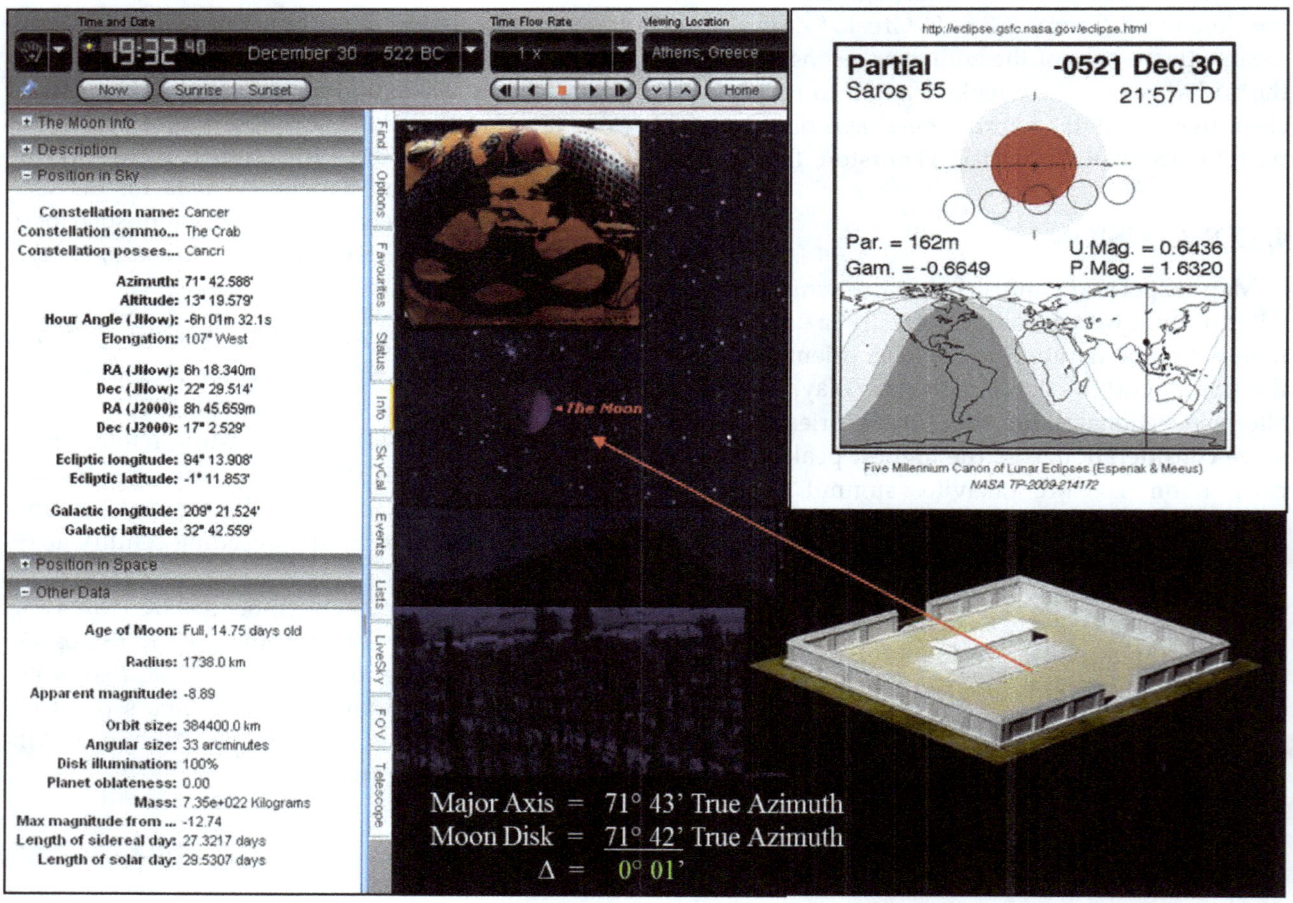

Figure 6. Typhon as Partial Lunar Eclipse at Midwinter Full Moon over NE peribolus opening, 30 December 522 BC. (Starry Night Pro Plus-6; https://eclipse.gsfc.nasa.gov/LEcat5/LE-0599--0500.html) Greek vase from Etruria, Italy, ca. 525 B.C., Staatliche Antikensammlungen Nr. 596, Munich)

ACKNOWLEDGEMENTS

We thank the anonymous reviewers for their constructive comments. The author would especially like to thank the Ephorate of Antiquities for Athens, and American School for Classical Studies in Athens for their generous assistance and technical support.

REFERENCES (online citations accessed June 2018)

Boutsikas, Efrosyni (2008) Placing Greek Temples: An Archaeoastronomical Study of the Orientation of Ancient Greek Religious Structures, *Archaeoastronomy,* Vol. XXI, pp. 4-19. https://www.academia.edu/362755/Placing_Greek_Temples_An_archaeoastronomical_study_of_the_orientation_of_ancient_Greek_religious_structures

Camp, J. M. (2010) *The Athenian Agora Site Guide*, Athens: The American School of Classical Studies at Athens, Fig. 3.

(1986) *The Athenian Agora: Excavations in the Heart of Classical Athens*, New York: Thames & Hudson, p. 45.

Crosby, M. (1949) The Altar of the Twelve Gods in Athens, *Hesperia Supplements*, Vol. 8, pp. 82-103, 447-450. https://www.jstor.org/stable/1353886?seq=1#page_scan_tab_contents

Gadbery, L. M. (1992) The Sanctuary of the Twelve Gods in the Athenian Agora: A Revised View, *Hesperia,* Vol. 61, pp. 447-489.

Greswell, E. (1862) *Origines kalendariae Hellenicae : or, the history of the primitive calendar among the Greeks, before and after the legislation of Sōlon*, 2 vols., Oxford: Oxford University Press.

Hawkins, G. S. (1983) Program STONEHENGE, *Mindsteps to the Cosmos*, New York: Harper & Row, pp. 328-330.

(1973) *Beyond Stonehenge,* New York: Harper & Row, pp. 303-304.

(1968) Astro-Archaeology, *Vistas in Astronomy* (A. Beer, ed.), Vol. 10, pp. 45-88.

Kleiner, D.E. (1992) *Roman Sculpture*, New Haven and London: Yale University Press.

Mauzy, C. A. (2006) *Agora Excavations 1931-2006 A Pictorial History*, Athens: American School of Classical Studies at Athens., Fig. 10.

Meritt, B. D. (1966) *Inscriptions from the Athenian Agora,* American School of Classical Studies at Athens, Princeton, NJ, Fig. 1 (photo credit Alison Frantz).

National Air and Space Administration, *Index to Five Millennium Catalog of Lunar Eclipses,* Greenbelt, MD: https://eclipse.gsfc.nasa.gov/LEcat5/LEcatalog.html

National Oceanic and Atmospheric Administration (2018), Magnetic Field Calculator, Boulder, CO: NCEI, https://www.ngdc.noaa.gov/geomag-web/ .

Plutarch (ca. AD 120) ***De Iside et Osiride*** 30:20,44:14,55:18, [J. Gwyn Griffiths, trans.] Cardiff: University of Wales Press, 1970, pp. 165, 189 and 207.

Prichett, W.K. (2001) *Athenian Calendars and Ekklesias,* Amsterdam: J.C. Gieben.

Pritchett, W. K. (1947), Julian Dates and Greek Calendars, *Classical Philology*, Vol. 42, no. 4 (Oct.), pp. 235-243. https://doi.org/10.1086/363049

Pritchett, W. K. and O. Neugebauer (1947) *The Calendars of Athens*, Cambridge: Harvard University Press.

Sarton, G. (1952) *A History of Science: Ancient Science Through the Golden Age of Greece, Vol. I,* Cambridge: Harvard University Press, pp. 119-120.

Schaefer, B.E., The Utter Failure of the Lunar Standstill Myth in Archaeoastronomy, *Bulletin of the American Astronomical Society*, Vol. 39, p. 872. http://adsabs.harvard.edu/abs/2007AAS...211.8201S .

Thucydides of Athens (ca. 430 BC) *History of the Peloponnesian War*, 6.54, 6-7 (Benjamin Jowett trans.) Oxford: Clarendon Press, 1881, p. 6.54.

Thurston, H. (1994) *Early Astronomy,* Ann Arbor, MI: Edwards Brothers, Inc., pp. 66, 110-111.

Tiede, V.R. (2017) ΒΩΜΟΣ ΔΩΔΕΚΑ ΘΕΩΝ/Altar of the Twelve Gods: An Astro-Archaeological Analysis (PowerPoint slides), 2017 SEAC conference, Santiago de Compostela, Spain, 18-22 September 2017. http://yale.academia.edu/VanceTiede

(2016) ARA · PACIS · AVGVSTAE: An Astro-Archaeological Analysis, *Mediterranean Journal of Archaeology and Archaeometry*, Vol. 16, no. 4, pp. 267-274. http://maajournal.com/Issues/2016/Vol16-4/Full37.pdf

U.S. Naval Observatory (2018) Altitude and Azimuth of the Sun, Washington, DC: USNO. http://aa.usno.navy.mil/data/docs/AltAz.php .

Vitruvius (c. 25 BC), *De Architectura Libri Decem*, (M.H. Morgan, trans.), New York: Dover, 1960, pp.5-6 (I:i,3), pp.14-15 (I:ii,5)

11

THE ALMOGAREN OF RISCO CAÍDO: A SINGULAR ASTRONOMICAL SANCTUARY OF THE ANCIENT CANARIANS

Julio Cuenca Sanabria[1], José de León[2], Cipriano Marín[3], José Carlos Gil[1,4], Juan Antonio Belmonte*[5], Carlos Gil Sarmiento[1] and José Miguel Márquez Zárate[6]

[1] *PROPAC and Cabildo de Gran Canaria, Spain*
[2]*Heritage Service, Cabildo de Gran Canaria, Spain*
[3]*Unesco Center of the Canary Islands and Cabildo de Gran Canaria, Spain*
[4] *Instituto Canario de Estadística (ISTAC), Gran Canaria, Spain*
[5] *Instituto de Astrofísica de Canarias and Universidad de La Laguna, Tenerife, Spain*
[6] *Architect, Tenerife, Spain*

**Corresponding author: Julio Cuenca Sanabria (juliocuenca@gmail.com)*

ABSTRACT

The almogaren (rock-cut sanctuary) of Risco Caído was discovered in 1996 in the Canary island of Gran Canaria. It is a paradigmatic example of a complex where light and shadow effects of an astronomical character have been found within the recent archaeological discoveries of a religious and ritual character in the Canaries. The main artificially excavated camera of the cultural complex Cave 6 takes the form of a cylinder, topped with a dome in the form of paraboloid. In this dome, a 2m long tunnel is excavated by which the light of the Sun penetrates at dawn, from spring to autumn equinoxes. The entering light projects enigmatic images on the western wall of the sanctuary, where numerous pubic triangles (vulvae – the universal symbol of fertility –) are recorded in low relief. Two dots of light of the sun first illuminate the decorated wall in March 19th (and september 25th) in the proleptic Gregorian Calendar, colliding and forming a single image for the time of the equinox, thus allowing the determination of a rough midpoint in time between the solstices. The rising and ascending sun then penetrates the cave during the spring and summer months, reaching its extreme at the moment of the summer solstice when the light takes a form roughly resembling a phallus illuminating the vulvae. This paper will desscribe the site and these series of illuminating effects and will discuss how this could have been interpreted by the ancient inhabitants of the island within the context of a lunar-solar calendar related to the cycle of fertility and permanent regeneration of life.

KEYWORDS: UNESCO, Archaeoastronomy, Archaeology, ancient Canarians, Gran Canaria, Risco Caído

1. INTRODUCTION

In 1996, the first author discovered the almogaren – or ceremonial centre –at Risco Caído in an archaeological prospect campaign in the mountains of Gran Canaria (Cuenca Sanabria, 2008 and 2012). This sanctuary is a unique, outstanding, religious archaeological site of the ancient Canarians, where astronomical relationship could have played a most relevant role. This was actually the rediscovery of a place of exceptional symbolic significance for the indigenous people.

The archaeological complex is situated on the edge of the Caldera de Tejeda, in the highlands of the north-western slopes of the island of Gran Canaria, 960 metres above sea level (see Figure 1). It is located in a remote and secluded part of Barranco Hondo ravine. Archaeological research indicates that the almogaren at Risco Caído was located at a strategic point on one of the main 'sacred routes' used by the indigenous people of the northern lowlands (area of Galdar, capital of the island in ancient times) to reach their main mountain sanctuaries situated in the Caldera de Tejeda and the surrounding mountains, such as Roque Bentayga (Esteban et al. 1996-7).

In these lands, where legends and actual facts mix and occupied by the early inhabitants of the island, archaeological studies have recovered a series of sites that have been identified as places of worship and ritual. They are found either on the top of prominent rock tors, where structures were carved out of the surface bedrock or, more frequently, in hollowed-out caves located in the most inaccessible or secluded parts of certain mountains, as is the case here (see e.g. Belmonte et al. 2018, this volume).

Figure 1. Panoramic view of the archaeological site of Risco Caído © Julio Cuenca

The two main and most remarkable neighbouring caves of the Risco Caído complex (caves C6 and C7, see Figure 2) comprise what the ancient Canarians called an 'almogaren', that is, a temple, a place where rituals are performed, where people congregated at certain times of the year or when the rains were scarce and rituals were needed to pray for them so as to ensure the cereal harvests. As ethnohistorical sources relate: *The houses of religious women were sacred for delinquents, they called them Tamogante en Acorán, which means the house of God. They had another house on a high cliff called Almogarén, which is a sacred house; there they invoked the deities and made sacrifices, sprinkling milk every day, which their God above looked down upon and they kept livestock for this purpose. They also went to two very high crags: Tirmah in the district of Gáldar, and another in Tirahana called Humiaia at White Cliffs. They swore by these two crags with utmost solemnity, they came to them in procession with branches and palm fronds, and the Maguas or virgins with their pitchers of milk to sprinkle; they called out and raised their hand and faces to the heavens and circled the rocky tor and descended, from there, they went to the sea to thrash it with their branches* (Gómez Escudero, 1682).

Archaeological evidence reaffirms the sacred nature of the site as one of the most important almogarenes of the aboriginal Canarians (see, de León and Marín, 2018 for a comparative). The first striking fact is that these troglodyte constructions are located on top of a mountain that had previously been covered by a laurel forest. Thus, this was a hidden, isolated place with abundant water, far from human settlement. Consequently, this is a very significant natural landscape, containing the mountain, rain forest, the caves, water sources and even plant fossils. Another archaeological indicator is the architecture itself. In this case, we have two unique hollowed-out caves that have been built differently from the caves used as dwellings or for economic use.

However, two other features unequivocally confirm the sanctuary-like character of this site. One is the presence of symbols or bas-relief rock engravings on the inside walls, in the form of inverted triangles. These are clear representations of pubic triangles, a universal sign of fertility (López Peña et al., 2002; see Figure 3). The other is that there are numerous circular cup-marks and small niches chiseled into the inside walls and floors of both chambers. The presence of these cultural expressions is considered a definite indicator of a place of worship and ritual (Cuenca Sanabria, 1992).

Cave 6, the most relevant of the Risco Caído complex, has an original and completely unique layout. The floor is practically circular. The cave has curved walls and a hollowed-out vault rising almost five metres from floor level, forming an almost perfect parabolic dome (see Figure 4). This aspect is extremely important: since no other example of an artificial cave with a domed roof of this complexity and shape is known in the aboriginal world of the Canary Islands. On the eastern side of the dome, almost coinciding with the highest point is a conduit or oculus facing east, allowing light to flow in through it (Fig. 4).

The most appealing feature of this remarkable site is the lighting effect representation produced by the sacred sun and moon inside Cave C6, in the form of images projected by the light passing across a light tunnel or oculus. This image changes shape as the days and months pass and as they move along the wall with its altar-like representations of triangular engravings and cup-marks. These may eventually have functioned as reference points in a lunar-solar calendar. This visual calendar would have started at a date very close to the spring equinox when the projected solar images first appear (equinoctial phenomenology is frequent in sacred sites of Gran Canaria, see Esteban et al. 1996-7), continuing until the autumn equinox. From then until the next spring equinox, either by chance or by deliberated desire of the designers, the light of the full moon between the months of October and February would have illuminated the engravings inside the sanctuary.

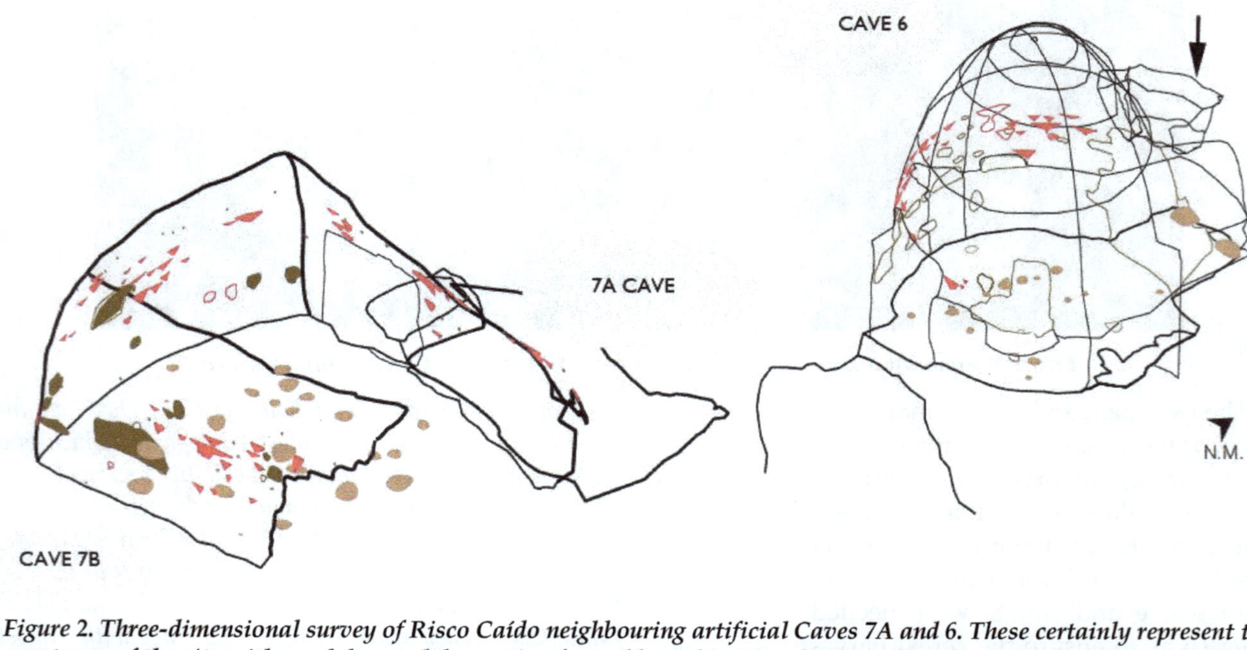

Figure 2. Three-dimensional survey of Risco Caído neighbouring artificial Caves 7A and 6. These certainly represent the sanctuary of the site with an elaborated decoration formed by pubic triangles (in red) and cup-marks (in brown). Notice the dome aperture or oculus in the upper right extreme of Cave 6. © Carlos Gil Sarmiento, PROPAC.

Amazingly, it is still possible to watch this visual tale unfold; one that has been projected inside this temple for countless centuries. It is thought to relate to fertility rituals, the fertility of Mother Earth being represented here by the carved ideogram for the female pubic triangle. Hence, this cave site is an ingenious creation that functions both as a sacred place and perhaps as an astronomical marker, where certain events such as the equinoxes and the summer solstice (when the moving image will stop and turn back) would be visually highlighted to coincide with the aboriginal Canarians' rituals, while eventually allowing them to keep a calendar, which they may have used to regulate farming and productive activities.

Archaeological research is still ongoing at Risco Caído. Two C14 radiocarbon dates have been obtained so far, one in 2013 from a sample of wood from a Canary Guelder Rose shrub (Viburnum rigidum; 1415-1450 A.D.), and the other in 2014 from an older organic sediment (1295 ± 25 A.D.), both collected from the inside walls of Cave C6. A thermally altered floor was discovered inside this cave during 2015: a sample taken from this still awaits paleomagnetic dating (de León and Marín, 2018).

Figure 3. Close-up of the series of the engravings of pubic triangles, a symbol of fertility, and the cup-marks scattered over the west wall of Cave 6. © Julio Cuenca.

The incoming light hits the mural of rock engravings on the back wall, opposite the light tunnel or oculus. For six months of the year, sunlight illuminates this wall creating different shapes depending on the time of day and the season of the year (see Figure 5). The floor of the cave has also been worked and levelled, and a whole set of unconnected circular cup-marks have been chiseled into it. The largest concentration of these cup-marks lies at the base of the rock-engraving mural. Halfway up the south wall is what may have been a silo. The entrance is rectangular and it has opposing orifices on the floor and lintel for fitting the closing system.

Perhaps most strikingly, halfway up the west wall is a composition of thirty or so engravings of inverted equilateral triangle-shaped motifs, forming a decorative frieze of two parallel rows (Figs. 2 and 3). Associated with these triangular motifs and forming part of the same panel, are numerous cup-marks and generally circular orifices. There are also two large niches within the same wall panel, the larger of which is rectangular in shape (see Figure 6).

With its unique architecture and the interesting design of the oculus, together with its remarkable rock engravings, Risco Caído Cave 6 is a religious-ritualistic structure that is unparalleled in pre-Hispanic Canary Island archaeology. As we shall soon show in greater detail, its design may also incorporate probable astronomical and calendar-related relationships. The only cave sanctuary anywhere in Gran Canaria, that bears any similarity is the Tara sanctuary, in Telde, which is currently under study, although Risco Caído indeed is a more evolved and sophisticated version.

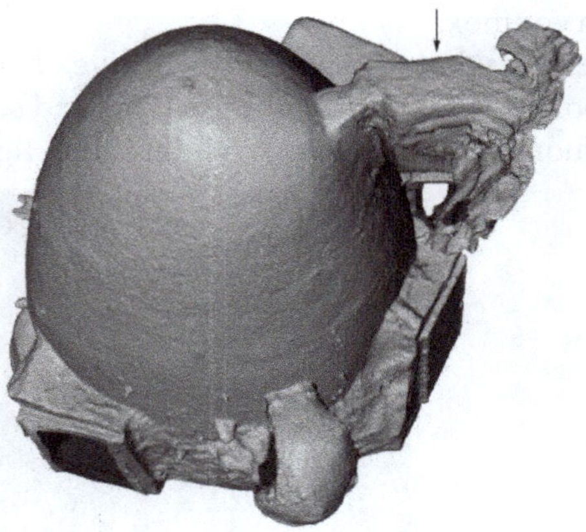

Figure 4. Digital model of Cave 6 using photogrammetry techniques and three-dimensional laser scanner. The paraboloid shape of the dome and the oculus (signalled by an arrow) are emphasized. © Carlos Gil Sarmiento.

2. RISCO CAÍDO, ARCHAEOSTRONOMY IN ACTION

It might be the case that Risco Caído Cave 6 acted as an ingenious astronomical marker. The light of the rising sun illuminating the interior of the sanctuary possibly signaled the arrival of the equinoxes and the summer solstice (Fig. 6), while the light of the full moon perhaps marked the passing of the 'winter' half' months after the autumn equinox until the next spring equinox, if this would not happen just by chance, which is unlikely since the ancient Canarians measured time through the year by 'moons' (Belmonte and Hoskin, 2002). These together would provide a calendar based on a lunar-solar year. This fact would indeed have allowed these aboriginal people to govern the vital yearly farming cycle.

The cave's significance as a possible astronomical marker was suggestive from the moment of its discovery. However, the fact that the light of the sun's rays enters two days before the spring equinox (a key moment of the ancient Canarian calendar; Esteban et al., 1996-7; Belmonte and Hoskin, 2002) and that the phenomenon continues until the autumn equinox was confirmed and documented in 2012, as part of the research work carried out during the first phase of the 'Archaeological Conservation, Protection and Research of the Cultural Complex' project. Every day at sunrise during the 'summer' half of the year, the cave is illuminated by a beam of light. This is projected onto the west wall and changes shape and direction as the sun rises and climbs into the sky. Each day, the descending beam of light illuminates the pubic triangles and niches located in the middle part of the mural, changing shape until it fades out in its path across the panel (see Figure 7). After six months of illuminating the cave every day, these effects disappear altogether at nearly the autumn equinox.

Hence, what is highly suggestive, and directly supported by empirical observations and subsequent astronomical calculations, is that the light enters through an aperture in the dome that could be described as an oculus, striking the dome for the first team each year almost at a time close to the spring equinox, dividing the year into two halves and allowing the determination of an approximate midpoint in time between the solstices. The recurring effect of the sunlight on the engravings would have progressed from two days before the spring equinox until the summer solstice, when the sunrise reaches its most northerly point on the local horizon of Risco Caído and its most southerly position on the panel of engravings (Fig. 7), before turning back along the same track until two days after the autumn equinox, when the phenomenon of the sunlight entering the cave comes to an end.

The chronicles of the Conquest reiterate the importance of the equinoxes (and particularly the spring equinox, as shown by the chronicles of Sedeño) and of the solstices, which are implicitly mentioned with the sun entering the sign of Cancer. Both are considered key moments in the annual cycle, as shown in the chronicle of Marín de Cubas (Esteban et al. 1996-7; Belmonte and Hoskin, 2002; Belmonte, 2015).

Figure 5. The diagram – a superposition of several photographs – shows the image projected by sunlight onto the panel of engravings at different moments of the year once the typical image of each epoch is clearly visualized. The sun's rays enter every day within the seasonal cycle between nearly the spring and autumn equinoxes. The sun's rays penetrate the chamber for the first time two days ahead the spring equinox and for the last time two days after the autumn equinox. The two separate dots on the extreme right mark these dates, close to the two equinoxes, moments when the two dots fused into a single image. In the intervening months of spring and summer, the image takes different forms that illustrate a rich visual language ranging from a 'primeval seed' to a 'fertilising phallus' (at the time of the solstice, time of harvest). In the middle, the light forms an image that resembles the Palaeolithic Venuses which, in turn, are reminiscent of certain female idols found in Gran Canaria. This, together with the presence of pubic triangles, cup-marks and niches along the light's path –as illustrated in Fig. 7 – suggests that we are dealing with a suggestive astronomical and symbolic phenomenology related to fertility cycles and perhaps timekeeping. © Julio Cuenca and José Carlos Gil.

Figure 6. The west wall of Cave C6, with the shapes of the extant pubic triangles superimposed for emphasis, viewed on the morning of the spring equinox. The patch of projected sunlight can be observed as it starts to descend the wall, moving down and right. The arrows indicate the most singular triangle of the panel (right) and the triangle touched by the path of sunlight at the summer solstice. Most of the central engravings on the panel (c. 65%.of the pubic triangles, including several of those to the left and right of the arrows as shown by our calculations; de León and Marín, 2018) are covered by sunlight or moonlight at different epochs of the year when a node regression lunar cycle is eventually considered. © Julio Cuenca

Figure 7. Photographic overlay showing consecutive images of the path of sunlight at the summer solstice (declination +23½°). The image, starting in the form of a phallus or shield, falls upon or fills several pubic triangles during its sweep down the wall, eventually becoming a dot that disappears into a niche, the lowest one illuminated by the rays of the sun or moon. This cycle of images is little different from the one that occurred in aboriginal times and serves as an example of what could be seen inside the sanctuary. This nice phenomenology is perhaps related to the fecundity rituals performed at the sanctuaries of Gran Canaria, at the great feasts celebrated in the lunation after the summer solstice. © Julio Cuenca and José Carlos Gil.

Further research work and the evidence collected over almost a decade have enabled us to gain a more accurate understanding of the probable astronomical function of the site. As already discussed, it is not only sunlight that penetrates Cave 6. The light of the full moon as it rises and climbs during the winter months from October to March (the time of the highest rainfall when the moon 'signals' the seasons in traditional Canarian agriculture; Belmonte and Sanz de Lara, 2001) also illuminates the main panel of engravings. The full moon also has the power and capability to illuminate the interior of the cave. This phenomenology happens from around the time of the autumn equinox and at each subsequent full moon until around the time of the vernal equinox. Although this effect could be for granted due to the architectural design and orientation of the sanctuary, we believe that it was certainly observed in ancient times.

The moon's movement is regulated by the 18.6-year lunar node cycle. This, together with the fact that full moons fall on different dates in different calendar years, mean that moonlight will not always fall on the same engravings from one year to another. During every 18th or 19th year, the moon's follows its most northerly path through the sky (the 'northern major lunar standstill limit'), which is beyond where the sun ever reaches. Interestingly, at these times, full moonlight can illuminate triangles beyond those illuminated by sunlight even at the summer solstice, lighting up some of the triangles situated to the left of the sweeps during the seasonal cycle between the spring and the autumn equinox (see Fig. 6).

3. FINAL REMARKS AND CONCLUSIONS

The structural profile of the ceiling of Cave 6 at Risco Caído is a parabola (Fig. 4), unlike all the other known caves on the island. Some have flat and some have domed ceilings, but their perimeters are generally rectangular and they usually have intentionally rounded corners, so as to reduce the maximum concentration of tangential forces where the ceiling meets the vertical faces of the walls. These, in turn, are vertical or lean in slightly towards the inside of the cave (e.g. Cave 7A, Fig. 2). From the point of view of its size, Cave 6 could perfectly well have been built like all the others, in a perceptibly prismatic shape. However, this is not the case!

Creatively and intentionally, the excavation and final ascending shape of the cave is a vaulted dome with a variable parabolic profile, with the main focus at the height of the window or oculus, facing east, through which natural light penetrates. In this way, the cave harnesses the geometrical qualities of a parabola not only to create a space that protects and embraces but also to obtain a uniform and concentrated distribution of the light that it captures, as well as the best diffusion of sound. These conditions enhance its quality as a space of sensations and beliefs.

Natural sunlight and moonlight – directed and, in turn, modified by the conduit that channels it like a spotlight into the cave – generates a series of images projected on to the opposite (west) wall that functions as a stone altar and potential astral marker. In addition to this, the beam of light that crosses the focus of the parabola creates indirect general lighting because its reflection is intensified by the alternating white rings of salts adhering to the surface of the upper laminar strata of the cave. These irradiate most of the complementary diffuse light that is concentrated downwards towards the base or inhabitable surface of this architectural masterpiece. Furthermore, thanks to its shape, the cave is evenly lit, without any extensive shadows or dazzling, apart from on the aforementioned stone altar where the solar and lunar phenomenology unfolds.

The construction process, assembly or system is one that, at the very least, materializes significant functional, structural, indeed artistic and symbolic and possible astronomical knowledge. It is a creation that has never been seen before in the islands; a work of great value for an isolated, unfamiliar with metal, culture like the one that produced this design of overall, integral conception. The architectural expression of the sacred site of Risco Caído, and the way it astronomical phenomenology might function, is outstanding for that historical period on the Island of Gran Canaria, because the inhabitants eventually applied what they learned from constantly observing the skies, because of the lack of auxiliary resources and, above all, because it is an eminently social creation conceived in abstraction.

The synchrony that apparently connects the Risco Caído sanctuary to the cosmos itself requires a level of perfection in form, proportion and execution that speaks of the singular technical achievements of the island society that designed and built it, that might be comparable with other ancient cultures around the world (Belmonte et al., 2016). The layout of the different architectural elements is based upon structural and compositional laws that transcend time, which are still present and which speak of the eternal and the unchanging. It would indeed be a perfect example of cultural astronomy in action.

Risco Caído is the most outstanding and enigmatic manifestation in the proposed Cultural Landscape of 'Risco Caído and the sacred mountains of Gran Canaria' (de León and Marín, 2018; Belmonte et al, 2018, this volume). It is in in certain aspects representative of the astronomical culture and knowledge of the ancient Canarians. This space would eventually have an interesting coherency as a sacred site closely related to celestial events – the skyscape – for measuring time, for marking commemorative dates, and for holding rituals. In this context, Risco Caído and the rest of the sacred sites in the neighborhood, such as the almogaren at Roque Bentayga (Esteban et al. 1996-7; Belmonte, 2015), are outstanding attributes because of their potential astronomical significance in a culture that evolved in isolation from the ancient know-how imported from the pre-Islamic Maghreb, and which developed as a genuine expression in this new island environment. Cave 6 at Risco Caído is a complex artificial indoor sanctuary with plausible astronomical connotations, where patterns of sunlight and moonlight interact with a series of engravings, possibly marking the passage of days and lunar months between the solstices and the equinoxes. It is an excellent example of a sacred site where an interesting astronomical phenomenology would be manifested. Indeed, its enigmatic anthropological perspective is still a clue to be disentangled and understood.

ACKNOWLEDGEMENTS

This work is supported by the Cabildo de Gran Canaria. The authors would especially like to thank all the people who have been involved in the different phases of study of the archaeological site for their support and cooperation and the two anonymous referees for their smart suggestions to improve the paper. JAB acknowledges the projects P/310793 'Arqueoastronomía' of the IAC, and AYA2015-66787 'Orientatio ad Sidera IV' of the Spanish MINECO.

REFERENCES

Belmonte, J.A. (2015) Pre-Hispanic sanctuaries in the Canary Islands. In: *Handbook of Archaeoastronomy and Ethnoastronomy*, C. Ruggles (ed.), Springer, Heidelberg, pp. 1115-24.

Belmonte, J.A., Sanz de Lara M. (2001) *El Cielo de los Magos*, La Marea, la Laguna.

Belmonte, J.A., Hoskin M. (2002) Reflejo del Cosmos. Atlas de Arqueoastronomía en el Mediterráneo Antiguo, Equipo Sirius, Madrid.

Belmonte, J.A., González-García, A.C., Rodríguez Antón, A.; Shaltout, M. (2016) Orientatio ad Sidera (OAS): highlights of a decade of archaeoastronomical research in the Mediterranean region and beyond, *Mediterranean Archaeology and Archaeometry*, Vol. 16, 4 (2016), pp. 93-101.

Cuenca Sanabria, J. (1996). Las manifestaciones rupestres de Gran Canaria, In: *Manifestaciones rupestres de las Islas Canarias*, Gobierno de Canarias, Las Palmas, pp. 133-222

Cuenca Sanabria, J. (2008). El Culto a las cuevas entre los aborígenes canarios: el almogarén de Risco Caído. In: *El Patrimonio troglodítico de Gran Canaria,* Asociación Insular de Desarrollo Rural de Gran Canaria, Las Palmas de Gran Canaria, pp. 77-95.

Cuenca Sanabria J. (2012) La arquitectura de lo sagrado de los antiguos canarios. In: *VIII Congreso de Patrimonio Histórico: Arquitectura indígena.* Cabildo de Lanzarote, Arrecife.

de León, J., Marín, C., eds. (2018) *Risco Caído and the sacred mountains of Gran Canaria Cultural Landscape*, Cabildo de Gran Canaria, Las Palmas.

Esteban, C., Belmonte J.A., Schlueter R., González O. (1996-7) Equinoctial markers in Gran Canaria Island, Parts I & II, *Archaeoastronomy*, Vol. 21, pp. S73-S79, and Vol. 22, pp. S51-S56.

Gómez Escudero, P. (1682): in Morales Padrón, F. (1993) *Canarias. Crónica de su conquista. Transcripción, estudio y notas.* Cabildo de Gran Canaria, Las Palmas de Gran Canaria, XIX, p. 440.

López Peña, F., Cuenca Sanabria, J., Guillén Medina, J. (2002) El triángulo púbico en la prehistoria de Gran Canaria. Nuevas aportaciones, In: *XV Coloquio de Historia Canario Americana,* Cabildo de Gran Canaria, Las Palmas de Gran Canaria, pp. 2243-64

12

DE QUATTUOR PARTIBUS MUNDI MEDIEVAL SACRED BUILDINGS ON THE VIA FRANCIGENA IN NORTHERN AND MIDDLE ITALY: SOLSTICE ALIGNMENTS AND ORIENTATIONS

Eva Spinazzè

Ca' Foscari University of Venice, IUAV University of Architecture Venice, University of Zurich, Independent Researcher

Corresponding author: aquadelph@gmail.com

ABSTRACT

This paper discusses different aspects of sun light linked to a religious rite and combined with ecclesiastical architecture. Our focus is on the four points of the solstices which form two sectors, one towards the sunrise and the other to the sunset. This examination is mainly based on the study of the ancient thought of scholars and Fathers of the Church. Despite the deep significance of the solstices (the Births of Jesus Christ and John the Baptist, His forerunner), only few of the over two hundreds medieval churches, situated in the North of Italy and long the old pilgrimage route, the Via Francigena, are oriented towards the solstice and equinox rise/set points on the horizon.

Georeferenced surveys on field with astronomical calculations were carried out on these sacred buildings by the author and the results were strengthened with primary and secondary written sources.

Throughout medieval times on path of faith a lot of hostels and sacred buildings were founded for pilgrims and travellers and a great deal of diaries written by the pilgrims survived so far. These travel reports record important building descriptions and knowledge of that time. In detail we will discuss a medieval drawing, one of the few we can admire and which represents the Holy Sepulchre of Jerusalem. It is particularly interesting to note, in this accurate drawing, the position of the doors of the building which appear oriented to the eight Winds Rose and the summer/winter solstices; the niches which show the cardinal points; and the *Sepulchri* that incorporates an Easter alignment. Finally, we discuss solsticial alignments of medieval church axes and of the position of sunlight inside sacred buildings among the summer solstice, when the sun is at its highest and simultaneously begins to appear lower.

KEYWORDS: sacred medieval buildings, orientation, light incidence, solstices, objective and written evidence, georeferenced surveys, the Via Francigena.

1. INTRODUCTION

For the last ten years the author has studied over 200 medieval sacred buildings regarding their orientation and their references, thus revealing an ancient building tradition. For each church she made a georeferenced survey with astronomical calculations and then cross-checked the results with primary and secondary sources. The analysed churches are situated in northern and central Italy and in the Swiss Alps. One third of them are along the old pilgrimage route, the Via Francigena (Spinazzè, 2016a). This was an important route for economic and religious reasons which facilitated cultural exchange between Northern Europe and Italy, just as the pilgrimage route to Santiago de Compostela that connected central Europe to the Atlantic coast. Along these paths of faith, numerous sacred buildings and structures, providing hospitality to travellers and pilgrims, were built about a day's walk one from another. Thanks to many medieval travel diaries written by pilgrims (Donner, 2002, *Pereginatio Egeriae; Itinerarium Burdigalense. Breviarius de Hierosolyma; Adamnani de locis sanctis libri tres. The Epitome of S. Eucherius about certain Holy Places,* 1890. Procopius, 1896. *Citez de Jherusalem,* 1896. Hodoeporicon S. Willibaldi, 1880/1881. Eusebius of Caesarea, *Life of Constantine*), scholars have primary evidence that construction knowledge travelled from one place to another. Furthermore, the beginning of every single stage of the journey began on a special day, e.g. with the sunrise on a saint's feast or on a holy day, sometimes illustrated with a drawing. In this section, we are going to discuss a medieval drawing (9th c.) which represents the *Sancti Sepulchri* of Jerusalem, one of the oldest surviving architectural drawings (Adamnanus, *De locis sanctis*, Ms. Rh. 73).

Cultural knowledge is essential for interpretation. In medieval times the horizon was divided into four parts, delimitated by the four solstice sunrise/set points, as we learn from the many treatises written by the Fathers of the Church and the medieval scholars. The solstices were related to significant religious feasts: the summer solstice with John the Baptist and the winter solstice with Jesus Christ. Moreover, the orientation of most medieval churches are delimited between these two points, formed by the solstices, with the highest and lowest light, and almost all of them show an alignment laid down by its history and dedi-cation (Spinazzè, 2007/2008; 2015a; 2016a; 2016c; 2017).

2. LITURGICAL ASPECTS THAT DETERMINED THE ORIENTATION

In the Christian World the solar theological concept for orientation was surely the most used (Podossinov, 1991). This theory reflects in:

- *the orientation – direction of the prayer (Figure 1):* at first pagans, and then Christians turned to the risen Sun and lifted up their hands to greet and pray the brightest star, seen as their divinity or God (Dölger, 1925. Vogel, 1964. De Blaauw, 2010, 2008, 1994. Piva 2010. Wallraff, 2001. Plato, *Symposium*, 220d. Cicero, *De natura deorum*, I.28.79. Virgil, *Aeneid,* XII.172-176);

Figure 1. Celebration of the mass, in Décrétales de Grégoire IX, Ms. 105, f. 133r, 13th c., Archives du Chapitre de Sion (CH).

- *the orientation of the sacred building:* thus, the medieval church took that orientation, especially looking to the point on the horizon where the Sun rose or set on a special day, an important day for the constructor. The orientation could be determined during the foundation ceremony of the church (empirical method) or by astronomical calculations, as it is indicated in treatises for an equinox and solstice alignment (Vitruvius, *De Architecture*, IX.7. References in: Spinazzè, 2016a; 2016b, 2-3);
- *the orientation of the baptistery:* The catechesis of the Fathers of the Church describes the baptismal rite in detail (Cyril of Jerusalem, *Mystagogic Catechesis*, c.313-387. John Chrysostom, *Homilies,* c.349-407. Theodore of Mopsuestia, *Baptismal Homilies,* c.350-428. Narsai c.399-502. Tertullian, *On Baptism*, c.160-c.220. St. Ambrose, *On the Mysteries*, c.339-397. Chromatius bishop of Aquileia, *Sermons,* c.335-408). The catechumen had to turn first to the West to renounce Satan, then he/she entered into the piscina where he/she turned to the rising Sun to obtain salvation. This turning from the West to the East, towards the region of light, symbolized the conversion to the Christian religion and a new re-birth. So, the architectural components of the baptistery and its church have a precise direction in correspondence with the symbolic path of the catechumen, performed inside the sacred building during the rite of baptism; e.g. the basilica of Saint Mary of the Assumption in Aquileia, Italy (Spinazzè, 2016c);

- *the orientation of the sepulture / burial:* archaeological excavation shows that often Early Christian burials goes along the church axes and the head of the dead faces the rising Sun waiting the Final Judgment (*Matthew* 25:31-46); e.g. the Romanesque church of St. Maurice d'Agaune, Switzerland (Spinazzè, 2015b; 2016a);
- *The light incidence inside the sacred building:* in some churches the light which passes the openings of the façade and/or the apse illuminates the altar or frescos on holy days thereby engendering a religious sentiment among believers (Incerti, 1999. Spinazzè, 2016a).

The orientation of a sacred building therefore emphasizes a deep symbolic meaning of its space.

3. THE FOUR SECTORS DEFINED BY THE SOLSTICES

We have the written evidence of numerous scholars and Fathers of the Church describing the path of the Sun and the Moon in their treatises and homilies. Such evidence provides us with a vision of the geographic, astronomical and liturgical concepts existing during the Middle Ages. These various descriptions divide the year into 8 moments and 4 sectors defined by the solstices. One indication can be found in the book *Etymologiae* (III. 51; V.34) by Bishop Isidoro of Seville (c. 560-636). The Bishop states the Sun stations like this: the sunrise and sunset on the winter solstice, 25th December; the sunrise and sunset on the equinoxes, 25th March and 24th September; the sunrise and sunset on the summer solstice, 24th June. He also associates the winter and summer solstices with the births of Jesus Christ and John the Baptist. We find the same information, also in the Roman calendar of the 5th century, which is the official Christian calendar, elaborated by Polemic Silvio, where the summer solstice (*solstitium*) occurs on June 24th and December 25th is memorized with these words *natalis domini corporalis* (Mortem, 1893, 279). Moreover, in the *martyrologium hieronymianum* (4th/5th century) on this day is written *nativitas Salvatoris Domini Nostri Jesu Christi,* and on June 24th appears the name of *sancti Iohannis baptis,* and on March 25th is recorded *Dominis nostris Iesus Christus crucifixus est et est conceptio sanctae Mariae et passio* (Mortem, 1893, 266-267. *Martyrologium Hieronymianum* 1894, II.1, 82, 36).

Also Bede, the Anglo-Saxon monk and scholar (c. 672-735), recommended in his work *De temporum ratione* (*caput* XXX, *De aequinoctiis et solstitiis)* to observe the equinoxes on the 8th *calendas* of April (25 March) and the 8th *calendas* of October (24 September) and the solstices on the 8th *calendas* of July (24 June) and the 8th *calendas* of January (25 December). In another passage of this work (*caput* XXX), Bede states that Christ was conceived and then died on 25th March (on this day the world was created; it is the day of the spring equinox) and Jesus was born on the winter solstice, December 25th (see also Tertullian, *De adversus Judaeos*, VIII. St. Augustine of Hippo, *De civitate Dei*, XVIII.54). Conversely, his forerunner, John the Baptist, was conceived in the autumn equinox, 24th September and was born on the summer solstice the 24th June. Bede explained, as the Creator was seen as the eternal light, the Sun of Righteousness, that's why He was born with the rising of the light, that is 25th December. Instead John the Baptist, the Messenger of Penance, had to diminish and so he was born when the light went shortening, 24th June, leaving his place to Christ.

Bede refers to the lines in the Gospel of John (3:28-30), when Saint John the Baptist exalts Christ: "You yourselves bear me witness, that I said, 'I am not the Christ, but I have been sent before him.' The one who has the bride is the bridegroom. The friend of the bridegroom, who stands and hears him, rejoices greatly at the bridegroom's voice. Therefore this joy of mine is now complete. He (Christ) must increase, but I must decrease."

It is important to distinguish the "liturgical equinox" fixed on March 25th from the astronomical equinox on March 21st [1]. This fundamental aspect has to be recognized, because most medieval churches are not aligned with the Sun on the astronomical equinox, but with the Julian calendar Equinox of March 25th, as the author demonstrated by her scientific studies (Spinazzè, 2016a, c). In the days following the Vernal Equinox, the darkness is completely defeated by the light, relating to the Bible (*Psalm* 18.29); it was believed that the world was created on March 25th, it is the remembrance of the Annunciation to the Holy Mary and the day of the Incarnation, feasts that had already been fixed in the 4th/5th century (Duchesne, 1903, 271-273. Strozzi, 1766, XIV. Brubaker, 2011, 19-22); it is a date with a deep religious meaning.

Throughout the whole Middle Ages these four solsticial points have been connected to the four fundamental dates: the summer solstice reveals the day of John the Baptist (June 24th), the winter solstice of Christ's Birth (December 25th) and the two close fests saints' days, i.e., protomartyr Saint Stephen (December 26th) and Saint John the Evangelist (December 27th); while Vernal Equinox marks the Annunciation to Mary (March 25th); while the Autumnal Equinox commemorates her Assumption (September 25th); and the conception of John the Baptist (September 24th), with the close feast of the Archangel Saint Michael (September 29th) (Hrabanus Maurus, *Liber de computo*, Ms I-27, I, 9th c.). The above four days are the memorials of the feasts of Christ, the Virgin and John the Baptist to whom some of the main saints were set beside.

Great importance was given to the winter solstice when the Sun, in the constellation of Capricorn, rises at its most southern point on the horizon and moves slowly over the horizon describing its lowest course in the sky. Then the Sun begins to move the ever-increasing

[1] As the Julian calendar bears an error, the astronomical equinox during the Middle Ages happened some days earlier (1 day approx. every 128 years).

orbit and goes further northward until it reaches its northern most point at the summer solstice. The increase of the hours of light are perceived by naked eye a few days after the solstice, exactly at Christmas, the day when the Renewed Sun, Jesus Christ is reborn. Roman Emperor Aurelian (r.d. 270-275), as it is said, set the birth of Christ on December 25th, on the same day of the pagans' feast when they worshipped their Sun god Mithra (*Sol Invictus*). His reform was intended to bring peace religeous unity to the Roman Empire (Usener, 1905. Hijmans, 2003. Berrens, 2004). The patriarch of Jerusalem Sophronius (6th-7th century) in his *Homilies* praised Christ's Christmas, seen as the only Sun of Righteousness. He invited the Christians to bring lit candles during the Lord's presentation to the Temple (40 days after His birth), associating their light with the divine light of Christ, in front of Him everything shines and darkness vanishes: "Everyone should be eager to join the procession and to carry a light. Our lighted candles are a sign of the divine splendor of the one who comes to expel the dark shadows of evil and to make the whole universe radiant with the brilliance of his eternal light. Our candles also show how bright our souls should be when we go to meet Christ… So let us all hasten together to meet our God" (Sophronius, *Homilies,* I. *Luke* 2.32).

The image of Christ seen as the Sun of Righteousness and bearer of victory was also based on a passage of the prophet Malachi (3.20 or 4.2): "But for you who revere my name, the Sun of Righteousness will rise with healing in its rays". Furthermore, one of the seven Antiphons which was sung on the vesper before Christmas, it exalts Jesus Christ as the true Sun of Righteousness and seen as the glory of eternal light:

"*O Oriens, splendor lucis aeternae, et sol justitiae: veni, et illumina sedentes in tenebris, et umbra mortis*" (Gauthier, Valentano, 1786, 185-227. Guillois, 1856, IV, 436-438).

"*O Morning Star, splendor of light eternal and sun of Righteousness: Come and enlighten those who dwell in darkness and the shadow of death*".

4. *DE QUATTUOR PARTIBUS MUNDI*

The heaven was also seen with "gates" of the Sun, so called *Solis portas*. Macrobius Ambrosius Theodosius (5th c.), in his treaty *Commentary upon Scipio's Dream* (I.12.1-3) said that there are two gates of heaven. He has an allegorical view of these *portas*: one is called the "gate of men", which is the Cancer from where the souls descend. The other is called the "gate of the gods", which is the Capricornus from where the souls ascend and become immortal. Through these gates it is believed that deserving souls / human beings reach and stay in heaven and are blessed with eternal happiness:

"*Per has portas animae de caelo in terras meare et de terris in caelum remeare creduntur. Ideo hominum una, altera deorum vocatur: hominum, Cancer, quia per hunc in inferiora descensus est; Capricornus, deorum, quia per illum animae in propriae inmortalitatis sedem et in deorum numerum revertuntur*".

The ancient scholars described the travel of the Sun in this way: after sunset, it is reborn in the morning at sunrise. It is believed that "the sun goes to the west and plunges into the Ocean, and it travels unknown paths under the earth, and once again runs back to the east" (Isidore, *The Etymologies*, III.52). So, Christ, seen as the Sun, descends in the evening through the "western gate" (seen also as the Passion) and reappears in his rising every morning from the "eastern gate" (seen as the Resurrection). Entering the church is a sacred gesture, one passes from this world to the world of God, which leads to eternal life.

We can find some other similar evidences in another treaties about astronomical topics (Anonymous author *Interrogationes et responsa varie*, Ms. I-27, IV, 9th c., Antonian Library of Padua). In a dialogue between a disciple and a magister, the disciple asks: "How many are the gates of heaven?" and the magister answers: "They are two, rising and setting, in fact, a door opens to the sun, the other closes to it" (Isidore, *The Etymologies*, XIII.1.7).

The summer-solstice-oriented architecture represents the descending phase of the solar cycle, whereas the one oriented to the winter solstice represents the ascending phase, called the door of the gods, of the Sun, of Heaven. Indeed, since that day in the northern hemisphere, the days are becoming increasingly longer and this moment is considered "the rebirth of the Sun", so it is written in the *Scriptures*: "This is the house of God and this is the gate of heaven" (*Genesis* 28:17).

This ancient thought represents the astronomical and liturgical concept of medieval scholars and shows that heaven is symbolically seen with two doors - *Solis portis* - which are facing the Sun: one door towards the sunset and the other one towards the sunrise. In this way, these two gates divide the heaven into four parts, formed by the solstices. In the *Munich computes* we have a clear understanding and explanation of it: "The Eastern sector is so called because the sun rises there every day between the extreme points of the solstices, where the Sun rises on June 24th and on December 25th. The South is measured from the place where the sun rises on December 25th to the place where it sets on the same day. The West is so called because it ends the day and is measured in relation to the space between the two extreme points of the solstices, where the Sun sets on December 25th and on June 24th. The North is measured from the place where the Sun sets on June 24th to the place where it rises on the same day" (*Computus*, Clm 14456, Bayerische Staatsbibliothek, 8th/9th c., ff. 25v, 26r. Warntjes 2010. Isidore, *The Etymologies*, III.42. Spinazzè 2016a).

This theory combined with the deep theological significance of the divine light associated with Christ and the Virgin, is not at all astonishing to get this kind of pattern (Figure 2. Spinazzè 2016a, 2017): most part of the medieval sacred buildings analysed, fall inside these two sectors defined by the solstices (182 of 203: 89,7%), (and nearly all of them if we consider also the lunistice arc, 197 of 203: 97,1%); plus most of them indicate a deliberate and proper alignment according to their history. Allegorically these churches are protected by the Sun which starts to decrease (John the Baptist) and the Sun which starts to increase (Jesus Christ), but also by the Moon which represents the Virgin Mary (*Apocalipse of John*, 12.1. *Song of Songs* 6.10. Liguori, 1836, II, 316.) who symbolizes the Church too (Gaudentius of Brescia, *Sermons*, III.1. Pope Innocent, *Sermon* II. Liguori, *The glories of Mary* (lived in 17th-18th c.) Briffault 1931), while Christ represents the newly risen light, the Sun of Justice (*Malachi* 3:20. Clement of Alexandria, *Christ the Educator*, IX.88. Gaudentius of Brescia, *Sermons*, I. Sophronius, *Homilies*, I. Bernard of Clairvaux, *Sermons on the song of songs*, XXI.4, XXVIII.13, XXXI.2).

The buildings themselves provide evidence of their orientation and the existence of the link between the church and the heaven.

Moreover, cardinal points and solstices were associated with the winds, so the rose of winds was born. Aristotle in his work *Meteorologica* (II.5-6. See also: Pliny, *Historia Naturalis*, II.45-46), mentioned and described eight principal winds which fell together with the astronomical main points: Zephyrus blows from the equinoxial West; Apeliotes, meaning Sun, comes from the Equinox Orient; Aparctias, also called Borea, is the Northern Wind; Notos blows from the South; Caecias (Volturna) blows on the Sunrise on the summer solstice; while Euros blows on the Sunrise on the winter solstice; Argestes, Olimpia or Scirone, blows on the Sunset on the Summer solstice and the wind Lips blows on the Sunset on the winter solstice.

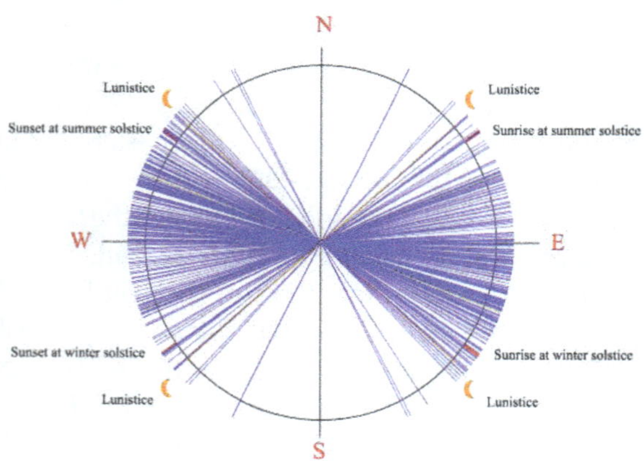

Figure 2. The above diagram shows the 203 alignments of the medieval sacred buildings' axis analysed. E. Spinazzè.

Objective testimonies are for example marble discs of Roman times, also called "wind tables" or anemoscopes, on which the directions of the winds are engraved (Rome, the Pio Clementino Museum, *Vestibulum*).

Coming back to medieval travel diaries it is significant to focus on the work *De locis sanctis* written by Adamnan (Adamnanus Hiensis), an Irish monk, who lived in the 7th century. He wrote down the journey of the Gaulish monk and bishop Arculf who visited the Holy Land. He described in particular the church of the Holy Sepulchre in Jerusalem. In a version dating back to 9th century, a drawing of the Sepulchre is also included beside the text. It is one of the few oldest drawings of buildings that has reached us, giving remarkable information about this sacred building. The Holy Sepulchre has eight doors, four facing the wind Volturno (summer solstice) and four facing the wind Euro (winter solstice), creating a real representation of the solstice section designed and planned for this church (Adamnanus, *De locis sanctis*, Ms. Rh.73, II.1-14):

"De eclesia rotundae formulae super sepulchrum domini aedificata et de ipsius sepulchri figura et eius tegorioli. "Haec bis quaternales portas habet, ... ex quibus iiii ad ulturnum spectant, qui et caecias dicitur ventus, ali vero iiii ad eurum respiciunt".

According to the Venerable Bede's description, Adamnano gave his book to King Aldfrith of Northumbria in 698 to be widely spread all over the Western Europe and therefore it had to be written in such a way that everyone could read and understand it (BEDE, *Ecclestiastical History of the English People*, V.15).

Observing this drawing (Figure 3), we can recognize sectors and lines like the ones termed by the medieval scholars who divided the heaven into four parts. The cardinal points (North, South. Isidore, *Etymologies* XIII.1.7: "*Cardines autem mundi due, Septentrio et Meridies; in ipsis enim volvitur caelum*") and the West point correspond to the directions in the drawing on the manuscript. Furthermore, we can notice an Easter sector which is similar to the solstice sector. If we calculate the solsticial arc for the city of Jerusalem (latitude 31°46' North), the azimuth between the astronomical North and the point where the Sun rises on the summer solstice is about 62° and on the winter solstice it is about 118°, so the solstice arc corresponds to 56°[2]. If we include these values in this medieval drawing, we note that the sector beginning of the doors corresponds, roughly, to the solstice arc for the city of Jerusalem, and the *sepulchrum domini*, which shows another orientation, (4th century. Azimuth of the *Sancti Sepulchri* approx. 104°-284°) to the sunset on the local horizon around 20th April, which could be Easter, highlighting the Passion of Christ (*Psalm* 104 (103). Isidore, *Etymolo-*

[2] Geographic coordinates: lat. 31°46′ N and long. 35°13′ E. Approx. azimuth of the *Sancti Sepulchri* in the direction of the risen Sun 104°, in the direction of the setting Sun approx. 284°. Astronomical horizon = local horizon. Altitude of Jerusalem approx. 750 m. Declination of the celestial body on the mountainous profile on the rising: -12°12′ and on the setting: 11°33′. Corresponding days on the sunrise on the mountainous profile for the first half of 4th century: 17th February, 25th October; on the sunset: 20th April, 23rd August.

gie, VI, cap. XVII, XIX. Saint Augustine, *Enarrationes in psalmos*, XXVI. 54). Eusebius of Caesarea (c. 260-c. 340), a Christian theologian, in his work of the *Life of Constantine* (III.31) states that the construction of the *Sancti Sepulchri* was ordered by the Emperor Constantine in the year 325 (in this year Easter fell on 18th April). Also in the *Itinerarium burdigalense* (Donner, 2002, n. 594), dating back to 333, written by an anonymous pilgrim, it is recorded that this basilica was built by Constantine. This church is the place where Christ was buried and it was called the church of Resurrection. Furthermore, the *Sancti Sepulchri*, "enclosed" by the two solstices, between the feast of John the Baptist when the light starts to diminish and the feast of Jesus Christ when the Sun starts to rise again, incorporates an Easter alignment, underlining His birth, death and Resurrection.

This church conveys a deep message to the Christians: the doors oriented towards the summer and winter solstice highlight the passage expressed by John the Baptist in the Gospel of John (3:28-30). The designation *Sepulchri* underlines Easter, emphasizing the death and Resurrection of Christ, eternally memorialized by the Paschal orientation of the sacred building[3].

5. SOLSTICIAL LIGHT INSIDE SACRED BUILDINGS

It was astonishing to observe that only 14 of 203 (7%) medieval churches were oriented towards the solstice points, either on local or astronomical horizons (Figure 4). Even if the summer solstice is seen as the highest light (Kreuser, 1851, I, 38-48), compared with the magnificence of Christ, and the winter solstice seen as the "new" Sun and the Nativity of Christ since ancient times, medieval builders chose in most cases another orientation for a sacred building: on one of the four Mary feasts commemorated during the Medieval age (Annunciation, March 25th; Assumption, August 15th / January 18th; Nativity, September 8th; Purification, February 2nd) or on the patron saint's day (Spinazzè, 2016a, c).

We wonder why there are so few alignments towards the summer solstice and the winter solstice even their deep symbolic significance. One reason for the first case could be represented by the symbolism of light, which begins to decrease after the summer solstice. In the second case, one might think of a practical reason linked to climate, for the difficulty of tracing the foundations during the winter especially in a mountainous setting.

On the other hand, we could notice in many more medieval churches an interior summer solstice alignment:

- Light which passes through the original openings (sometimes rebuilt on the original position) such as a rose window on the façade, crosses the whole church and goes out from the windows of the apsis with the setting Sun around the summer solstice. As we can see in the churches: St. Mary, Piacenza; St. Peter in ciel d'oro, Pavia. In the church of Piacenza, around the summer solstice the "highest light" on its setting passes through the large rose on the façade going through the whole nave and out the apsis window showing even today this fundamental moment, the magnificence of Christ. On this particular morning, this "highest light" brightens also the altar. A similar case: church of St. Martin, Siccomario, Pavia (Spinazzè, 2016a, n. 35, 24, 22; 2016b).
- Or the summer solstice light signs the length of the whole church touching the apsis area with the sunset. As we can see in the medieval churches in Italy: St. Mary, Gambassi Terme, Chianni; St. Stefan, Sorano, Filatiera; St. Theodor, Pavia; Chiaravalle Milanese, Milan; Chiaravalle della Colomba, Alseno (Spinazzè, 2016a, n. 56, 43, 27, 20, 38. Incerti, 1999).
- Also frescos sometimes had a special place in the church and were illuminated on most significant days and hours of the year. In the Cistercian church of Saint Mary near Milan the fresco over the top of the lantern, which depicts the Marriage of the Virgin, is illuminated on the summer solstice when the Sun reaches its highest light and at the same time it starts to diminish (Spinazzè, 2016a, n. 20).

6. CONCLUSION

In medieval times the horizon was divided into four parts, delimitated by the four points of the solstices. In this way two sectors were formed facing the Sun, one sector in direction of the sunrise and the other in direction of the sunset. The orientation of most medieval churches up to the Romanesque age are enclosed between these two liturgically and astronomically important days when the Sun begins to diminish (summer solstice), the feast of John the Baptist and when the Sun begins to rise again (winter solstice), the birthday of Jesus Christ. This representation recalls a deep message written in the line of the Gospel of John the Apostle (3:28-30) when John Baptist says: "I have to diminish and He (Jesus Christ) has to increase".

But there are only few medieval churches with major axes oriented to the solsticial light. On the other hand, the medieval builders preferred to align a sacred building between the solstice points of minimum and maximum sunlight: viz. mostly towards the Sun on the Virgin's feast (Annunciation, Assumption, Nativity, Purification) or on the patron saint's day. However, the solsticial light often strikes significant architectural feature inside the church. Either the length of the entire church or the altar is fully illuminated among the summer solstice, or a significant image, fresco is illuminated around these days. Or we can

---

[3] Another sacred building in the Middle East, which contains an orientation with the solsticial light: the church of St. Sophia, Constantinople, built in 6th c. In this church the axis has a perfect winter solstice alignment (Schibille 2009).

observe how the solsticial light passes through the openings on the façade and goes out from the apsis window.

One significant medieval drawing, dating back to the 9th c. of a sacred building which reached us through the work *De locis sanctis* of Adamnano (7th c.), represents the *Sancti Sepulchri* of Jerusalem. In this drawing the doors were aligned in the direction of the wind Volturno (summer solstice) and the wind Euro (winter solstice). It shows us a real representation of the solsticial arc, with the cardinal points through the niches and with a Pasqual alignment of the *Sancti Sepulchri* on the sunset recalling with its dedication the birth, death and Resurrection of Jesus Christ.

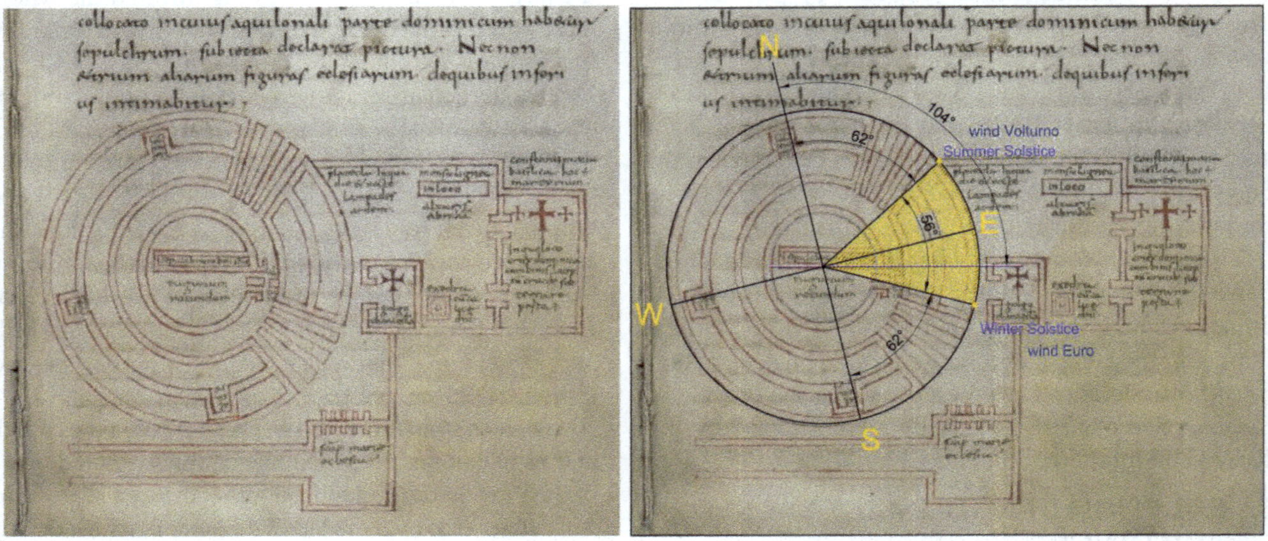

Figure 3. Left: Drawing of the Ecclesia Sancti Sepulchri in Jerusalem, Church of the Resurrection, in Adamnanus of Iona (lived in 7th c.), De locis sanctis, Zurich, Zentralbibliothek, Ms. Rh. 73, 9th c., ff. 2r-28r, drawing f 5r. Right: Reconstruction by E. Spinazzè.

Sacred building Place in Italy	Century of construction	Geographic coordinates Lat. N. Long. E.	Azimuth true rising setting	Decl. on a.h. rising setting	Error of azimuth <	Days corresp. to foundation age rising setting	Horizon Altitude rising setting	Decl. on l.h. rising setting	Days corresp. to foundation age rising setting	Link with the history of the sacred building interpretation
San Giacomo di Rialto Venice	5th	45°26'18'' 12°20'06''	123°31'38'' 303°31'38''	-23°15' 22°21'	0,4°	10 Dec - 29 Dec 10 July - 1 June	--- ---	-23°15' 22°21'	10 Dec - 29 Dec 10 July - 1 June	Christmas, winter solstice
San Fermo e Rustico Verona	8th	45°26'23'' 11°00'01''	55°45'03'' 235°45'03''	23°28' -23°42'	1,2°	summer solstice winter solstice	7°58' ---	28°34' -23°42'	lunistice winter solstice	lunistice winter solstice
San Pietro Camaiore (Lucca)	8th	43°56'18'' 10°18'36''	123°18'16'' 303°18'26''	-23°44' 22°51'	0,6°	winter solstice 5 June - 4 July	4°44' 6°43'	-20°07' 27°33'	18 Jan - 18 Nov ---	winter solstice on a.h. St Peter, 18 January, on l.h.
San Marziale Venice	9th	45°26'37'' 12°19'57''	124°43'59'' 304°43'59''	-24°00' 23°06'	0,4°	winter solstice 6 June - 1 July	--- ---	-24°00' 23°06'	winter solstice 6 June - 1 July	winter solstice St Marziale, Limoges, 30 June
Santa Maria Follina (Treviso)	10th	45°57'16'' 12°07'05''	60°06'50'' 240°06'50''	19°49' -20°42'	0,4°	15 May - 21 July 13 Jan - 19 Nov	4°48' 5°51'	23°29' -15°37'	11 June - 21 June 1 Feb - 1 Nov	summer solstice Purification of Mary, 2 Feb
Abbadia Isola Monteriggioni (Siena)	10th	43°23'15'' 11°11'41''	127°01'47'' 307°01'47''	-26°23' 25°30'	0,6°	--- ---	3°51' --	-23°25' 25°30'	10 Dec - 23 Dec ---	Christmas, winter solstice
San Stae (Eustachio) ribuilt 17th c. , Venice	10th	45°26'29'' 12°19'50''	124°35'25'' 304°35'25''	-23°55' 23°01'	0,4°	winter solstice 4 June - 28 June	--- ---	-23°55' 23°01'	winter solstice 4 June - 28 June	winter solstice
San Simeone Profeta (grande) Venice	10th	45°26'26'' 12°19'27''	54°08'01'' 234°08'01''	23°49' -24°43'	0,4°	summer solstice ---	--- ---	23°49' -24°43'	summer solstice ---	summer solstice
Sant'Aponal Venice	11th	45°26'16'' 12°19'56''	55°02'26'' 235°02'26''	23°15' -24°09'	0,4°	7 June - 26 June winter solstice	--- ---	23°15' -24°09'	7 June - 26 June winter solstice	approx. summer solstice winter solstice
San Martino Arnad (Aosta)	11th	45°38'42'' 7°43'09''	126°00'32'' 306°00'32''	-24°43' 23°48'	0,5°	approx. winter solstice summer solstice	15°32' 9°57'	-12°16' 31°27'	12 Feb - 17 Oct ---	summer solstice on a.h.
Eremo di Sant'Alberto Butrio (Pavia)	11th	44°51'18'' 9°08'59''	124°51'22'' 304°51'22''	-24°20' 23°27'	0,9°	approx. winter solstice summer solstice	10°39' ---	-15°59' 23°27'	31 Jan - 30 Oct summer solstice	Purification of Mary, 2 Feb summer solstice
Santa Maria Carità Acc. Venice	12th	45°25'53'' 12°19'41''	123°21'13'' 303°21'13''	-23°08' 22°14'	0,4°	5 Dec - 25 Dec 27 May - 5 July	--- ---	-23°08' 22°14'	5 Dec - 25 Dec 27 May - 5 July	Christmas, approx. winter solstice
Santo Stefano Venice	13th	45°26'01'' 12°19'50''	125°19'56'' 305°19'56''	-24°23' 23°29'	0,4°	--- summer solstice	--- ---	-24°23' 23°29'	--- summer solstice	summer solstice
Sant'Andrea Apostolo Venice	13th-14th	45°26'18'' 12°18'57''	125°53'32'' 305°53'32''	-24°44' 23°50'	0,4°	--- summer solstice	--- ---	-24°44' 23°50'	--- summer solstice	summer solstice

Figure 4. Alignments on the solstices. 14 medieval sacred buildings (in Italy) of 203 (6,9%).
a.h. = astronomical horizon; l.h. = local horizon. For each sacred building an accurate georeferenced survey together with astronomical and trigonometrical calculations was carried out. (For detailed information see the publications of the author, with the diagrams, drawings, surveys and calculations).

REFERENCES

Adamnanus of Iona (9th c.) *De locis sanctis.* Ms. Rh. 73, manuscript originally from Reichenau, Zentralbibliothek, Zurich.

Anonyimous author (9th c.) *Interrogationes et responsa varia.* Ms. I-27 – Miscellanea, Biblioteca Antoniana, Padova, part IV.

Aristotele (2003) *Meteorologia*, by Lucio Pepe, Bompiani, Milano.

Berrens, S. (2004) *Sonnenkult und Kaisertum von den Severern bis zu Constantin I.* F. Steiner, Stuttgart.

Briffault, R. (1931) *The Mothers.* New York.

Brubaker, L.; Cunningham, M. B. (2011) *The Cult of the Mother of God in Byzantium.* Ashgate, Surrey GB.

Cicerone (2000) *La natura divina,* by Cesare Marco Calcante, Bur, Milano.

Citez de Jherusalem (1896) by C.R. Conder, *City of Jerusalem.* Palestine Pilgrims'Text Society, Adelphi, London.

Computus (8th/9th c.) Clm 14456. Bayerische Staatsbibliothek. Manoscript transcipted and translated in English by Immo Warntjes (2010) *The Munich Computus: Text and Translation, Irish computistics between Isidore of Seville and the Venerable Bede and its reception in Carolingian times.* Franz Steiner Verlag, Stuttgart.

De Blaauw, S. (2010) In vista della luce. Un principio dimenticato nell'orientamento dell'edificio di culto paleocristiano. In Paolo Piva, *Arte medievale, le vie dello spazio liturgico.* Jaca Book, Milano, pp. 15-45.

De Blaauw, S. (2008) Orientierung. Architektonische Ausrichtung. *Reallexikon für Antike und Christentum.* A. Hiersemann, Stuttgart, Vol. XXII, pp. 278-282.

De Blaauw, S. (1994) *Cultus et decor, liturgia e architettura nella Roma tardoantica e medievale.* Biblioteca Apostolica Vaticana, Città del Vaticano, Vol. I e II.

Dölger, F.J. (1925) *Sol Salutis, Gebet und Gesang im christlichen Altertum.* Aschendorff Verlagsbuchh., Münster in Westfalen.

Donner, H. (2002) *Pilgerfahrt ins Heilige Land, die ältesten Berichte christlicher Palästinapilger (4.-7. Jahrhundert).* Verlag Katholisches Bibelwerk, Stuttgart: *Pereginatio Egeriae*, pp. 68-133; *Itinerarium Burdigalense. Breviarius de Hierosolyma*, pp. 214-225; *Adamnani de locis sanctis libri tres*, pp. 296-394.

Duchesne, L. (1903) *Origines du culte chrétien*, ed. Albert Fontemoing, Paris.

Gauthier, F.L. and Valentano, N. (1786) *Riflessioni sopra gli O, le antifone dell'Avvento, in forma di omilie,* ed. S. Occhi, Venezia.

Guillois, A. (1856) *Explication historique, dogmatique, morale, liturgique, canonique du catéchisme*, ed. C. Monnoyer, Paris, *tome* IV.

Hijmans, S. (2003) Sol Invictus. The Winter Solstice and the origins of Cristmas. *Mouseion*, Series III, Vol. 3, pp. 377-398.

Hodoeporicon, S. Willibaldi (1880/1881) *S. Willibalds Pilgerreise, geschrieben von der Heidenheimer Nonne*, by J. Brückl, Druck Däntler, Eichstätt.

Hrabanus Maurus (9th c.) *Liber de computo.* Ms. I-27 – Miscellanea, Biblioteca Antoniana, Padova, part I.

Incerti, M. (1999) *Il disegno della luce nell'architettura cistercense.* Certosa Cultura, Firenze.

Kreuser, J.P.B. (1851) *Der christliche Kirchenbau, seine Geschichte, Symbolik, Bildnerei.* Bonn.

Liguori, Alphonsus (1888) *The glories of Mary.* Kenedy & Sons, New York.

Macrobio (2007) *Commento al sogno di Scipione*, by M. Neri, Bompiani, Milano.

Martyrologium Hieronymianum (1894) by G. B. de Rossi, L. Duchesne, *Acta Sanctorum Novembris.* Société de Librairie, Bruxelles.

Mortem, Caesaris (1893) *Inscriptiones Latinae Antiquissimae*, ed. Reimerum, Berlin.

Piva, P. (2010) Le vie dello spazio liturgico. In Paolo Piva, *Arte medievale, le vie dello spazio liturgico*, Jaca Book, Milano, pp. 7-13.

Platone (1992) *Il Simposio*, by Carlo Diano, Marsilio, Venezia.

Plinio Gaio Secondo (1982) *Historia Naturalis*, by Alessandro Barchiesi, Roberto Centi, Einaudi, Torino.

Podossinov, A. (1991) Himmelsrichtung. *Reallexikon für Antike und Christentum.* Anton Hiersemann, Stuttgart, Vol. XV, pp. 233-286.

Pope Innocent (2006) *I Sermoni,* by Stanislao Fioramonti, Libreria editrice Vaticana, Città del Vaticano.

Procopius (1896) *Buildings of Justinian*, by Aubrey Stewart, Palestine Pilgrims' Text Society, Adelphi, London.

Publio Virgilio Marone (2012) *Eneide*, by Alessandro Fo, Einaudi, Torino.

Schibille, N. (2009) Astronomical and Optical Principles in the Architecture of Hagia Sophia in Constantinople. *Science in Context*, March, n. 22, pp. 27-46.

Spinazzè, E. (2007/2008, 2015a) *Luce ed Orientazione nelle Abbazie Benedettine Altomedioevali e Medioevali nel Veneto*. Università Ca'Foscari Venezia, Magister Thesis in Medieval Archaeology, year 2007/2008; published in 2015, title: *Luce e canto incisi nelle pietre. Allineamenti astronomici delle chiese monastiche benedettine medioevali nel Veneto*, Cleup, Padova.

Spinazzè, E (2015b) Un santo, una Pasqua e un lunistizio a Saint-Maurice d'Agaune. *Atti del XIV Convegno nazionale annuale S.I.A. Padova, 17-18 ottobre 2014*, Padova University Press, pp. 307-314.

Spinazzè, E. (2016a) *La luce nell'architettura sacra: spazio e orientazione nelle chiese del X-XII secolo tra Romandie e Toscana*. Mediaevistik, Peter Lang Verlag, Frankfurt am Main (Ph.D. Thesis, 2013/2014, Department of Philosophy and Cultural Heritage, Ca' Foscari University of Venice, IUAV Venice in joint supervision program with University of Zurich).

Spinazzè, E. (2016b) The alignments of sacred medieval buildings and the path of light inside the church on the Patron Saint's day. *SEAC-proceedings, Societé Européenne pour l'Astronomie dans la Culture, International Conference, Rome, 9-13 November 2015, Astronomy in Past and Present Cultures, Mediterranean Archaeology and Archaeometry*, Vol. 16, no. 4, 455-463.

Spinazzè, E. (2016c) Il battistero e il fonte: un'interpretazione sull'orientazione degli edifici battesimali medioevali nel medio-basso Friuli. Il percorso 'simbolico' del catecumeno da Occidente verso Oriente. *Atti dell'Accademia 'San Marco' di Pordenone*, Pordenone, Vol. n. 18, pp. 491-550, 967.

Spinazzè, E. (2017) Gli allineamenti negli edifici sacri medioevali situati nel Nord e Centro Italia. Risultati e interpretazioni. *Atti dell'Accademia 'San Marco' di Pordenone*, Pordenone, Vol. n. 19, pp. 379-436.

Strozzi, Pier Filippo (1766) *L'assunzione al cielo della Madre di Dio*. Stamperia Generoso Salomoni, Roma.

The Epitome of S. Eucherius about certain Holy Places (1890) by A. Stewart, Palestine Pilgrims' Text Society, Adelphi, London.

Usener, H. (1905) Sol invictus. *Rheinisches Museum für Philologie*, n. 60, pp. 465-491.

Vitruvio Pollione Marco (1990) *De Architectura*, by di Luciano Migotto, Edizioni Studio Tesi, Padova.

Vogel, C. (1964) L'orientation vers l'est du célébrant et des fidèles pendant la célébration eucharistique. *L'orient Syrien*, Vernon, IX, pp. 3-37.

Wallraff, M. (2001) *Christus verus sol: Sonnenverehrung und Christentum in der Spätantike*. Jahrbuch für Antike und Christentum, Aschendorff Verlagsbuchhandlung, Münster in Westfalen, Ergänzungsband XXXII.

Works by the Fathers of the Church are cited in the text *supra*.

13

CONSOLIDATION OF THE RAPANUI ASTRONOMY CONCEPT INVENTORY AND RE-APPRAISAL OF APPLIED ASTRONOMIC OBSERVATION AT PAPA UI HETU'U, RAPA NUI

Alexandra Edwards*[1,2], Barthelemy d'Ans[2,3] and Edmundo Edwards[1,2]

[1] *Pacific Islands Research Institute*
[2] *Rapa Nui Planetarium*
[3]*Universidad Peruana de Arte Orval*

Corresponding author: Alexandra Edwards (alexandraedwards@yahoo.com.au)

ABSTRACT

Rapa Nui, the easternmost settlement and most isolated trade outpost in ancient Polynesia, represents one of the most ambitious feats in Polynesian wayfinding, and by default reflects the high level of competence in navigation and interpretation of astronomical and natural phenomena, currently a popular field of inquiry in Polynesian ethnoastronomy. However, as skywatching was not limited to navigation, the full scope of Rapanui applied astronomical observation and complexity of astronomical devices was much broader. Observatories, calendric calibration systems, "star-maps," and structures with astronomical alignments, have all been identified for Rapa Nui, yet the two main challenges confronting researchers is the scarcity of extant information, as well as degradation and site location due to errors in the official mapping of the island. The present investigation re-examines and extends the Astronomy concept inventory of the ancient Rapanui, with regards to the toponymy of local stars and asterisms, and an inquiry into the stars and methods used by traditional navigators to locate Rapa Nui in past times. The results have been analyzed within both a local and regional context. In addition, our team did a re-appraisal of the astronomical functions of the Matariki (Pleiades) stone and related petroglyphs in the area of Papa Ui Hetu'u (star-gazing rock) on Poike Peninsula, which E. Edwards and Antonio Belmonte had investigated in 2004. For the first time, this collected data was evaluated within the context of a digital planetarium, permitting the review and identification of alignments between astronomical phenomena and sites of observation as if viewed in situ and in the pertinent chronology; together with photogrammetry/3D-modelling of astronomical devices at Papa Ui Hetu'u, these techniques offer a new and more precise line of research, rendering intriguing results.

KEYWORDS: archaeoastronomy, ethnoastronomy, toponymy, Easter Island, Rapa Nui, Polynesia, planetarium, 3D modelling

1. INTRODUCTION

The purpose of this paper is to consolidate the Rapanui Astronomy Concept Inventory mainly obtained from the works of Routledge (1911-1923; 1919) Englert (1974), and Métraux (n.d.; 1960), as well as the previous ethnographic research of E. Edwards (2010, 2013). Previous studies have built upon the works of Routledge and Métraux, which are invaluable but brief, and contain several inconsistencies due to the inaccurate interpretation of data. Routledge and Métraux collected ethnographic information from direct sources, however, subsequent scholars without access to the original material built upon the errors of their previous colleagues, incrementing and confirming fallacies, which upon closer inspection yielded an assembly of unreliable results. Based upon the field notes of both Routledge and Métraux, identifying errors, offering a few amendments, and hypothesizing on the correct interpretation of specific elements we were able to form a compendium of the terms used to designate different types of heavenly bodies (planets, asterisms, and individual stars), ultimately offering an updated and exhaustive list of Rapanui star names, as well as their meaning and cultural significance. Regarding Papa Ui Hetu'u, our team re-appraised the cultural features and landscape context of this observatory, adding alignments to the Pleiades (or June solstice) and the path of the stars that Polynesian navigators would have used to locate Rapa Nui.

2. PREVIOUS STUDIES AND METHODOLOGY

Of the many scholars that have worked on Rapa Nui only a few have covered the subject of ethnoastronomy. Ferdon (1961), Mulloy (1975), and Liller (1986, 1989, 1993, 2000, 2004) worked on the island between 1955-2004, however most of their work focused on *archaeo*astronomy, recording alignments between the famous Rapanui ruins and celestial phenomena, and less on the cultural aspect regarding the purpose behind these orientations beyond basic time-keeping. Excluding E. Edwards and his team, the few scholars that collected information about the names of stars and their significance in a cultural context are Routledge (1911-1923; 1919), followed by Englert (1974), and Métraux (n.d.; 1960). However, all of their collected data does not amount to much if presented individually. Routlege worked on Rapa Nui between 1914 and 1915. Her principal informants were Kapiera Maherenga and Parlemón Te Haha. Maherenga, once a competitor (*hopu manu*) in the Bird-man cult competitions was born ca. 1846 and was a member of the Ure o Hei clan via paternal descent and the Hamea Clan from his maternal side. Maherenga provided most of the star names recorded by Routledge. Te Haha was born near Maunga Kou'a in approximately 1832. He was an *ariki paka* nobleman of the prestigious Miru clan and was a member of the noble entourage of paramount chief Ngaara. His nephew Charlie Teao was one of Métraux's principal informants. Another of Routledge's informant was José Fati (Hati) Rongo Pua Tea, Fati was the father of Isaías Fati, one of the main informants of Englert, Métraux, and E. Edwards.

Englert's main informants were the previously mentioned I. Fati, as well as Rafael Haoa, Mateo Hereveri, and Santiago Pakarati. Haoa was the informant for a host of present-day ethnographers working on Rapa Nui. Hereveri, a nobleman of the Miru clan was considered an authority regarding Rapanui lore. Hereveri was also an informant to E. Edwards. Pakarati was the son of Rapanui catechist Nicolás Pakarati Ure a Potahi (also an informant of Routledge and Métraux). Métraux's informants were Victoria Rapahango Tepuko and Charlie Teao. Rapahango was also a Miru Clan member and she had learned much from Mateo Hereveri and her uncle Parlemon Te Haha who had lived with her in her youth; both men had been informants of Routledge. Teao, was a nephew of Te Haha and a grandson of Tomenika Vaka Tuku Onge, also Routledge´s informant.

For this paper we reviewed the original and published works of Routledge, Métraux, Englert, and E. Edwards, as well as the dictionaries of Tregear (1891) and Churchill (1912)

3. CULTURAL CONTEXT

The same as other Polynesians, the Rapanui visualized the celestial sphere as several superimposed heavens, which together with the stars and constellations were believed to be the abode of the gods. The careful observation of the sky was useful for navigation but also to determine the time of the year, establishing a ritual calendar and regulating farming and fishing activities. Astronomer priests or skywatchers studied the Moon and stars, and less often the Sun, from special observatories built in places with the best vantage point for each astronomic event, announcing when festivities, ceremonies, prohibitions, and the different seasons started and ended. A total of 58 alignments have been registered on Rapa Nui, with observatories (*tupa*) oriented to significant celestial markers, most to the Pleiades and Orion's Belt, but also Vega, Alpha Beta Centauri, in a North–South axis, etc.…(Edwards and Belmonte 2004; Edwards and Edwards 2010). These constructions reflect a shared bank of knowledge that dates back to the Austronesian roots of the Rapanui people, and which evolved for very practical reasons:

subsistence, immigration and trade. Clark has calculated that "the probability of these alignments being accidental is 10,000,000 to 1." (Edwards and Clark 1984).

Makemson (1941) recorded 772 star names elsewhere in Polynesia, and skywatching was taught in special schools. In Rapa Nui unfortunately much of this information did not survive the great cultural and social upheaval as a result of outside contact, colonization, and missionary proselytization.

Like other Polynesians, the Rapanui had an annual cycle of activities that determined what sort of chores and rituals people carried out during each month of the year, evidently linked to the weather and the seasons. Astronomical phenomena helped astronomer priests (skywatchers) determine the different times of the year, divided between periods of rain and drought, a season of plenty and of want. Since the chiefs were the medium by which deified ancestors provided for their descendants by making the plants grow and animals multiply, the entire clan ensured the success of their subsistence activities by making offerings and sacrifices to honor the gods in an agricultural cycle that in other Polynesian islands was called Te Fekai Nga Atua (the gods produce food; Edwards and Edwards 2013). The Rapanui diet depended mainly on agricultural products, chicken, and fish, in that order. The Rapanui carried out several rituals related to the planting and harvest of their most important crops. The harvest was important not only as related to food production and surplus, but through feasting and ceremonies it was also of great social and symbolic significance. Food-sharing expressed kinship ties and solidarity, and validated a set of principles, social codes, rights and duties. The opening of the deep-sea fishing season in October/November was also widely celebrated. It was imperative that the seasonal activities and feasts should be performed at the right and auspicious times, and that was the jurisdiction of skywatchers. Aside from the lunar months, specific annual activities and events were marked by the first appearance of specific stars at dusk and dawn. Since there was a fair amount of prediction involved in their work, these specialized priests also delved into the world of prophecy and magic.

4. RESULTS

Polynesians had names for over 700 stars, and it was believed that many of them were placed in the sky by the gods to simply humor humans while others were thought to be the eyes of deified ancestors (Makemson). This explains why several Polynesian star names include the term *mata,* meaning "eye" (Ibid). The generic Rapanui name for stars, asterisms, and constellations is *hetu'u,* a term with a root common to all Polynesian languages, however, Eastern Polynesians grouped stars to form figures that differed from those of Western Polynesia with only the names of the most important calendar stars remaining constant, specifically the Pleiades (Matariki) and Orion (Tautoru; Makemson, Tregear, Churchill).

We were able to record 25 star names (three of which remain undetermined in relation to classic astronomy terms (Mere, Ehuo/Ehua, and Pau), in addition to the names for the Milky Way (Te Ngo´e), and Venus (various terms) and Mars (Matamea). We amended the designation of several Rapanui star names in relation to the previous concept inventory. We also added Tau Ehu for Sirius, and Tauhoru and Hetu'u Tea in addition to the existing names for Venus. Four "new" star names were added: Atutahi, He Mata/Mataa Te Tautoru, Ehuo/Ehua, and Pipiri, while two were deleted as a duplicates (Vari Hariu/Horia and Hare a Tanga). Sirius and Vega had multiple Rapanui names, which seems accurate for the latter, but may be a source error for the former. Orion's Belt on the other hand is both a constellation of six stars (Etui = Orion´s Belt + Sheath), but is called Tautoru when referring only to the belt (Routledge n.d., Metraux n.d.). We also revised the translation of Rapanui stars names, some of which have more than one interpretation. A list of the complete and revised Rapanui astronomy concept inventory appears in Appendix 1. The table in Appendix 1 starts with the terms for the Milky Way, followed by the planets Mars and Venus, continuing with the names of fourteen asterisms and constellations, and ending with eleven single star names. These are presented in alphabetical order within each of their categories.

For the Rapanui, the Milky Way was a mythical marine creature, a non-fearsome sea monster of sorts. Regarding the planets, the Rapanui realized that Venus and Mars were different, that they do not twinkle, and move unlike stars, however the distinctions stopped there, with bright stars and planets both being called *hetu'u purapura* (Churchill).

As for the planets, only the Rapanui names for Venus and Mars survived to the present day, although it is likely that like other Polynesians they would have also recognized the other visible planets, namely Mercury, Saturn, and Jupiter (Makemson, Tregear).

Important time-keeping stars and asterisms were: Matariki (the Pleiades), Nga Vaka (Alpha and Beta Centauri), Mere (possibly Vega), Tau Ehu or Te Pou o Te Rangi (Sirius), and Atutahi (Canopus), as well as the constellation Orion, which the Rapanui divided into six separate parts: E Tui (Orion's Belt and sheath - 6 stars), He Mata/Mataa te Tautoru (Bellatrix and probably Meissa and φ Ori 1 and 2), Tautoru (Orion's Belt), Po'o or Po Roroa (Saiph), Rei a Tanga (Betelgeuse), and Vari Koreha (Rigel). Matamea (Mars) was also used as a calendar marker together with the Pleiades (Matariki), Orion's Belt (Tautoru), and Sirius (Tau Ehu or Te Pou o Te Rangi). The events that each of the above marked can be seen in Appendix 1. Other stars that stand out are Nga Vaka (Alpha and Beta Centauri), which in the Southern Hemisphere may be used to find the South cardinal position in relation to Crux (Tatauro), and Antares (Te Hau Vaero), the traditional star used to mark the latitude position of Rapa Nui in traditional Polynesian navigation when at its zenith; islands were marked by the brightest stars that passed directly above them and Antares may be seen at 89°19'07.5" over Rapa Nui.

Three out of 25 collected Rapanui star names include the word *mata*, which as mentioned previously, is characteristic of the Polynesians, who considered that some stars were the watchful eyes of important ancestors (Edwards and Edwards 2013). In addition, the planet Mars was called Matamea, but in this case it is probably best to interpret it in reference to color and the custom of visualizing stars as eyes, rather than it being the eye of a particular person.

5. OTHERS CONSIDERATIONS

In addition to star names we also collected a list of terms that are related to astronomical concepts, (from Churchill, Métraux n.d., and Englert). Many of these had not been recorded together before (see Table 1).

Table 1. Additional Rapanui Astronomy Terms

Rapanui Name	Translation	Comments	Source
Api te Mahina e te Ra'a	The Sun presses the Moon	A lunar eclipse?	3
Hetu'u Kokiri	Shooting star		1
Hetu'u Pupura	1. Planet or very bright star 2. *-mai*. To ogle, stare	Pupura is to shine bright	1 3
Hetu'u Rere	Meteor		1
Hetu'u Ave	Comet		2
Hetu'u Viri	Shooting star, meteor		2
Hiti	To rise (regarding the light of the moon, sun, stars)		2
Homo	Meteorite, meteoroid crater		2
Kai te Mahina e te Ra'a	The Sun eats the Moon	A lunar eclipse?	3
Kai te Ra'a e te Mahina	The Moon eats the Sun	A solar eclipse?	3
Kihi Ra'a	Ray of sunlight		3
Kihi Mahina	Moonbeam		3
Me'e Nui Te Hetu'u	Sky with many stars		3
Ruaroa	Tropic of Capricorn		1
Uero	Comet, twilight, a ray, radiant star		1
Verovero	To twinkle like the stars		1

Of the new terms listed above, there are a few that stand out because of there telling implications. First of all, there are several terms for different kinds of eclipses, all of them indicating that the Rapanui theorized that it was the sun and moon that caused eclipses. In addition, the word "*homo*" or meteoroid crater indicates that this event must have occurred on the island in past times, and perhaps more than once.

6. THE PAPA UI HETU'U OBSERVATORY.

In 1914, K. Routledge's informants told her that astronomer priests climbed Poike peninsula, on the Eastern headland of the island, during the lunar month of Tuaharo (February/March) to watch the rising of the Pleiades from Papa Ui Hetu'u (Figure 1), which in the Rapanui language means "rock to watch the stars from". Santiago Pakarati showed this boulder to E. Edwards in 1960, consisting of a basalt block of greyish color (Munsell chart 2.5Y-7/1) that probably originated in one of the lava flows located in the surrounding area that was purposely moved to its present location. The upper surface of the boulder has 13 cupule-like depressions of different sizes and depths that most probably are a consequence of the natural cooling process of the lava and later erosion, but which locals have long considered to be akin to the Pleiades. In 2005, E. Edwards and A. Belmonte found that this spot was ideal for observing the heliacal rise of many important local calendric stars, as it is completely open towards the Eastern quadrant of the island.

Figure 1. Papa Ui Hetu´u

The boulder is located 178 m above sea level at 109°14'31. 80" W, 27°5'44. 21 S. It lies east of Pua Ka Tiki, the peak of Poike volcano. As it has an irregular shape, its approximate dimensions are: 1 m high by 1.5 m long and 0.80 m in width. It weighs approximately 300 kgs. The photogrammetric mould that we made of it appears in Figure 2.

Figure 2. Comparative images of the boulder known as Papa Ui Het'u--photograph on the left, and photogrammetry mould on the right, which highlights the distribution and depth of the cupules said to represent the Pleiades.

We proceeded to register the landscape's profile surrounding the Papa Ui Hetu'u site with a fisheye lens, and established several checkpoints upon the terrain, so we could later use it to correct the planetarium digital view (Stellarium, Figure 3).

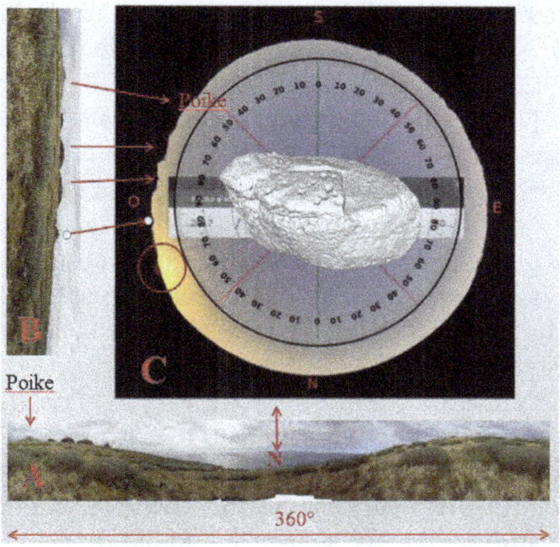

Figure 3. (A) development of the 360° panorama. (B) Important elements in its landscape. (C) Landscape profile with fisheye lens and mask used in the stellarium format. Installing the landscape profile in Stellarium and reproducing the observed event. Field check of the landscape profile altitude, plus the time. azimuth, and sunset on July 20, 2017.

This image was later edited in Photoshop and we then proceeded to erase completely the sky leaving only the profile and a transparent background. We then used this masked image in the profile of the digital planetarium and inserted into it the coordinates obtained from our GPS and reoriented the profile to show the most prominent elements of the landscape with the correct azimuth. Finally we verified the model to see if it adjusted to reality, by superposing an astronomic event, which in this case was the sunset of July 20, 2017, and check that the hour, elevation, and it profile, together with the height and azimuth of the Sun, were correct with the model. Using a Google Earth map, we proceeded to verify the area of the horizon that can be observed from there, and also to make sure that this adjusted to reality. This method served to locate in a more exact manner all the cartographic elements contained in our profile.

We were interested in measuring the intervals between the angles and the azimuths observed on the nearly 180° wide landscape view from the stone, paying special attention to azimuths 302 and116, while also checking if there was any relationship to the heliacal rising or setting of any important Rapanui calendar stars or the Pleiades as as suggested in oral tradition (Figure 4).

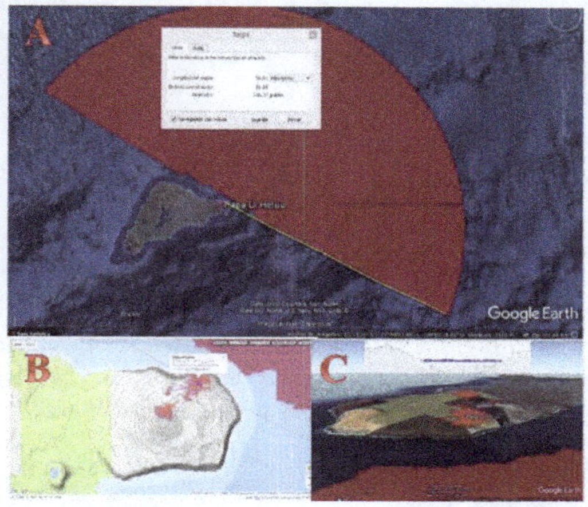

Figure 4. (A) Ocean view azimuths and intervals upon an oceanic horizon (B) Detail of the visible area of land and its geographic markers (C) Perspective view of the area that can be observed from Papa Ui hetu´u.

7. ANALYSIS OF THE CULTURAL EVENTS RELATED TO THIS SITE AND ITS PHYSICAL LOCATION

We carried out a basic survey of all the surrounding area to locate any contiguous archaeological remains, as well as specific places that could be associated with Rapanui cultural landscape.

On the East side there are three trachyte domes that are the most conspicuous elements in the landscape (Figure 5). On the northernmost dome, lie the remains of a disfigured statue (moai) with no platform or ahu, which is clearly visible from Papa Ui Hetu'u and may have bee placed there intentionally. The southernmost dome is associated with several important oral traditions: on its southern foothill there is cave that was the abode of a group skywatchers who lived permanently with their acolytes, while above and to the left of the cave there is a large rock carving called Vai a Heva (water of Heva) that consists of a face with a large protruding eyes and a gaping mouth that served as a basin to collect rainwater. To the East, at a distance of approximately 200 m in a E-SE direction, lies a small basalt outcrop with several petroglyphs of fishhooks of different sizes and shape. These petroglyphs were previously registered by E. Edwards and G. Lee, and now we added a photogrametric record of them (Figure 6).

Significantly, petroglyphs of fishhooks and deep sea creatures, such as tuna and shark, are also present in several Rapanui sites where skywatchers were known to observe the night sky.

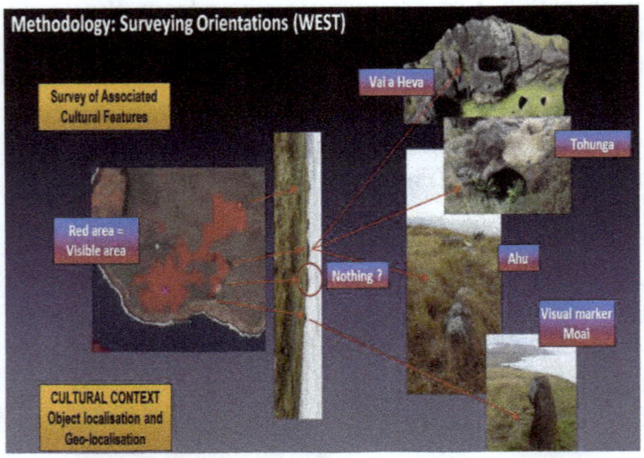

Figure 5. View from Papa Ui Hetu'u to the three trachyte domes in Poike; these were also exploited for trachyte and several small statues raised on platforms on the eastern headland of the island were sourced there.

Figure 6. View towards Papa Ui Hetu'u as seen from the unnamed outcrop with fishhook petroglyphs. The cave where the skywatchers are said to have lived with their acolytes and the stone water basin known as Vai a Heva are also visible from this spot.

Lastly, to the Northeast there is a ceremonial platform that lies in a perfect East-West position. The relationship between the sites and/or significance of this emplacement remains to be determined.

8. ANALYSIS OF THE CULTURAL EVENTS RELATED TO THIS SITE AND ITS PHYSICAL LOCATION

By inputting our collected data into Stellarium we were able to determine the following:

There are two zenithal celestial objects that pass directly over Rapa Nui and Papa Ui Hetu'u. One is Te Hau Vaero (Antares), and the other is Ko Tau Erua Ehu, which in our consolidation of star names we found to possibly represent Castor and Pollux or alternatively Wezen and Adhara, which are two bright stars in the Canis Major constellation. In past times, zenithal stars were known as pito (umbilicus) in many parts of Polynesia, and their path was considered to mark the umbilicus of the sky for the specific island that these stars passed directly over. Zenithal stars were of prime importance in traditional Polynesian navigation, serving to exactly locate each of the islands these stars marked. Interestingly, Antares is visible during the winter months, while Wezen and Adhara may be observed in the Summer, thus if Ko Tau Erua Ehu indeed refers to Wezen and Adhara it is possible to consider that traditional navigators would have been able to more easily locate Rapa Nui year-round. Now, determining what are the most propitious times of the year to travel to Rapa Nui from elsewhere in Polynesia is a subject of another study.

Most significantly, from Papa Ui Hetu'u it is possible to observe the entire path of the Pleiades from heliacal rising to heliacal setting. Such observations were very important to the ancient Rapanui as the first full Moon after the heliacal rising of the Pleiades marked the first night of the year and opened a time of festivities and island-wide rejoicing. This was also true in most of Polynesia. Due to the latitude of Rapa Nui, the Pleiades follow nearly the same trajectory of the path of the Sun on the June solstice. To determine approximately when the Rapanui new year began depends very much on the visible altitude of the Pleiades at the time of observation, which would have been in the very early morn, before sunrise. On Rapa Nui, the Pleiades disappear for a few months, becoming visible once again before sunrise in the last two weeks of June, considering that it is best observed the higher it is in the sky, 20-30 degrees degrees above the horizon. The Rapa Nui new year would have then begun on the first full Moon after this celestial event. Conversely, the observation of the Pleiades high in the sky shortly after sunset in November, marked the opening of the deep sea fishing season on Rapa Nui.

9. CONCLUSIONS

Although much has been lost regarding the astronomical knowledge of the ancient Rapanui, the extant names recorded here adeptly convey the breadth and richness of Rapanui mythology and calendrics. The stars that were remembered up to modern times were the most important ones, part of a complex annual cycle of activities, both ritual and practical. They also point to a knowledge base that hails back to the Lapita ancestors of the Rapanui, as is the case with Matariki and Tautoru (i.e. the Pleiades and Orion's Belt), but also that places Rapa Nui within a contact network extending to the Tuamotu Islands and Tahiti (i.e. the moral tale regarding the repentant jealous husband and/or the selfish parents). It seems significant that an important number of Rapanui stars refer back to the legendary homeland of Hiva, and to its infamous persons and notable culture heroes. The same as elsewhere in Polynesia, Rapanui ethnoastronomy went beyond navigation, serving a religious purpose associated to a belief system that on Rapa Nui survived until the mid-nineteenth century. Evidently, the present Rapanui star concept inventory is incomplete as there must have been many more star names than those presented here. Nevertheless, we have compiled an exhaustive list, benefitting from a comparative analysis of the various ethnographic sources. The number of clarifications and amendments to the previous concept inventory serves as a reminder of the pitfalls of relying on secondary source materials when carrying out this sort of work. Considering the total number of compiled star names, it seems significant that we were able to add two entries and provide a much more accurate index of star names, even if a few items remain undetermined. Lastly, it is important to note that the multiplicity of particular star names is simply a sign of a healthy evolution of concepts, in accordance with cultural growth and in response to societal changes. In the case of dual meanings, oftentimes this is purposeful, in much the same way that the connotations of words in a verse, lend themselves to a poetic layering of meanings. Ethnoastronomy and the unique Rapanui idiosyncrasy deserve as much.

As the Poike peninsula forms the eastern headland of the island, it's location is a priviledged place to observe the heliacal rise of certain celestial objects from the ocean. The rock known as Papa Ui Hetu'u undoubtely was intentionaly placed in that position and although its cupules are of natural origin, the islanders identified them with the Pleiades (Matariki) and believed they held some magical power (*mana*). This was most probably due to the fact that from that precise location it was possible to observe the rising and setting of the Pleiades, the foremost calendar stars of the Rapanui and all Polynesians. Diverse cultural remains located in the vecinity of this site, testify to the presence of specialized observers or skywatchers as mentioned in their oral traditions.

Appendix 1. The Rapanui astronomy concept inventory.

Rapanui Name	Translation	Star / Asterism	Comments	Source
Te Ngo'e	Fabulous Marine Creature	The Milky Way	A fabulous marine creature	5
Matamea	Red Eye	Mars	One "star" Observed from Poike Announced: start of Koro festival; egg hunt in Dec. with Matariki, Tau Ehu, & Tautoru Ominous with Pau and Tautoru appearing in Oct./Nov.	2, 4
Taurua *or* Hetu'u Tea, *or* Tauhoru *and* Hetu'u Ahi-ahi	Beautiful Two (Venus) *and* Evening Star	Venus	Tuaurua refers to Venus (archaic) Presently Hetu'u Ahi-ahi	3, 6, 7
Taurua *or* Hetu'u Tea, *or* Tauhoru *and* Hetu'u Popohanga	Beautiful Two (Venus) *and* Beautiful Bucking to & fro *or* Morning Star	Venus	Tuaurua refers to Venus (archaic) Tauhoru is a historical name Presently Hetu'u Popohanga	3, 6, 7
Ehuo *or* Ehua	-	– Undetermined –	A large constellation	3
E Tui	The Expelled	Orion's belt and sheath	Six stars	2, 3
He Mata Te Tautoru	The Eyes of the Beautiful Three	Bellatrix and possibly Meissa and φ Ori 1 and 2	Two "stars"	2, 6
He Kete	Basket	Possibly Aldebaran and Δ,Θ,Ɛ Tauri *or* Pegasus	Four stars	2, 6
Ko Tau Erua Ehu	Two Beautiful Firebrands	Possibly Castor and Pollux or Wezen and Adhara	Two stars near Canis Major W&A: Antagonic to Antares, zenithal star?	2, 3, 6
Ko Toe Ko Peu Renga	Wake of the Fine Pickaxe/Energy	Menkalinan and Capella	Rnga is energy, but also to paint something yellow or red (sacredness)	2,, 5, 7
Matariki	The Small Eyes *or* The Eyes of the Chief	The Pleiades	Six stars Announced: New Year; first harvests; bountiful season; Bird-man races; egg hunt in Dec. with Matamea, Tau Ehu, and Tautoru	2
Pipiri	The Deprived	Clustered star pair on the tail end of Scorpius	May have marked the December season	1, 2, 4, 5, 8, 9

Nga Rau Hiva	The Sprout from Hiva *or* Twins *or* The Weak	The Hyades	Hiva = Rapanui homeland	6, 3, 5
Nga Toa Rere	The Flying Sugarcane	Stars in Ursa Major	-	6
Nga Vaka	The Canoes	Alpha & Beta Centauri	Important for navigation Canoes of founders	2, 5
Taura Nukunuku	The *Nukunuku* Rope	Procyon and Gomeisa	Two stars	2, 6
Tautoru	The Three Handsome Ones	Orion's Belt	Three stars Announced: colder rainy season (with Tau Ehu); start of Paina festivals; egg hunt (with Matariki and Tau Ehu); time for planting Legendary father and 2 sons (from Hiva)	2
Tatauro	Cross	Crux	Te Tatauro (modern name) Four to five stars	2
Atutahi	The Wayward One *or* The First Apprentice	Probably Canopus	End of Paina (with Tautoru) Start of the planting season The first of all stars, chief of the others	1, 3, 3, 6
Ko Para Tahiri	Tahiri's Club *or* Para´s Fan, *or* Decayed Fan	Probably Arcturus	-	3, 6
Ko Te Mata Pu/Po Nui	Eye With a Big Hole/Darkness	Spica	-	6
Mere	Voice of Joy	–Undetermined– Some possibility it was Vega or Corvus	If Vega, marked: Opening of the eel fishing season and ritual tattooing at Orongo	1, 3 6, 7
Pau	Open Piercing	–Undetermined–	Probably one star Announced: Ill-omen when appearing in Oct./Nov. with Tautoru and Matamea	3, 4, 6
Po'o or Po Rongo	O Rongo's Darkness	Probably Achernar	-	6
Po'o or Po Roroa	The Great Darkness/Infinity	Saiph	-	2
Rei a Tanga	Tanga's Breastplate	Betelgeuse	Wife and mother of men of Tautoru	2
Tau Ehu *or* Te Pou o Te Rangi	Beautiful Firebrand *or* The Post of the Sky	Sirius	Announced: colder rainy season with Tautoru; Dec. egg hunt with Matariki and Tautoru	2, 3, 6
Te Hau Vaero	Roostertail Feather Headdress	Antares	Rapanui zenithal star?	2, 6
Vari Koreha	Fearsome Eel/Centipede Curling Into a Circle	Rigel	-	2, 6

ACKNOWLEDGEMENTS

Would like to thank our informants Valentino Riroroko and Alberto Hotus, who both put in an extra amount of effort to share their knowledge with us, as elders and leaders of the Rapanui people. We would also like to extend our thanks to Christopher Browder who secured the copy of the unpublished field notes of Alfred Métraux, which proved invaluable to our studies.

REFERENCES

Belmonte, J. A., and Edwards, E. (2007) Astronomy and landscape in Easter Island: new hints at the light of ethnographic sources. *Ethnography Papers for the Annual Meeting of the SEAC (European Society for Astronomy in Culture).* XII Conference of the SEAC, held in Kecskemét, August 26-28, 2004. E. Pásztor (ed.). Strasbourg: Institute of Ethnology at Strasbourg University, 2007, pp. 79-85.

Belmonte, J. A., and Edwards, E. (2011) Archaeoastronomy: Archaeology, topography and celestial landscape from the Nile to Rapa Nui. In *Highlights of Spanish Astrophysics VI: Proceedings of the IX Scientific Meeting of the Spanish Astronomical Society (SEA).* M. R. Zapatero Osorio, J. Gorgas, J. M. Apellániz, J. R. Pardo, and A. Gil de Paz (ed.). Madrid: Sociedad Española de Astronomía, pp. 786-796.

Cristino, C., Vargas, P., and Izaurrieta R. (1981) *Atlas Arqueológico de la Isla de Pascua.* Universidad de Chile, Santiago, Editorial Universitaria.

d'Ans, B. (2016) Moldes digitales de estructuras complejas y generación digital de paisajes culturales para estudios arqueo-astronómicos. R. Yachachiy (ed.). II Semester 2016. Available: http://www.alberdi.de/MOLDIESCODIGPAIARQABARTIIS-16.pdf, retrieved February 4th 2017.

d'Ans, B. (2014) Generación de paisajes culturales tipo fisheye para uso en planetarios digitales. Scientific paper delivered at the IV Annual Meeting of the Interamerican Society for Astronomy in Culture (SIAC), Instituto Peruano de Astronomía (I.P.A.), Planetarium Maria Reiche, Universidad Peruana de Arte ORVAL. Rio de Janeiro, October 16.

Edwards, E., and Edwards's A. (2010) Rapanui astronomy and ethnoastronomy: Flag 83 Expedition Report. Expedition Report, The Explorers Club, October.

Edwards, E., and Edwards A. (2013) *When the Universe Was an Island.* Santiago, Hanga Roa Press, 2010.

Edwards, E., and Belmonte J. A. (2004) The Megalithic astronomy of Easter Island: A reassessment. *Journal For The History of Astronomy*, Vol. 35, Part 4, N° 121, pp. 421-433.

Edwards, E., and Clark, M. (n.d.) Preliminary report of possible astronomical relationships in Quadrangle 31, Ahu Ra'ai, Easter Island, South Pacific. Unpublished.

Edwards, E., and Clark, M. (1984) Sun, moon, and volcanoes on Easter Island. Scientific paper delivered at the First International Congress: Easter Island and Eastern Polynesia, Centro de Estudios de Isla de Pascua, University of Chile. Hanga Roa, September 6-11.

Ferdon Jr., E. N. (1961) The Ceremonial site of Orongo. In *Archaeology of Easter Island: Reports of the Norwegian Archaeological Expedition to Easter Island and the East Pacific.* T. Heyerdahl and E. N. Ferdon Jr. (ed.). Santa Fe, School of American Research and Museum of New Mexico, Vol. I, pp. 221-72.

Kirch, P. V. (2004) Solstice observation in Mangareva, French Polynesia: New perspectives from archaeology. Archives of the JAC, N° 18.

Lee, G. (1992). *The Rock Art of Easter Island: Symbols of Power, Prayers to the Gods.* The Institute of Archaeology, UCLA, Monumenta Archaeologica, N° 17.

Lee, G., and Liller, W. (1987) "The sun stones of Easter Island: A re-evaluation." Archaeoastronomy, N° 11, pp. 1-11.

Lewis, D. (1974) "Voyaging stars: aspects of Polynesian and Micronesian astronomy." In *Philosophical Transactions for the Royal Society of London.* Series A, Mathematical, Physical, and Engineering Sciences, Vol. 276, Issue 1257, May, pp. 133-148.

Liller, W. (1989) "The megalithic astronomy of Easter Island: Orientations of *ahu* and *moai.*" Archaeoastronomy, N° 13, 1989, pp. 21-48.

Liller, W. (1993) *The Ancient Solar Observatories of Rapanui: the Archaeoastronomy of Easter Island.* Woodland, The Easter Island Foundation.

Liller, W. (2000) Ancient Astronomical Monuments in Polynesia. In *Astronomy Across Cultures: The History of Non-western Astronomy.* H. Selin (ed.). Dordrecht, Kluwer Academic Publications, pp. 127-160.

Liller, W. (2004) Hetu'u Rapanui: The Archaeoastronomy of Easter Island. In *Asteroids to Quasars: A Symposium Honouring William Liller."* P. M. Lugger and W. Liller (ed.). Cambridge, Cambridge University Press, pp. 267-286.

Liller, W., and Duarte, J. (1986) Easter Island's Solar ranging device, Ahu Huri A Urenga, and vicinity. *Archaeoastronomy*, N° 9, pp. 39-51.

Makemson, M. (1941) *The Morning Star Rises: An Account of Polynesian Astronomy.* New Haven, Yale University Press.

Métraux, A. (1960) *Ethnology of Easter Island.* Honolulu, Bernice P. Bishop Museum Press.

Mulloy, W. (1975) A solstice-oriented *ahu* on Easter Island. In *Archaeology and Physical Anthropology in Oceania*, Vol. X, N° 1, pp. 13-18.

Routledge, K. (1919) *The Mystery of Easter Island: The story of an Expedition.* London, Hazell, Watson, and Viney.

Routledge, K. (1911-1923) Papers of Mrs. Katherine Scoresby Routledge relating Chiefly to Easter Island. Unpublished manuscript. Reels 1, 2, 3, 4, Pacific Manuscripts Bureau, Research School of Pacific Studies, The Australian National University, Canberra.

WAS THE PARTA NEOLITHIC SANCTUARY IN ROMANIA ASTRONOMICALLY ALIGNED?

Marc Frincu

Romanian Society for Cultural Astronomy and West University of Timisoara Faculty of Mathematics and Computer Science

**Corresponding author: Marc Frincu (marc.frincu@e-uvt.ro)*

ABSTRACT

Since its discovery, the Neolithic sanctuary from Parta, Timis county, Romania has been the subject of many archeoastronomical and ethnoastronomical studies. While interesting, the sanctuary itself is no longer visible in situ, with a scaled replica, based on original materials, accessible inside the National Museum of Banat in Timisoara. Studies have focused on its solar alignment, lunar and stellar symbolism, eclipses, and horizon astronomy. The lack of actual azimuth readings of the original sanctuary make any astronomical alignment studies challenging if not impossible. The only evidence lies in the original experiment performed in situ during the autumnal equinox sunset on 23 September, 1982, and on maps showing the direction of the North. Regarding eclipses, the high ΔT uncertainty in Neolithic times makes any eclipse study questionable. In this paper we critically review prior work and introduce our own hypotheses regarding some interesting aspects of the sanctuary. We also identify possible horizon markers for the WSSR and equinoxes.

KEYWORDS: Parta, Neolithic, Romania, solar and lunar alignments, eclipses, constellations, horizon astronomy.

1. INTRODUCTION

The purpose of this paper is to examine the proposition formulated by many researchers over the past 30 years that the Neolithic Sanctuary from Parta Romania was astronomically oriented. The site of the sanctuary was first excavated in 1931. Based on archaeological findings and carbon dating the site was determined to be part of Banat culture itself derived from the local evolution of the Starčevo-Criş over which elements of Vinča A are intermixed. According to Lazarovici (2006), the earliest carbon dating for the Parta settlement are from the Parta 7a layer (5600 CAL BCE) while the latest are from grain seeds found in the Parta 6a layer (4800 CAL BCE). Most settlements in the perimeter of Parta are placed at an elevation of 90-95 meters in floodable areas near the Timis River. The main settlement, tell 1, was placed on a former river meander which during floods could have been islanded. The architecture of the settlement cannot be generalized, Parta, presumably being a religious, economic, cultural, and sociopolitical center. Over time the settlement has been intensely occupied with seasonal character at times. Overall the definitions given to a medieval commercial center can be without doubt applied to Parta (Lazarovici, 2006).

In the center of the settlement there was a ceremonial building called Sanctuary 2 (Parta 6 layer, 5200-5000 CAL BCE). Over time the building was reconstructed and destroyed repeatedly. Evidence suggests an earlier Sanctuary 1 (Parta 7c layer, 5300 CAL BCE). This was also predated by another building with an unknown purpose. This older building had an N-S orientation of the longer axis whereas the two sanctuaries were E-W oriented based on their entrances and longer axes. The entire settlement was found burned in layer 6 a result that has been attributed to conflict. The function of Sanctuary 2 as a temple is motivated according to Lichter (2014) by: (1) the existence of clay boxes with incised decoration interpreted as cult or libation tables, (2) the existence of bull skulls and horns, and (3) raised applications of a stylized human face and a bull skull as well as a sickle-shaped clay application around a hole in the wall (Sun-Moon window).

Over the years many archaeoastronomical researches on Sanctuary 2 were conducted (Lazarovici et al., 2002; Szűcs-Csillik, 2013; Szűcs-Csillik, 2015; Lazarovici, 2009; Szűcs-Csillik, 2017). We summarize and analyze here their relevance from an astronomical and ethno-astronomical context. Our objective is not to dismiss them entirely but to raise new questions about the uncertainties regarding the accuracy of astronomical algorithms and actual orientation of the sanctuary which make any theory regarding its orientation rather probabilistic. Any solid archaeoastronomical study needs to consider both archaeological and astronomical uncertainties as well as (where possible) cultural evidence. Recent examples include the debate around the meaning of the animals depicted on pillars at Göbekli Tepe (Notroff et al., 2017). In case of the Parta sanctuary we cannot dismiss the fact that the building was excavated and reimagined based on several incomplete or uncertain components. While its importance in the Neolithic period is unquestionable uncertainties about its orientation, placement and existence of some of its components, as well as the accuracy of numerical algorithms make it hard to prove any archaeoastronomical study on the building itself.

Our analysis is based on the following data (also see Figure 1) from Lazarovici et al. (2002):

- GPS coordinates: 45°37′ N, 21°06′ E.
- Dimensions: 11.6m x 6m.
- Distance of the Sun-Moon window (on the west wall) from the north wall: 1.5m.
- *Presumed* height of the Sun-Moon window from the ground: 1.35m.
- Diameter of the Sun-Moon window: 0.35m.
- Distance of the interior window from the north wall: 2.25m.
- *Presumed* height of the interior window from the ground: 1m.
- Diameter of the interior window: 0.3m.
- Distance of the bull-female twin statue from the north wall: 3.05m.
- Distance of the bull-female twin statue from the east wall: 1m.
- Width of the bull-female twin statue: 0.5m (*approximation*)
- Distance of the interior wall from the east wall: 6.65m (*determined from the scaled model*).
- Thickness of walls: 0.3m (*approximation*)

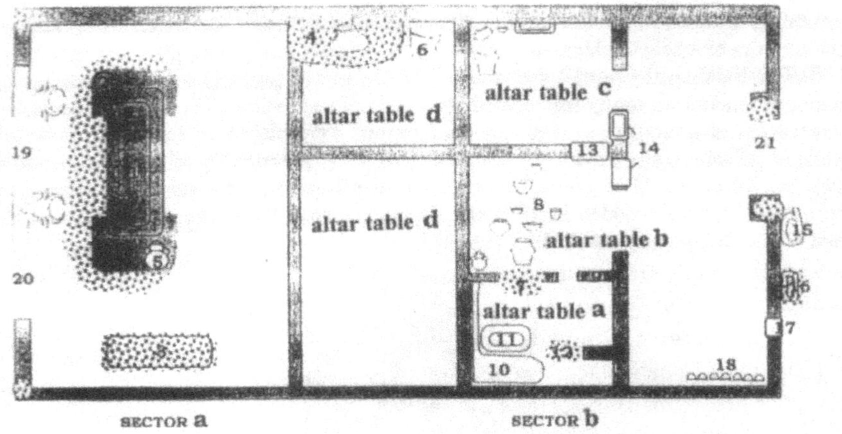

Plan of the second shrine. 1: the double idol; 2: pedestal; 3: adobe trough; 4: portable fireplace; 5: amphora; 6: great cup; 7: taurian skulls; 8: vessel; 9: mounting for the taurian skull; 10: cassette; 11: 'man-faced vessel'; 12: mounting for the taurian skull; 13: taurian skulls; 14: mounting for the taurian skull; 15: hand-mill (mortar); 16: adobe cup; 17: Sun–Moon couple; 18: vertical loom; 19: window; 20: eastern entrance; 21: western entrance

Figure 1. Plan of Sanctuary 2 (from Lazarovici et al., 2002). East is left.

2. EXISTING HYPOTHESES

We have identified five main studies addressing the following problems: equinoctial alignment of the sanctuary (Lazarovici et al., 2002), eclipses (Szűcs-Csillik, 2013), Moon symbolism (Szűcs-Csillik, 2015), and stellar symbolism (Lazarovici, 2009), and alignments (Szűcs-Csillik, 2017).

2.1. *Solar alignments*

On September 23, 1982 Lazarovici performed an in situ experiment during excavations at Sanctuary 2 in which it was determined the path of the light through the Sun-Moon window on the *reconstructed* western wall. Assuming that the dividing wall had a similar window the light would have hit the back of the bull-female twin statue[1] approximately at the point where the amphora was found. To represent the window Lazarovici used a cardboard disk stuck on a wooden stick. Right before sunset he noticed the shadow (in the original setting it was the light) touching the division of the statue, the floor, the idol, the fireplace and the man-faced vessel. In Lazarovici et al. (2002) it is speculated on the light path at SSSS (Summer Solstice Sun Set) and WSSS (Winter Solstice Sun Set) as well. At SSSS the light would have hit the dividing wall and illuminated the southern wall where the vessel with corn was found. Correspondingly, at winter solstice, the light would have touched the hand-loom. In both cases the authors identify ritualistic symbols associated with the corresponding moments. Unfortunately, no exact measurements of the orientation of Sanctuary 2 exist, and only approximate indications of an E-W oriented long axis are given. For this reason we attempted to demonstrate the illumination effect on scaled models. Our own experiments on a scaled gypsum model and computed based simulations showed this to be true in both cases but only if the sanctuary is shifted to N-E (see Figure 2 and Section 2.1). If an exact E-W alignment is used the Sun does not touch the southern wall at SSSS as in Lazarovici's experiment. We will discuss the alignment problem later in this section.

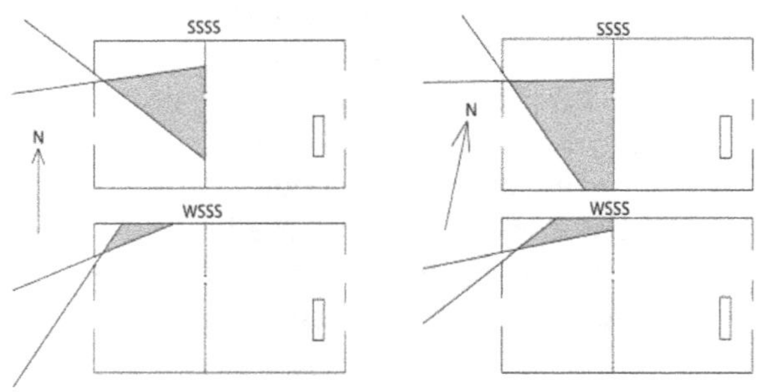

Figure 2. Simulation of sunlight (shaded area) entering in 5000 BCE inside the sanctuary at solstices (E-W aligned – left; shifted to N-E – right).

An interesting aspect is given by the orientation of the sanctuary and the path of the beam of light through the two windows. From the dimensions of the sanctuary and the placement of various elements we can determine that the beam of light is 8.82° deviated from the E-W axis. Other sources indicate a deviation of 9°-10° (Szűcs-Csillik et al., 2017). Considering that many sources indicate an *exact* E-W orientation of the sanctuary (Szűcs-Csillik et al., 2013; Szűcs-Csillik et al., 2015) while others just mention an E-W orientation (Lazarovici et al., 2002;

[1] The female was interpreted as the Great Mother Goddess and the bull as the Bull God (Lazarovici et al., 1994).

Szűcs-Csillik et al., 2017) then the beam of light had an azimuth of 278.82° and an altitude of 7.18° at the moment the first beam of light would enter both windows (considering that light first touches the upper limb of the Sun-Moon window and the lower limb of the inner window). Szűcs-Csillik et al. (2017) give a value of 20° but that would place the inner window at only 1.94m from the Sun-Moon window making the west room too little compared to known plans. In our computations we used a distance of 4.95m plus 0.6m to account for the thickness of both walls. The last beam would enter from an altitude of 0.51° (about one solar diameter). It is clear that it points to a position of the Sun in the sky prior to the autumnal equinox and after the spring equinox. By using Stellarium 0.15 with DE431 algorithm for the year 5200 BCE we checked the date for the position of the Sun at the computed azimuth and inclination and found that the first light would enter about 30 days before the autumnal equinox and about 30 days after the spring equinox. Hence there is an apparent contradiction between the results of the 1982 experiments and the orientation of the path between the two windows which would have caused light to illuminate the idol about a month before the fall equinox and a month after the spring equinox. This gives approximately 8 months between the late summer and early spring illumination. Currently we have no means of knowing whether or not this orientation was intentional or not.

A major problem with the solar alignment hypothesis is represented by the actual orientation of the sanctuary. There are no known azimuth readings or any accurate maps and the sanctuary has been reconstructed at the National Museum of Banat with the original location buried after the excavations. Most maps given in papers have a North arrow roughly parallel with the N-S axis of the sanctuary. However, prof. Lazarovici provided for this article another map (see Figure 3) which gives a calculated offset of about 38-40° N-W (for the North arrow compared to the initial maps). Lazarovici mentioned in a discussion that he sometimes used a centesimal compass so the difference may be due to an error in plotting centesimal readings using a sexadecimal protractor. Sadly, the actual azimuths are missing and without additional information it is impossible to rely on it. Prof. Drasovean from the National Museum of Banat recalls the long axis of the sanctuary to be S-E oriented but does not remember any azimuth readings. The hard question in our case is: *why was the experiment on September 23, 1982 a success if the sanctuary as cited in most papers is E-W aligned but the light path is off by 8.82°?* A possible answer is that the experiment was conducted on site with no outer walls and an approximately placed cardboard on a wooden stick as seen in Figure 4 (left). Another reason could be that an error was incorporated in the first reports stating the E-W *exact* orientation of the sanctuary when in fact this is not the case. Or maybe the term *exact* refers to the equinoctial alignment? In this case the sanctuary may be off by at most 16.18° to the N-E. Given the inclination of the beam through the two windows (7.18°) the Sun would start to illuminate the idol at equinox sunset from an azimuth of 262.64°. Further considering the azimuth offset of the light beam from the longitudinal axis of the sanctuary (8.82°) this gives an orientation of the sanctuary of 253.82°, or a total of 16.18° from the true north. In this case the Sun would illuminate the back of the idol at equinox through the two windows starting about one hour before sunset up until an altitude of 0.51°, validating the experiment performed by Lazarovici. Interestingly, this N-E deviation would also validate Lazarovici's initial statement that during SSSS the southern wall is illuminated (Figure 2 – right image). Common sense dictates that the offset range is too large to go unnoticed, the experiment was probably erroneous, and that initial maps depict a realistic image: a slightly S-E orientation. If so, the idol would have been illuminated about 30 days before fall equinox.

Lazarovici has also showed a connection between the sanctuary and the Neolithic fertility and fecundity cult so perhaps the light entering through the two presumed windows at equinoxes had some sort of ritualistic role when hitting the twin idol's back. Interestingly, the female idol on the pedestal seems to be pregnant (see Figure 4, right side). We will discuss this aspect in greater detail in Section 2.4.

But how could have they marked the equinox date? Ruggles (1997) gives four possible methods: (1) finding the spatial midpoint between the rising/setting at two solstices, (2) finding the half-way point in time between two solstices, (3) finding the day on which the sunrise is exactly opposite of the sunset, and (4) finding the day when the length of time from sunrise to sunset is the same as from sunset to sunrise. Method (4) is inapplicable as we do not know of any exact time keeping methods in Neolithic. Method (3) is inapplicable to Parta as we have the Pades Mountains rising due east making the Sun visible to the south of true east on the equinox day. This also makes method (1) unreliable *except* if they watched the *sunset horizon* since it lacks any horizon markers. Method (2) as explained by Ruggles always gives a positive declination as the Earth does not have a constant speed around the Sun and the Sanctuary would have been N-E aligned.

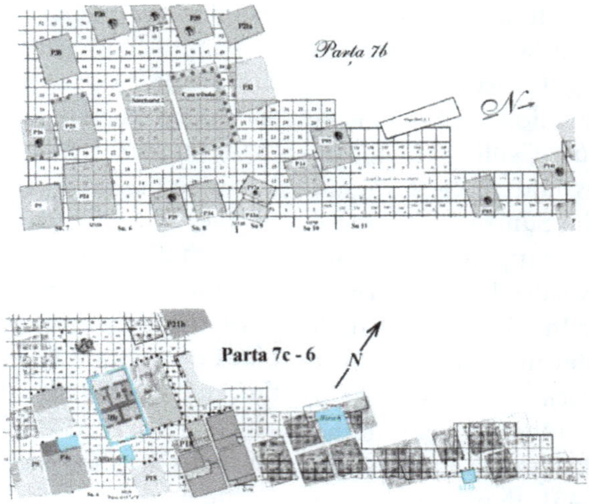

Figure 3. Orientation of Sanctuary 2 by Lazarovici (top: orientation from the 2006 book; bottom: map possibly wrongly showing north by reading the centesimal azimuth not the converted sexadecimal).

Figure 4. Left: partially reconstructed wall and the cardboard used to test the alignment hypothesis (half of the cardboard can be seen on top of a wooden staff in the upper part of the picture). Right: twin bull-female idol from Parta. There is a prominence on the belly of the female which would indicate her being pregnant.

2.2. *Eclipses*

While the subject of identifying eclipses in Neolithic times is known to be virtually impossible due to the ΔT problem this has not stopped researchers from associating ancient sites, including Parta, with the observation of eclipses. Simply put, the ΔT problem arises from the variable rotation rate of Earth which is unknown in the past and cannot be estimated for the future accurately. ΔT is the difference between the Uniform Terrestrial Time used to compute the positions of the planetary bodies and the Universal Time which is linked to the Earth rotation speed which varies (with a tendency of slowing down) over time due to many factors (e.g., ice ages, lunar gravity). ΔT can be approximated and correlated based on observed historical eclipses. Presently, the oldest identifiable eclipse has taken place on October 30, 1207 BCE (Humphreys, 2017). Numerical algorithms such as JPL DE (Jet Propulsion Laboratory Development Ephemeris) (Folkner et al., 2008) and VSOP (Variations Séculaires des Orbites Planétaires) (Bretagnon and Francou, 1988) as well as methods based on the Saros cycle can predict eclipses but their accuracy in determining *where* they would be visible from decreases drastically the further one goes in the past or future. If the oldest mentioned eclipse indeed took place at the indicated date then it means that the currently used expressions for ΔT can be extended up to 1200 BCE which is not enough for Neolithic times. Figure 5 shows the increase in ΔT uncertainty over years. Even if an eclipse took place around the time Sanctuary 2 was in use the uncertainty area is of about 70° in longitude meaning that the eclipse could have been visible also from either the East coast of Canada or China.

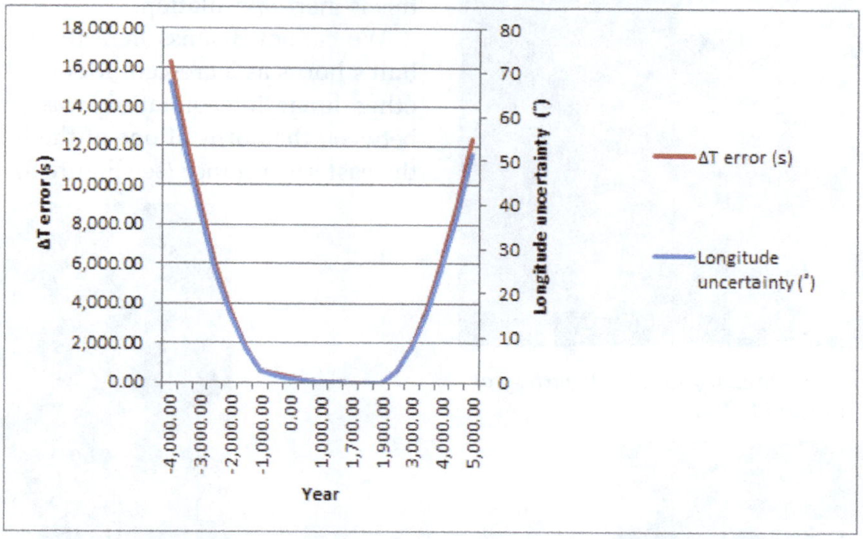

Figure 5. ΔT uncertainty as a function of time as derived from https://eclipse.gsfc.nasa.gov/SEhelp/deltaT.html.

Szűcs-Csillik et al., (2013) theorized that eclipses could have played a major role in the local society having a deep impact on the lives of the people of the era. The occurrence of an eclipse could have triggered wars or a migration of the people after the great fire that destroyed the settlement. A debatable account which has survived to this day is that of Herodotus in his work *The Histories* about an eclipse that stopped the war between Medians and Lydians. While their eclipse theory is interesting, Szűcs-Csillik et al. fall (as we also had initially) in the trap of actually trying to predict the date for a Neolithic eclipse. They identify using a custom made Matlab code based on Saros cycles, two solar and two lunar eclipses that might have been visible on 4974 BCE and 4563 BCE (solar), respectively on 4825 BCE and 4631 (lunar). Furthermore, in a different article Szűcs-Csillik et al., (2015) mention (based on Stellarium) a solar eclipse that presumably took place on September 18, 5401 BCE (Szűcs-Csillik et al. mention 5400 however Stellarium uses astronomical year numbering which includes year 0).

Assuming the eclipses had been visible from the sanctuary, we analyse the actual possibility of them being *observable* by the inhabitants and focus on the one in September 18, 5401 BCE and the one on April 30, 4593 BCE (Szűcs-Csillik et al. mention 4592 for the same reason as depicted above) which we could precisely identify ourselves using Stellarium 0.15 with DE431 algorithm installed. The 5401 BCE eclipse despite being too early for the Sanctuary 2 period covered approximately 6.7% (Szűcs-Csillik identify it as an annular eclipse using an older version of Stellarium 0.12 with the default VSOP87 algorithm) around noon making it impossible to be seen with the naked eye due to the intensity of the Sun. On the other hand, the one in 4593 presumably took place right before sunset starting with about 10% coverage at an altitude of 1° (roughly two solar diameters). While such eclipses are clearly visible (as seen in Figure 6 depicting the rising of a partially eclipsed Sun as seen with the naked eye) people must actually know where to look. It is not unconceivable that ancient man looked at the sun at sunset or sunrise and that they observed partial eclipses with at least 5% coverage as witnessed by today's eclipse hunters. The possible 4593 BCE eclipse taking place around the sunset was however after the period of Sanctuary 2. Furthermore, the ΔT uncertainty makes it rather improbable that the eclipse would have been visible from Parta.

An interesting fact we mention here with regard to Sanctuary 2 is the existence of a crescent Moon shaped ornament on the outer western wall window as depicted in Figure 7. This may hint of a possible eclipse observed at some point by the local inhabitants. This so called Sun-Moon window is traditionally interpreted as follows: the crescent shape object is the Moon and the window is the Sun (Lazarovici et al., 2002; Szűcs-Csillik et al., 2015). However, astronomers know that this is an impossible interpretation of a solar eclipse since it is the Sun who has a crescent during the event. Hence, if the solar eclipse representation theory is correct then the window would symbolize the Moon and the crescent the Sun. Following the same reasoning if the window depicts a lunar eclipse than the crescent represents the unshaded part of the Moon while the window represents the shadow of the Earth.

Figure 6. Solar eclipse on January 15, 2010 (courtesy of Cătălin Beldea).

Figure 7. The Sun-Moon window on the western wall at Parta (reconstruction from the original).

2.3. *Lunar significance*

According to researches the sanctuary holds significant lunar symbolism probably most obvious through its Sun-Moon window (as seen in Figure 7). Probably the most detailed work done on the subject is that of Szűcs-Csillik et al. (2015). We focus here on the symbolism alone and ignore any references to eclipses as we have discussed them previously. Under the Sun-Moon window there was a rimmed hand-mill and a cup in which remains of seeds were found. All these together form a complex picture depicting an agricultural culture and possible offerings to the Sun and Moon both symbols of fertility and procreation. In the ancient world the light of the Moon was regarded as favourable to vegetal grow and animal reproduction. According to Eliade (1976), the connection between the Moon, rain and plant life was realized before the discovery of agriculture.

An interesting interpretation given by Szűcs-Csillik about the pregnant female on the twin idol is that it is in fact a symbol for the Full Moon or a total lunar eclipse when the Moon looks reddish (Blood Moon). This can be further correlated with the light entering about a month before/after the fall/spring equinox and hitting the bull's back (if the sanctuary was indeed exactly E-W aligned) but without definitive proof on the actual orientation of the sanctuary this is mere speculation.

We cannot dismiss here the interpretation of the bull's horns as a crescent Moon (Drössler, 1986). Another lunar interpretation was given to the bulge between the horns of one of the bull heads guarding the eastern entrance (see Figure 8).

Figure 8. Bulge between the horns of the bull interpreted as the Moon. Another one on the second column is interpreted as the Sun.

2.4. *Constellations*

During the period of Sanctuary 2 the sun rose from the constellation of Gemini at spring equinox and set in the constellation of Sagittarius during the autumnal equinox. This means that at the autumnal equinox sunset the constellation of Taurus together with Gemini would have been visible on the eastern horizon. The twin idol faces east through a door marked by the columns with bulls on their head. Only the female is placed in front of the door, the bull shape figure facing the wall. Symbolically, this might indicate some sort of fertility ritual with the pregnant woman facing the rising Bull constellation full of sexual energy. The symbolism could have been increased by the light of the Sun at sunset through the two windows hitting the bull idol's back.

Recently, a paper was published by Szűcs-Csillik et al. (2017) were it is hypothesized that through the line of sight defined by the inner and Sun-Moon windows priests could see the night sky. Their theory is that someone standing near the pedestal could have looked through the two windows at the starry sky at an angle of 20°. This is somewhat strange given that for that angle the distance between windows should have been of 1.94m whereas the dividing wall with the inner window is closer to the middle of the sanctuary. Figure 9 depicts the maximum angle of the line of sight. The angle is trivially computed from the dimensions of various sanctuary elements as tan(0.70/5.55)=7.18°. From the northern corner of the twin idol a viewer looking through the two windows would have seen a region of about 2.3° of the western night sky. While Szűcs-Csillik uses a date of 5400 BCE which may be 200 years earlier than the date Sanctuary 2 was constructed they identify several stars which might have been visible at least partially through the two windows: Castor and Pollux (Gemini), Capella (Auriga), Regulus (Leo), Antares (Scorpio), Altair (Aquila). Interestingly they do no mention Mirphak (Perseus) which also has a similar declination. Our analysis using Stellarium for 5200 BCE showed that among them the best candidate is Pollux with the others being outside the window due to precession. This is not surprising as it is one of the two brightest stars in the constellation that in 5200 BCE rose heliacally at spring equinox. Szűcs-Csillik et al. also link the twin idol with the constellation itself.

In the same study, Szűcs-Csillik et al. also attribute the V shape and zig-zag symbols found on the wall of the sanctuary to constellations such as Cassiopeia or other less noticeable constellations such as Serpens and Ophiucus but this aspect deserves more investigation as except few cases (e.g., Taurus, Ursa Major) we have no records of these constellations being known as they are today.

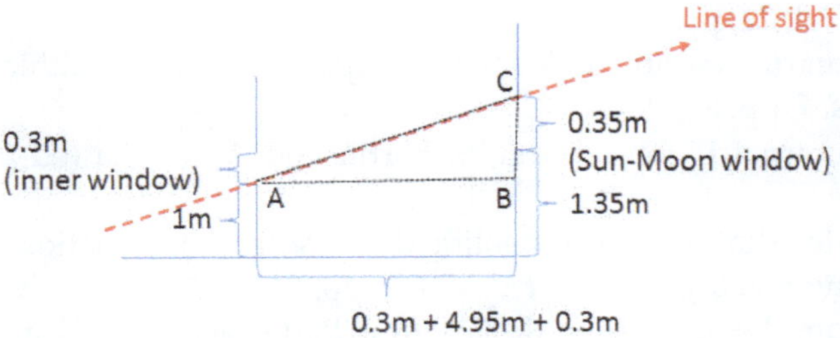

Figure 9. Maximum angle for the line of sight of a viewer placed inside the sanctuary. Conversely it is the maximum angle at which sunlight can enter through the two windows.

3. OUR DISCOVERIES

Parta is located on a plain with few horizon markers available. To the east there are the Pades Mountains at 94km and 87.34° azimuth (Pades peak, 1382m), to the S-W are the Semenic Mountains at 88km and 123.83° azimuth (Piatra Goznei peak, 1447m) and to the south are the Vršac Mountains at 59km and 159.10° azimuth. In 5200 BCE WSSR would have been visible from an azimuth of 125.20°. Semenic is a plateau with an altitude of 0.5° between azimuths 122-126°. Based on this data we went on the field to investigate the possible horizon alignments of visible from the site. On March 20, 2017 after three unsuccessful events due to bad weather we witnessed the sunrise from behind Pades Mountains. The alignment is almost perfect with the rising from the middle of the peak due to its elevation (Figure 10). In the same year we photographed after one unsuccessful attempt the WSSR from behind the highest peak in Semenic Mountains. The image was taken, due to bad weather, 10 days after the solstice on January 1, 2017 (Figure 11) when the Sun rises from behind the plateau and to the left side of the peak at an azimuth of 124°. On the solstice day the Sun would have risen from behind the peak and in 5200 BCE from the right side of the peak. While the alignment is not perfect Piatra Goznei (visible in Figure 11) is the most significant landmark on the plateau and could have been used as a marker for the *winter solstice period* (not date itself).

Figure 10. Sunrise at Spring equinox on March 20, 2017 from behind Pades Mountains.

Figure 11. Sunrise near the winter solstice on January 1, 2017 from behind Semenic Mountains.

4. DISCUSSION

Given the existing data it is impossible to determine the *exact* orientation of the sanctuary. However, we believe most maps to be fairly accurate and that the orientation was roughly E-W. This questions the accuracy of the 1982 fall equinox. Due to the offset of the light beam from the E-W axis of the sanctuary the twin idol would have been illuminated about a month before fall equinox. Despite uncertainties in the actual orientation, the geographical position however seems to indicate that the astronomy of the horizon may have played an important role. Simulated experiments on scaled models have shown that the handloom, is illuminated at winter solstice and that light touches the southern wall at summer solstice but only if the sanctuary is N-E oriented. We believe that the eclipse hypotheses regarding Parta are interesting but impossible to validate due to the ΔT problem. Finally, while a lot of lunar symbolism seems to be present the constellation alignments are hard to prove due to the sanctuary's long period of existence and the precession of equinoxes which may have *aligned* different stars in the sky with the two windows over time. Also, any association with present day constellations is purely speculative.

Concluding, while the sanctuary is a fantastic example of Neolithic craftsmanship uncertainties about its orientation (and even placement of some of its components) and numerical algorithms make it hard to prove any archaeoastronomical study on the building itself. Nevertheless, the sanctuary remains an interesting place full of (astronomical) symbolism in the Neolithic period.

ACKNOWLEDGEMENTS

We thank the delegates at the SEAC 2017 conference and the anonymous reviewers for their constructive comments on our original presentation. We also thank students in the Astronomy in Culture lecture for the gypsum model (Daniel Sîrbu, Anamaria Gödör, Teodora Marişescu, Luana Alexa) and computer based simulations (Casian Benga) of the Sanctuary.

REFERENCES

Bretagnon, P., Francou, G. (1988) Planetary theories in rectangular and spherical variables. VSOP87 solutions, *Astronomy & Astrophysics*, Vol. 202, pp. 309–315.

Davis, B.A.S., Brewer, S., Stevenson, A.C. and Guiot, J. (2003) The temperature of Europe during the Holocene reconstructed from pollen data, *Quaternary Science Reviews*, Vol. 22, No. 15, pp. 1701-1716.

Drössler, R. (1986) *Als die Sterne Götter waren*, Leipzig, Prisma.

Eliade, M. (1976) *Histoire des croyances et des idées religieuses vol 1 et 2*, Paris, Payot.

Folkner, W.M., Williams, J.G., Boggs, D.H. (2008) The planetary and lunar ephemeris DE 421, *JPL Interoffice Memorandum*.

Humphreys, C. and Waddington, G. (2017) Solar Eclipse for 1207 BC helps to date Pharaohs, *Astronomy and Geophysics*, Vol. 58, No. 5, pp. 39-42.

Ruggles, C. (1997) Whose Equinox?, *Archaeoastronomy*, Vol. 22, pp. 45-50.

Szűcs-Csillik, I. and Maxim, Z. (2013) Eclipsele si Sanctuarul Neolitic de la Parta (Eclipses and the Neolithic Shrine at Parta), A. Stavila et al. (ed.), *Arheovest I*, pp. 847-855.

Szűcs-Csillik, I. and Maxim, Z. (2015) Goddess of Nocturnal Light at Parta, S. Fortiu and A. Stavila (ed.), *Arheovest III*, Vol. 2, pp. 605-620.

Szűcs-Csillik, I. and Maxim, Z. (2017) Constelatii "vazute" din sanctuarul neolitic de la Parta (Constellations Seen from the Parta Neolithic Sanctuary), Fortiu S. (ed.), *Arheovest V*, Vol. 2, pp. 641-647.

Lazarovici, G., Maxim, Z. and Drasovean, F. (1994) Complexul Neolitic de la Parta (Parta Neolithic Complex), *Analele Banatului*, pp. 106-134.

Lazarovici, C. M. and Lazarovici G. (2006) *Arhitectura Neoliticului şi Epocii Cuprului din România (The architecture of Neolithic and Bronze Age in Romania)*, Iaşi, Trinitas.

Lazarovici, G., Chis, G., Oproiu, T. and Szűcs-Csillik, I. (2002) Neolithic Shrine at Parta. *Konkoly Monographs*, Vol. 4, pp. 7–17.

Lazarovici, G. and Lazarovici, C. (2009) The Architecture of Temples/Sanctuaries in the Banat and Transylvania during Neolithic and Copper Age Periods, *Arheologia Spiritualităţii preistorice în ţinuturile carpato-ponto-danubiene*, pp. 65-76.

Lichter, C. (2014) Temples in the Neolithic and Copper Age in Southeast Europe? *Documenta Praehistorica*, Vol. XLI, pp. 120-136.

Notroff, J., Dietrich, O., Dietrich, L., Tvetmarken, C.L., Kinzel, M., Schlindwein, J., Sönmez, D., Clare, L. (2017) More than a vulture: A response to Sweatman and Tsikritsis, *Mediterranean Archaeology and Archaeometry*, Vol. 17, No. 2, pp. 57-63.

15

THE ARCHAEOASTRONOMY OF HIGH ALTITUDE INCA CEREMONIALISM

J. McKim Malville

Department of Astrophysical and Planetary Sciences, University of Colorado
malville@colorado.edu

ABSTRACT

The Incas are renowned for their architecture, skillful masonry, complex political organization and their extensive system of roads. One of their most remarkable achievements was the ascent and the building of ceremonial structures on many of the highest peaks of the Andes, including Llullaillaco with an altitude of 22,110 feet, containing the world's highest archaeological site.

Offerings on the summits were made after state-supported pilgrimages, which often involved weeks or months of travel, covering distances of 1000 km or more. This paper discusses the sacrifices on the summit of Llullaillaco, their possible astronomical attributes and explores the meaning behind these challenging endeavors.

Llullaillaco can be climbed only during the southern summer, between November to March. The summit platform contained the bodies of three children, a 13-year-old girl and a boy and girl aged 4-5 years. Because the burials were in undisturbed conditions when excavated by Reinhard and his colleagues they also provided evidence for the role of astronomy in this ceremony. The platform has been rotated toward December solstice sunrise, which would have been an extraordinary endeavor, considering the difficulties of building and orienting stone structures at such an extreme altitude. Alignment of the young boy to sunrise on Capac Raymi may have been intended. The girl may have been oriented to June solstice sunrise. These ceremonies appear to have been a combination of imperial geopolitics and reciprocity between humans and mountain deities.

KEYWORDS: shrines on Andean peaks, capacocha ceremony, child sacrifice, reciprocity

1. INTRODUCTION

Evidence of the most remarkable example of solar ritual on a mountain summit is found on the summit of Llullaillaco with an altitude of 6739 meters (22,110 feet). It is the 7th highest peak of the Americas and contains on its summit the world's highest archaeological site. It was extensively and carefully excavated by Reinhard, Ceruti and assistants. They have published definitive description and analyses of the site (Reinhard and Ceruti, 2006, 2010). This paper addresses the possibility of astronomical alignments at the site. (Note: Reinhard comments that he suspected an astronomical alignment of the summit platform, but chose not to include that speculation in his publications).

The mountain is on the Chilean-Argentine border, east of Antofagasta, some 750 miles south of Cusco. It was climbed by the Inca sometime around 1500, close to December solstice, the start of the climbing season, and then not again until 1952 when two Chilean climbers, anticipating a first ascent, were disappointed to discover they were not the first. Ruins on this peak have been carefully excavated and mapped by Johan Reinhard and his colleagues (Reinhard and Ceruti, 2010). Other examples of similar sacrifices have been hard to find; most of these shrines, despite their high altitude, have been looted or excavated with little documentation.

Many of these mountain top shrines, of which where may be over 100, were the sites of *capacocha* (Quechua, *Qhapaq hucha*) ceremonies. (Beorchia, 2001; Reinhard, 2005; Reinhard and Ceruti, 2006, 2010). There are 16 shrines for which mummies have been documented and these almost entirely involved child sacrifice, boys of ages 4 to 10 and teen age girls chosen for their beauty and perfection, who were offered to the sun and mountain deities. Capacocha ceremonies were also performed, probably in Cusco, during critical events in the life of the Inca as well as to avert natural calamities such as volcanic eruptions. The most astronomically significant capacocha ceremonies were cyclical, some performed annually at the times of major solstice festivals of Inti Raymi (June) and Capac Raymi (December). The Inca empire was hierarchical and authoritative; little was done without permission of the authorities. Yet the struggle of climbing a 22,000 mountain was an intensely personal experience for which motivation had to be voluntary as well as imposed. Considering the extraordinary difficulty of climbing such a high mountain, the determination of the participants and victims to participate in this ritual is extraordinary, revealing something of the power of belief in Andean gods and associated ritual. According to Spanish chroniclers, offerings on the summits were made after state-supported pilgrimages, which often involved weeks or months of travel. The elaborate, state-supported pilgrimages started in Cusco and may have covered distances of 1000 km or more. The party included priests, officials, assistants, local inhabitants, the child to be sacrificed, and sometimes proud parents. (It is not entirely clear that all parents would have been proudly happy. Many may have feared the dire consequences of refusing to offer their children to the state for such an event). We can only speculate what the children felt. They would have been treated briefly like royalty. When passing through the mountains the pilgrims would keep as silent as possible to avoid angering the mountain gods who were not to be honored at that time. These were important imperial events, and when they passed through villages, the residents were not allowed to view them directly and had to prostrate themselves. We are encountering an ontology here in which the mountains and sky deities were understood to be living beings who could be actively involved in people's lives. In addition, a distinction between the living and dead, between animate and the inanimate, was seriously blurred. (Alberti and Bray, 2009; Allen, 2015; Malville, 2016) Death did not mean a permanent departure from this world. The children could thereby function as mediators between powerful deities and the living.

These ceremonies were apparently a combination of reciprocity involving mountain deities and state ordained geopolitics. In the Andean world humans, quite wisely, sought to achieve reciprocal relationships with the powerful forces of the land and sky by making offerings, which they hoped would require a substantial positive response. Mountains, as living beings capable of being alternately very dangerous or benign, provided water for agriculture or were responsible for droughts, avalanches, earthquakes, and fierce weather. The establishing of close relationships was vitally important: offerings to the sun were hopefully reciprocated with bountiful harvests and relief from natural disasters. Climbing to the summit of the mountain also brought the offering closer to the solar deity. Only the most perfect of offerings (such as unblemished children) were considered worthy offerings to insure the desired reciprocal response. The chosen child also had to be able to survive in the unpredictable and dangerous realm between the living and dead

2. PILGRIMAGE AND LIMINALITY

Liminality relates to passages, often transformative ones, from one realm to another, coming from the Latin word *līmen*, meaning a threshold (Turner, 2008). These passages may involve the ambiguity and disorientation that can occur when one crosses into a new and unfamiliar space or time. The idea of liminality was extensively developed by Turner in his discussion of pilgrimage and the unsettling experience of traveling into unfamiliar landscapes. While in the liminal state, human beings have a heightened awareness of their surroundings and are open to transformative suggestions from the environment or their companions. Liminality can involve places as well as experiences. Liminal places can include springs, caves, shores, rivers, crossroads, bridges, sacred spaces such as temples, and especially the cold and oxygen-deprived summits of high peaks in the Andes. In India, passages involving a "crossing-over" are known as *tirthas,* the most famous of all is on the banks of the Ganga in Varanasi. Tirtha has multiple meanings; it is a place for passing from one side of a river to the other, a place for meeting gods who have passed from their realm into ours, and a place to pass out of this life. The ascent of pyramids and high mountains in the Andes had a similar symbolism of moving across the three worlds into the upper realm where some of the gods resided.

3. TIMING OF THE CEREMONIES

There are two climbing seasons in Peru, Chile, and Argentina. In Peru, mountains are accessible during the dry cold winter months, while in the south, they can be climbed only during the southern summer between November to March. The most likely times for pilgrimage would be days around December solstice, the time of the Inca celebration of Capac Raymi, a major festival when young men from the aristocracy were initiated.

The timing of this celebration required incredible planning and precision. The pilgrims and priests spent weeks on the road before reaching the base of the mountain. Everything on the mountain had to be ready for the participants. Weather would have added additional uncertainty, but because the sun does not move appreciably for several days around solstice, the date of the celebration was slightly flexible.

4. ASCENT

The route upward to the summit of Llullaillaco contains a number of resting places for pilgrims, priests, and sacrificial victims. The largest of these way stations is a tambo at 17,000 feet, where most of the pilgrims and, perhaps, the parents would have stayed, which could have housed as many as 100 people. Ascent to the summit by a smaller party, which included the children, would have taken at least three days. They would have had to spend a night on the summit before the sunrise ceremony. The two room building on the summit may have been used by the priests and children (Figures 1 and 2). Other members of the party must have huddled during the night behind the wind break.

Whether it was in the rooms or behind the windbreak, spending the night at such a high altitude would have been an excruciating test of endurance, as any climber who has bivouacked at high altitudes can attest. Clearly they did not have down sleeping bags or down jackets, but some of the discomfort may have been alleviated by chewing coca leaves. When the first gleam of the sun appeared, its location would have been burned in the memories of those waiting for the dawn. Few sunrises could have been greeted more enthusiastically than by those who spent a frigid night on the summit of Llullaillaco.

Figure 1. Room at the Top (Graham Zimmerman)

Figure 2. Summit of Llullaillaco. *Note the rooms, wind break and trail to the platform (Graham Zimmerman)*

5. THE SUMMIT PLATFORM

The platform (Figure 3), 30 feet long, 20 feet wide, contained the bodies of three children, a 13-year-old girl and a boy and girl aged 4-5 years. These may have been interred on three different years. The summit rooms, windbreak, and summit platform appear too extensive for them to have been used only once. It has been suggested that the two younger children may serve as attendants to the older girl. She had a higher status based upon her clothing and her elaborately braided hair. The trail to the summit platform, in particular, is too well marked in the soil for only one visit. Because the burials were in undisturbed conditions when carefully excavated, there were three perfectly preserved mummies as well as information on their approximate orientation. The platform has been rotated approximately 10 degrees away from the natural ridge line suggesting an intention to orient it toward December the date of solstice sunrise (Figure 4). Capac Raymi, which was an important festival among the Inca when boys of the aristocracy were initiated. The long side of the platform is oriented to 30° magnetic north (Reinhard and Ceruti, 2010, p. 66). The perpendicular to the long side would then have a true azimuth of 118.4°, and the boy is approximately facing that direction. There are no reported measurements of the angle of depression of the horizon, but it can be estimated by the panorama generator of heywhatsthat.com. At an azimuth of 118° the horizon is depressed by approximately 1.3°. Measurements of refraction made from Cerro Tololo (2215m and 645 km to the south) (Schaefer and Liller, 1990), involving an average depression angle of -2.2°, give an average refraction of .7° with a standard deviation of .27°. Using this refraction angle, the azimuth of the upper limb of the sun would be 117.3°, which is within approximately a degree of the orientation of the platform. It thus appears possible that the young boy was intentionally placed facing the first gleam of the sun near the date of December solstice.

Figure 3. Summit Platform (Graham Zimmerman)

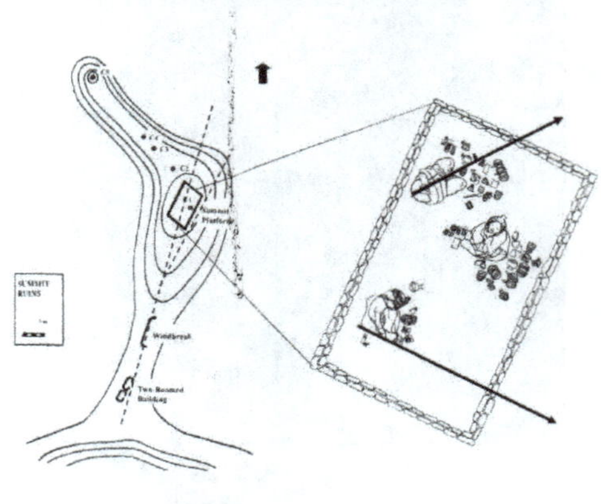

Figure 4. Summit Platform Showing Burials (after Reinhard and Ceruti, 2010)

The 13-year-old girl was facing northeast, approximately 53° north of the boy or to an azimuth 65.4°. (Note: personal communication from Reinhard expresses caution because of the difficulty of making such a measurement on the irregularly shaped mummy bundle *in situ*). That orientation is close to the direction of June solstice sunrise, the date of the Inca festival of Inti Raymi. The depressed horizon in that direction has an angle of -1.3°, giving an azimuth of the first gleam of sunrise of 66.1°, using a refraction angle of .7°. The orientation of the girl thus appears to have been within 0.7° of the first gleam of June sunrise. However, the mountain cannot be climbed during the winter month of June. There are two methods that could have been used to estimate the azimuth of June solstice sunrise. A ritual specialist could have noted the direction of the last gleam of the sun on December solstice. In that direction the depressed horizon had an angle of approximately -2.2°. Once that direction had been established, the priest could have turned himself 180° to establish the approximate direction of June solstice sunrise. That approach would have resulted in an orientation of the girl to an azimuth of 62.3° some 3° short of the direction toward which she was buried. Alternately, perhaps the more likely approach would have been to transfer the angular separation of June and December solstices measured somewhere else to the summit. That angle would have to be established at a site readily accessible in winter and close in latitude. The angular separation of the solstices could have been marked using a horizontal pole held at arm's length.

Where could such a measuring device have been calibrated? Because of the considerable importance of the solstices in the Inca ceremonial world, such a measurement may have had ritual importance and may not have been unusual. The modern town San Pedro de Atacama is 5 km south of the Inca center of Cartepe, which at an altitude of 8400 feet should have been occupied throughout the year. As observed from the town the two solstices occur 52° apart, within a degree of the angular separation of the northern and southern mummies. The orientation of two of the three children buried in the summit platform suggests an intention to mark the two major solar festivals of the Inca. The third child is approximately facing the major summit, perhaps honoring the mountain deity. Analysis of the hair of the older girl indicates that 12 months before her death her diet became rich in animal protein and maize, marking the time she became an Acllas or chosen women and entered elite lifeways. Her hair was cut 6 months later and then trimmed just before her death. Ritual hair cutting is often associated with a change of status, perhaps a liminal rite. Hair cutting at six month intervals may indicate synchronization with the solstices as well as a countdown to sacrifice (Wilson et al., 2007, 2013).

6. FINAL THOUGHTS

Children were chosen for these capacocha ceremonies because they were deemed to be the most perfect offering that could be made. Sacrificial techniques employed in Inca capacocha ceremonies avoided any overt display of cruelty. The practice of burying children alive or sometimes of suffocating them prior to burial apparently was based on the belief that only "complete" offerings were acceptable to the sun or mountain deity. A victim who had shed blood would have been considered to be an "incomplete" offering in the eyes of the Incas. All of the children had ingested coca and alcohol (Wilson et al., 2007, 2013). The older girl had markedly higher levels of coca and alcohol, perhaps indicating a greater need to sedate her. The boy's knees were tightly bound against his chest by a rope. He may have suffered from pulmonary edema and died during the ascent. Next to him was an aribalo, containing the residues of chicha, which may have been intended as an offering to the sun whom he faced in death.

Climbing a 22,100 foot mountain is not an easy task; one has to contend with physical exhaustion, cold, and oxygen deprivation. It would not have been an ordinary journey, but rather a challenge that required a willingness on the part of the children to reach the summit. It's hard to believe that anyone could be coerced into climbing such a mountain. Spending the night on top may have been terribly frightening for the children. A brilliant sunrise in the luminous atmosphere of the summit of the mountain may have been exciting. Being in the presence of gods of the sun and mountain could have been an epiphany. We shall never know.

ACKNOWLEDGEMENTS

I am greatly indebted to Johan Reinhard and Maria Constanza Ceruti for their careful field work carried out under extremely difficult conditions. I thank Graham Zimmerman for the use of his photographs.

REFERENCES

Alberti, B. and T. Bray. (2009) Animating Archaeology: of Subjects, Objects and Alternative Ontologies, *Cambridge Archeaological Journal* Vol. 19, pp. 337-343.

Allen, C. (2015) The Whole World is Watching: New Perspectives on Andean Animism, in Tamara Bray (ed) *The Archaeology of Wak'as. Exploration of the Sacred in the Pre-Columbian Andes.* Boulder, University Press of Colorado.

Beorchia, A. (ed.) (2001) *Revista del Centro de Investigaciones Arqueológicas de Alta Montaña,* Vol. 6.

Bray, T. (2009) An Archaeological Perspective on the Andean Concept of Camaquen: Thinking Through Late Pre-Columbian Ofrendas and Huacas. Cambridge Archaeological Journal 19: 357-366.

Malville, J. M. (2016). Animism, Reciprocity, and Entanglement. *Mediterranean Archaeology and Archaeometry,* Vol 16, pp. 51-58.

Reinhard, J. (2005) *The Ice Maiden: Inca Mummies, Mountain Gods, and Sacred Sites in the Andes.* Washington D C, National Geographic Society.

Reinhard, J. and M. C. Ceruti (2006) Sacred Mountains, Ceremonial Sites and Human Sacrifice Among the Incas. *Archaeoastronomy,* Vol. 19. pp. 1-43.

Reinhard, J. and M. C. Ceruti. (2010) *Inca Rituals and Sacred Mountains: A Study of the World's Highest Archaeological Sites.* Los Angeles: Cotsen Institute of Archaeology.

Schaefer, B.E. and W. Liller (1990) Refraction near the horizon. *Publications of the Astronomical Society of the Pacific,* Vol. 102, pp. 796-805.

Turner, V. (2008) Liminality and Communitas, in *The Ritual Process: Structure and Anti-Structure,* New Brunswick, Aldine Transaction Press.

Wilson A. et al. (2007) Stable isotope and DNA evidence for ritual sequences in Inca child sacrifice. *Proceedings of the National Academy of Sciences of the USA,* Vol. 104(42). pp 16456-16461.

Wilson A. et al. (2013) Archaeological, radiological, and biological evidence offer insight into Inca child sacrifice. *Proceedings of the National Academy of Sciences of the USA,* Vol 110(33). pp 13322-13327.

ARCHAEOASTRONOMICAL STUDY OF THE PROTOHISTORIC SHAFT TOMBS NECROPOLEIS OF THAPSOS (SICILY)

Andrea Orlando*[1], Davide Gori[2] and Carlo Veca[3]

[1]Institute of Sicilian Archaeoastronomy, Italy
Fondazione Floresta Longo, Italy
[2]Wood plc, Italy
Institute of Sicilian Archaeoastronomy, Italy
[3]Archaeologist, Superintendence of Cultural Heritage, Malta

**Corresponding author: Andrea Orlando (orlando@archeoastronomia.com)*

ABSTRACT

Thapsos is a settlement located on the Magnisi peninsula, a triangular strip of land between the gulfs of Syracuse and Augusta. Here it developed a remarkable Middle Bronze Culture (1440-1250 BC), with a village with proto-urban character for the influence of import and know-how from the Aegean, Cyprus and Malta. The necropoleis of Thapsos are arranged in 3 areas of the peninsula. For our analysis we considered the two groups of rectangular shaft tombs with underground chamber necropolis, widespread on the limestone plateau to the North and South of the peninsula.

A first partial study about the orientation of the Thapsos' shaft tombs was realized at the beginning of the third millennium on 25 tombs in the North necropolis; in this first analysis it was observed that there were two privileged orientations: one of astronomical character, linked to the sunrise at the summer solstice, and the other of topographical character, connected to Mount Etna (North direction).

The new archaeoastronomical analysis on the Thapsos' shaft tombs was started by the authors in 2016 and involved 70 tombs located on the peninsula. The opportunity to study the orientations of these tombs allowed to realize the first mapping of the necropoleis and the numbering of the tombs. The proposed study is therefore of great scientific importance, as it was previously thought that in the Bronze Age had lost cultic tradition of building shaft tombs necropoleis with astronomically oriented entrance. The study thus opens new scenarios as part of the funerary rituals and religious needs, showing a discontinuity with the previous phases.

KEYWORDS: Thapsos, shaft tomb necropoleis, Paolo Orsi, Middle Bronze Age, topographic orientation, Etna mount, summer solstice, astronomical orientation.

1. INTRODUCTION

In Sicily, the research on the astronomical orientation of monuments, megaliths and funerary hypogea during Prehistory and Protohistory still represents a poorly explored field, which is slow to take off. Excluding the most recent attempts to develop this interpretation in the island context (Orlando, 2015; 2016; 2017), the first and only study dedicated to the orientation of rock necropoleis in Sicily was realized at the end of the 20th century (Tusa and Foderà Serio, 2001). The archaeoastronomical study involved:

- the rock-cut and shaft tombs of the Copper Age (IV-III millennium BC);
- the rock-cut tombs of the Early Bronze Age (Castelluccio Culture, late III - mid II millennium BC);
- the chamber tombs with "dolmenic" corridor of south-western Sicily, also of the Early Bronze Age (end III - beginnings of the II millennium BC);
- "pseudo-dolmenic" evidence (end 3rd - beginnings of the II millennium BC).

This study did not take into consideration the shaft tomb necropoleis of the subsequent phase of the Middle Bronze Age.

The Eneolithic shaft tombs were generally built on flat rocky terraces. On these lithic tables a shaft was dug and then it was possible to direct the tomb towards a predetermined or desired point of the horizon, without any kind of conditioning. It has therefore been thought, therefore, that in the sepulchral orientation there could be the possible compliance with religious beliefs connected to the stars, first of all our star, the Sun.

As for the Eneolithic shaft tombs, in the study of Tusa and Foderà Serio (2001), the sites of *Roccazzo* (Mazara del Vallo) and *Tranchina* (Sciacca) were investigated to see if there was a preferential choice in the orientation of the cell.

In the Eneolithic settlement of *Roccazzo* the orientations of 34 rock-cut tombs were measured. Of these:

- 5 fall into the I quadrant (0°-90°);
- 22 in the II (90°-180°);
- 7 in the III (180°-270°);
- none in the IV (270°-360°).

While in *Tranchina*, on a total of 30 rock-cut tombs, it was possible to measure the orientations of 28 of them. Of these:

- 3 fall into the I quadrant;
- 22 in the II;
- 3 in the III;
- none in the IV.

The azimuth values found indicate that the orientation of the tombs of the two sites is framed in the typical pattern widespread in the Mediterranean, which often sees prehistoric sepulchres oriented between the rising of the sun and its maximum height (Hoskin et al., 1994; Hoskin and Zedda, 1997; Hoskin and Palomo y Perez, 1998; Hoskin, 1998). So in the hypogeal tombs examined in the sites of *Roccazzo* and *Tranchina* the cell was dug, in the great majority of cases, in such a way that the axis of the opening (from the inside to the outside) fell into the II quadrant (90°-180°).

2. HYPOGEAN TOMBS IN THE BRONZE AGE

With the beginning of the Early Bronze Age (end of the III millennium BC), the hypogeal tomb typology changes radically. The most macroscopic change is the disappearance of the shaft and the consequent position of the tomb no longer on a flat surface, but on a slope or in a wall. It is a typology with a very wide and widespread diffusion in the island territory, so much so that it became an intrinsic element of the landscape, especially in the countryside of the provinces of Syracuse, Ragusa, Enna, Caltanissetta and Agrigento. This is what the current archaeological bibliography called "tomba a forno" or "tomba a grotticella", due to the effective analogy with domestic ovens used in the countryside and in the farming villages of Sicily (Orsi, 1892).

The tomb with a small chamber is intrinsically linked to the spread of the most typical culture of prehistoric Sicily, that of Castelluccio from the Early Bronze Age (2200-1450 BC) (Alberti, 2013). The existence of intentionality in the positioning of the entrance of the tombs is very problematic. In fact, they are always placed on almost vertical walls on the sides of river valleys (the famous "Cave" of the Iblei Mountains) or of hills (Caltanissetta, Enna, Agrigento, Catania areas) receiving, therefore, a strong environmental conditioning. However, it cannot be excluded that certain intentionality has existed in the choice of the wall to be "drilled" according to its orientation. In this regard, the study by Tusa and Foderà Serio took into consideration two emblematic sites of the most typical Castelluccio culture of the Iblea area: *Cava Lazzaro* and *Castelluccio di Noto*.

In the tombs of *Cava Lazzaro* the homogeneous orientation (towards the North) is not enough to highlight a very strong environmental conditioning due to the orientation of the rock face where the caves were dug. However, it is interesting to note that on the opposite wall to that used for the necropolis there are no rock-cut tombs. If this clear intentionality is evident in *Cava Lazzaro*, this does not happen for *Castelluccio di Noto*, where the tombs are arranged indifferently on either side of the small valley. It would seem therefore that in Sicily, during the Early

Bronze Age, the Eneolithic tradition of building tombs with astronomical orientation had been lost.

On the other hand, new and significant indications could be drawn from the study of the orientations of the shaft tombs of the Thapsos necropoleis (Orlando and Veca, in press). Thapsos is the eponymous site that fully represents the events of the Middle Bronze Age (XV-XIII century BC), a period of Sicilian Protohistory not held in consideration from the aforementioned archaeoastronomical studies.

3. THE ARCHAEOLOGICAL SITE OF THAPSOS

Thapsos is a settlement located on the peninsula of Magnisi (lat. 37.15° N, long. 15.23° E), a strip of land with a triangular shape (2300 × 800 meters, about 20 masl) between Syracuse and Augusta, linked to the mainland of the eastern coast of Sicily by a narrow sandy isthmus (Figure 1).

Figure 1. The Magnisi Peninsula viewed from satellite (Google Earth).

The Magnisi peninsula represents a calcareous structural high (Horst) within the distensive tectonic (NW-SE orientation) of the Iblean Avampaese (Carta geologica, 1986). The main lithology is characterized by cenozoic limestone rocks of the Monti Climmiti Formation (Miocene middle-Inf.), and in particular by the Calcari di Siracusa Member composed of white-greyish calcarenites and algals calcirudites, irregularly layered and notably karst. There are also sand, calcarenites and fossiliferous conglomerates at Strombus bubonius constituting Tirreniano's terraced deposits (Pleistocene sup.).

In this settlement developed a very remarkable and articulated Middle Bronze culture (Veca, 2016b). The inhabited area has at least three development phases (Figure 2B):

- the first (oldest), characterized by circular-shaped huts;
- the next, characterized by a spatial distribution of a proto-urban type with quadrangular plan settings around a central courtyard, served by roads and wells;
- the most recent one from small quadrangular or rectangular-shaped buildings that interfere with the proto-urban installations of the previous phase.

The culture of Thapsos follows to that of Castelluccio and is contemporary with that of Milazzese of the Aeolian Islands. Its chronological limits are XV-XIII centuries BC. The influence of the Aegean civilization is particularly sensitive in this period (La Rosa, 2004), with numerous Mycenaean imports (Myc. III A-B). Gold, ivory, amber and precious stone are also imported; locally produced, instead swords and daggers (Veca, 2016a; 2017). More disputed is the Aegean influence exerted on architectural forms, both domestic and funerary although in this last case the ancient tradition of rock-cut tomb goes through a process that will lead to the creation of the chamber tomb with a tholoid profile (Tomasello, 1995-96, page 222).

Figure 2. Penisola Magnisi: (A) 1, protohistoric settlement; 2, distribution of the rock-cut and shaft tombs necropoleis; 3, area of the enchytrismòs tombs; 4, city walls. B) Thapsos, planimetry of the protohistoric settlement (from Voza, 1985).

The ceramic repertoire of the Thapsos culture is characterized by grey monochrome pottery with a brown surface. The shapes are new and varied with respect to the previous culture: cups and basins on high tubular feet with bifid plate handles, bowls on high feet and with long handles, globular orbs with cylindrical neck, globular pyxis on the foot or cylindrical apode. The decoration is plastic, with low horizontal cords; but it is also engraved raw, and presents simple geometric patterns (bundles of lines, zigzag, festoons), or more rarely phytomorphic and zoomorphic figures, especially volatiles. There is a case of depiction of a human figure standing on a boat (Orsi, 1895). Relations with the Maltese culture of Borġ in-Nadur, and with the Cypriot cultures of the same period are attested by numerous ceramic imports (Tanasi, 2011; 2015; Alberti, 2015).

4. THE PROTOHISTORIC NECROPOLEIS OF THAPSOS

The rock-cut and shaft tombs necropoleis of Thapsos are arranged in three areas of the peninsula, and consist, according to the literature (Orsi, 1895), in about 300 tombs (Figure 2A).

The largest group is located in the North and North-West, between the lighthouse of the Navy and the posts of the Second World War. The tombs are presented in two different types:

• tombs with quadrangular shaft entrance and hypogeum chamber, diffused on the calcareous plateau of the peninsula (Figure 3);

• rock-cut tombs with a pavilion entrance carved into the low cliff wall near the sea (Figure 3).

Figure 3. Types of tombs visible at Thapsos: shaft tomb (up) and rock-cut tomb (down).

The area of the prehistoric necropoleis of Thapsos has been destroyed in antiquity by metal seekers. The excavations of the Thapsos necropoleis were conducted by the archaeologist Paolo Orsi in 1894. In his work the scholar describes the architecture and the funeral goods of only five shaft tombs, marked by numbers (Orsi, 1895): 51, 52, 53, 65 and 66 (Figure 4). Orsi, even though he numbered some tombs, did not make a plan for the necropoleis. The archaeological site of Thapsos was investigated with the excavation of the settlement and the necropoleis (Bernabò Brea, 1970; Voza, 1972; 1973; 1980- 81; 1984-85). Most chamber tombs are characterized by a narrow entrance corridor (dromos) that gave monumentality to the tomb and allowed an easy flow of water. In both types, the funeral chamber, sometimes preceded by a small vestibule, has a circular plan with an arched or flat profile, and often presents a series of perimeter niches.

Another group of shaft tombs, with a smaller extension than the one in the North, is located in the southern portion of the peninsula, immediately south of the fortifications (Figure 2A).

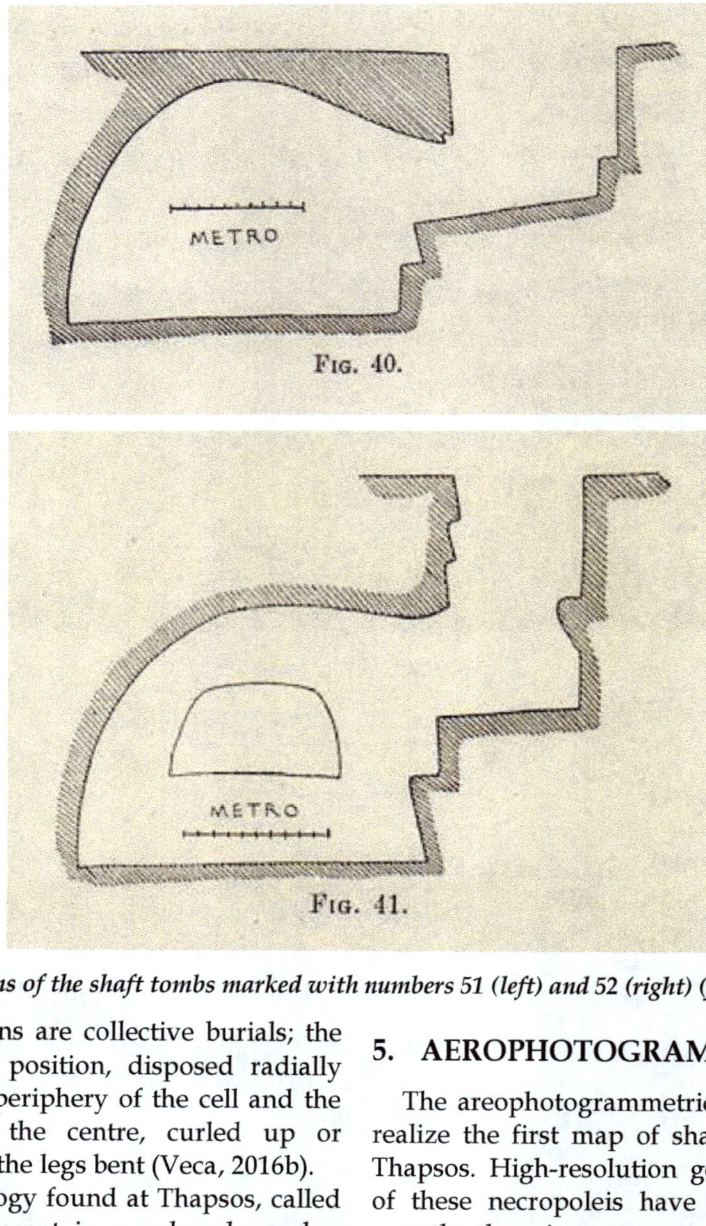

Figure 4. Sections of the shaft tombs marked with numbers 51 (left) and 52 (right) (from Orsi, 1895).

In general, depositions are collective burials; the bodies, in a crouched position, disposed radially with the skulls on the periphery of the cell and the lower limbs towards the centre, curled up or stretched out with only the legs bent (Veca, 2016b).

Another burial typology found at Thapsos, called "a enchytrismòs", is present in a reduced number com-pared to the tombs dug into the rock. It consists in the deposition of bodies within large "pithoi" deposited in small ravines or depressions of the rock (Veca, 2014). This burial ground is located in a small cave in the middle part of the peninsula (Figure 2A). Although it is interesting from a funeral archaeology point of view, the burial ground to enchytrismòs is not part of our treatment, and is mentioned here as additional information.

5. AEROPHOTOGRAMMETRIC ANALYSIS

The areophotogrammetric technique was used to realize the first map of shaft tombs necropoleis of Thapsos. High-resolution georeferenced ortophotos of these necropoleis have been produced. Subsequently, these images were digitized to produce distribution maps of shaft tombs (Figure 5). For the first time it was possible to enumerate the tombs, also identifying not yet discovered tombs (at least 4). It is not excluded, in fact, that there may be other tombs not yet identified. For the future, a not-invasive instrumental survey (e.g. georadar or other) could be performed.

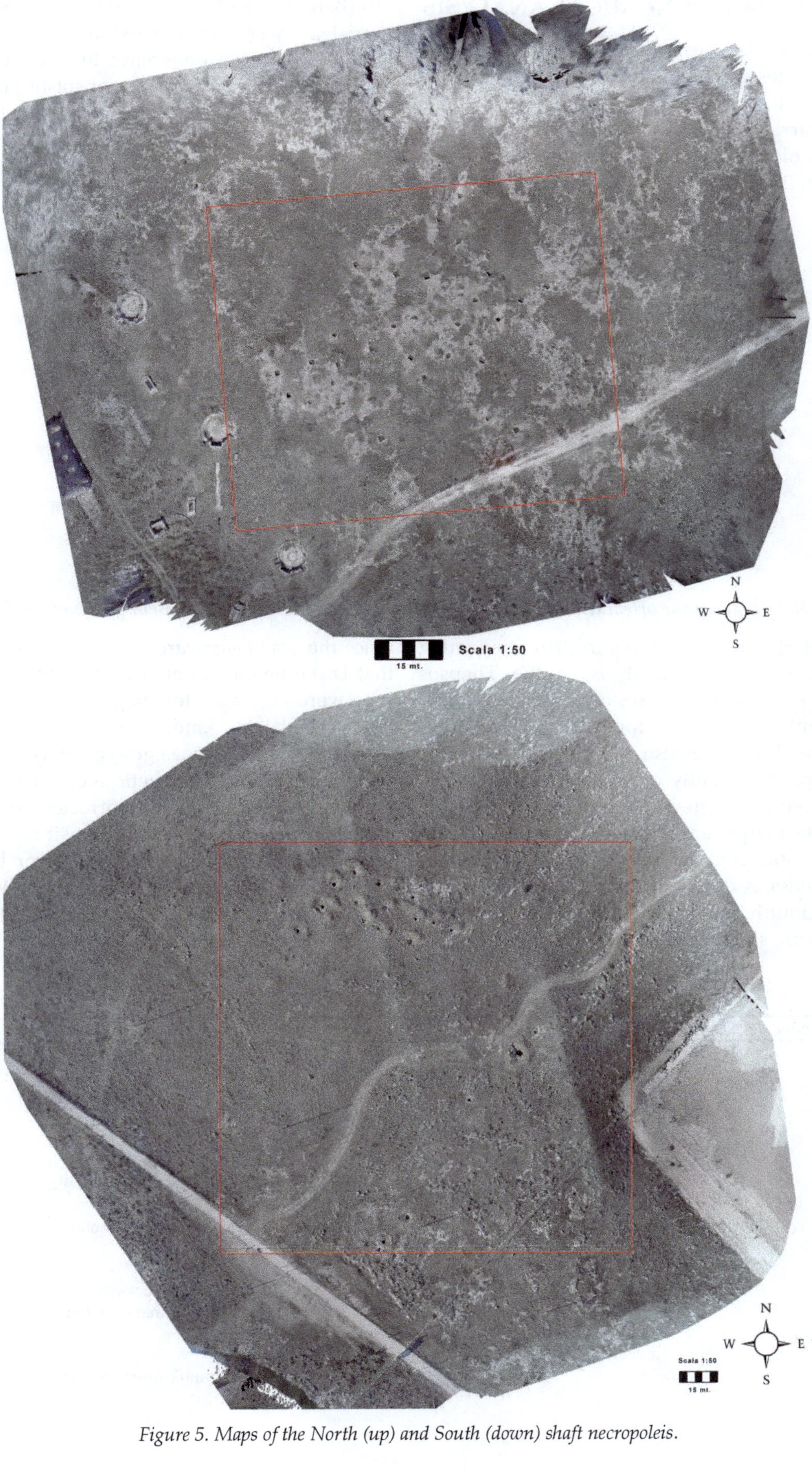

Figure 5. Maps of the North (up) and South (down) shaft necropoleis.

6. ARCHAEOASTRONOMICAL ANALYSIS

Very few scholars, especially in Italy, know the study that professors Belmonte and Hoskin realized at Thapsos following a brief survey on the Magnisi peninsula carried out in January 2001. During the visit, the scholars were able to measure the orientation of about 25 shaft tombs of the Northern necropolis (Belmonte and Hoskin, 2002). The results of their partial study are summarized in Figure 6. Data clearly show a double preference in orientation, that is one for the dawn at the summer solstice (SV) and the other for the silhouette of the volcano Etna, which stands out on the northern horizon.

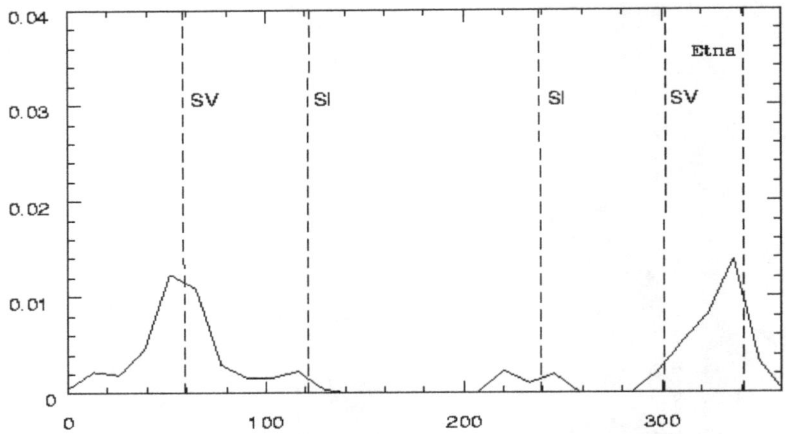

Figure 6. Histogram of the orientation of the 25 shaft tombs analysed in 2001 (from Belmonte and Hoskin, 2002).

The study started by the authors in 2016, concerning the archaeoastronomical analysis of the Thapsos shaft tombs necropoleis, focuses on the orientation of about 70 tombs. There are 2 tombs that have a double entrance. We have also identified 4 probable tombs still predominantly covered with soil. Therefore, compared to the previous study, it was possible to investigate a larger and complete sample. The first survey was conducted in situ with a Wilkie-type geologist compass. A double measurement was carried out. The azimuth measurements with heights and declinations are shown in Table 1. Since the sea is eastward, the eastern horizon is flat, that is the height is equal to 0°. The compass readings were corrected for magnetic declination using the calculator kindly provided by NOAA (http://www.ngdc.noaa.gov/geomag-web/#declination) (magnetic anomalies are not to be expected at Thapsos). Our corrected compass readings are fully consistent with transit readings, as the latter all fall within the nominal error band of the corresponding compass readings (±½°). We calculated declinations using the program GETDEC kindly provided by Clive Ruggles.

Table 1.

Tomb number	Azimuth (°) (average)	Height (°)	Declination (°)	Note
1	76,5	0	+10,35	-
2	68	0	+17	-
3	92	0	-0,05	solar orientation
4	12	0	+50,65	*partially buried tomb*
5	62	0	+21,6	solar orientation
6	60,5	0	+22,7	solar orientation
7	338	1,5	+48,6	topographic orientation (Etna Volcano)
8	35	0	+40,3	-
9	53	0	+28,25	lunar orientation
10	49	0	+31,1	-
11	24	0	+46,2	-
12	92	0	-0,05	solar orientation
13	52	0	+29	lunar orientation
14	43,5	0	+34,9	-
15	15	0	+49,8	-
16	357,5	0,5	+52,8	topographic orientation (Etna Volcano)
17	27	0	+44,75	-
18	-	-	-	*probable tomb*
19	82	0	+6	-
20	24	0	+46,2	-
21	62	0	+21,6	solar orientation

22	122	0	-24,5	solar orientation
23	34	0	+40,9	-
24	65	0	+19,3	lunar orientation
25	66	0	+18,5	lunar orientation
26	8,5	0	+51,45	-
27	74,5	0	+11,9	-
28	82	0	+6	-
29	26	0	+45,25	-
30	100,5	0	-7,3	-
31	73	0	+13,1	-
32	45	0	+38,9	-
33	62; 332	0; 1	+21,6; +45,2	tomb with double entrance/solar orientation and topographic orientation (Etna Volcano)
34	75,5	0	+11,15	-
35	72	0	+13,9	-
36	62	0	+21,6	solar orientation
37	62	0	+21,6	solar orientation
38	75	0	+11,5	-
39	108,5	0	-15	-
40	8	0	+51,5	-
41	83	0	+5,2	-
42	24	0	+46,2	-
43	65,5	0	+18,9	lunar orientation
44	76,5	0	+10,35	-
45	-	-	-	*totally covered tomb*
46	200,5	1	-46,25	topographic orientation (Pizzo Belvedere)
47	-	-	-	*partially collapsed tomb*
48	28	0	+44,25	-
49	70,5	0	+15,05	-
50	55	0	+26,8	lunar orientation
51	304,5	1	+26,95	lunar orientation
52	61; 320,5	0; 0	+22,35; +37,5	tomb with double entrance/solar orientation
53	58,5	0	+24,2	solar orientation
54	57	0	+25,3	-
55	74	0	+12,3	-
56	33	0	+41,5	-
57	62	0	+21,6	solar orientation
58	231	2,5	-27,5	lunar and/or topographic orientation (Hyblaean Mountains)
59	2	0	+52,2	-
60	104	0	-10,5	-
61	336,5	1	+47,5	topographic orientation (Etna Volcano)
62	335,5	1	+47	topographic orientation (Etna Volcano)
63	61	0	+22,35	solar orientation
64	41	0	+36,5	-
65	313	0	+32,5	-
66	122	0	-24,5	*tomb full of soil*/solar orientation
67	-	-	-	*probable tomb*
68	323	0	+39,1	-
69	324	0	+39,7	-
70	352	0,5	+52,1	topographic orientation (Etna Volcano)
71	233	2,5	-26,9	lunar and/or topographic orientation (Hyblaean Mountains)
72	44	0	+34,5	-
73	231	2,5	-27,5	lunar and/or topographic orientation (Hyblaean Mountains)
74	43	0	+35,2	-

The measures allow us to present the following first considerations. In this preliminary analysis we did not consider possible stellar orientations. This check will be carried out in the near future, when the study will be completed.

As we can see from Table 1 more than 70% of the azimuths falls in I (0°-90°) and in the II quadrant (90°-180°). So the eastern direction is therefore privileged. In that direction there are 14 shaft tombs that have a solar orientation, so the axis of the entrance of the shaft tombs (from the inside towards the exterior) is directed towards: a) summer solstice (10 tombs); b) winter solstice (2 tombs); c) equinoxes (2 tombs). So we have 20% of the shaft tombs that presents a solar orientation.

Another group of 10 tombs presents a topographic orientation, and in particular their entrance is oriented towards: a) Etna volcano (North, 6 tombs); Hyblaean Mountains (West, 3 tombs); Pizzo Belvedere (South-West; 1 tomb). We must remember that Etna dominates the North horizon and the northern landscape. In the southern part of the Etna volcano, the one visible from Thapsos, there are numerous important archaeological sites, such as that of the Sicel village of Mendolito (territory of Adrano). Similarly, the Iblei Mountains are an integral part of the western horizon. The Iblei preserve important archaeological sites, such as that of Pantalica, a UNESCO heritage site since 2005. So we have a 14% of the shaft tombs that presents a topographic orientation.

The three tombs whose entrance is directed towards a particular point on the western horizon, where the Iblei Mountains form a stairway, could however also have a lunar orientation; in fact, the calculated declinations suggest this possibility. However, the analysis shows that at least 7 tombs have a lunar orientation, since the declinations are close to those connected to the so-called lunistices. So we have a 10% of the shaft tombs that presents a lunar orientation.

In total, therefore, for this first analysis, 44% of the shaft tombs seem to have a non-random orientation.

7. CONCLUSIONS

This preliminary study about the orientation of the Thapsos' shaft rock necropoleis is quite interesting, as up until now it was assumed that during the Bronze Age, the cult and tradition of building rock-cut shaft tomb necropoleis with astronomically or topographical oriented entrance was lost. From the preliminary analysis we have found that almost half of the rock-cut shaft tombs are orientated, both astronomical (solar and lunar) and topographical. Particularly in this phase we have considered only the solar and lunar orientation, but the next step will be to check possible stellar orientations.

From our study it is clear that it is not possible to confirm the existence of only two privileged directions, an affirmation that we find in the work of Belmonte and Hoskin (2001), which, it should be remembered, considered only a group of 25 shaft tombs.

Thapsos' tombs, characterized by collective depositions, represent a further evolution of Sicilian funerary hypogeism, already of a clan type with the spread of the multi-chamber hypogeal sepulchre during the Copper Age (Malpasso's horizon, half of the III millennium BC). This study thus opens up new scenarios in the funerary rituals and needs of worship, highlighting, during the Middle Bronze Age, a discontinuity with previous Castelluccio culture of the Early Bronze Age.

ACKNOWLEDGEMENTS

We want to thank Dr. Alessandra Alberici for his valuable and tireless collaboration in IAS activities.

REFERENCES

Alberti G. (2013) - *A Bayesian 14C chronology of Early and Middle Bronze Age in Sicily. Towards an independent absolute dating*, Journal of Archaeological Science, 40(5), pp. 2502–2514.

Alberti G. (2015) - *Brocche Base Ring II da contesti del Bronzo Medio in Sicilia: produzione levantina, cipriota o locale? Alcune considerazioni*, Jasinks A.N., Bombardieri L. (a cura di), AKROTHINIA. *Contributi di giovani ricercatori italiani agli studi egei e ciprioti*, Firenze, pp. 1-18.

Amari, S. (2006) - *I materiali in esposizione nell'Antiquarium, Sale I-II-III*, Palermo.

Belmonte J.A. and Hoskin M. (2002) - *Astronomía y paisaje en Thapsos*, in Reflejo del Cosmos: Atlas de Arqueoastronomía del Mediterráneo antiguo, EQUIPO SIRIUS, pp. 217-221.

Bernabò Brea L. (1970) - *Thapsos. Primi indizi dell'abitato dell'età del bronzo*, Adriatica Praehistorica et Antiqua, Miscellanea Gregorio Novak dedicata, Zagreb, pp. 139-151.

Carta geologica del settore nord-orientale ibleo (1986) scala 1:50000. Università di Catania, S.E.L.C.A., Firenze.

Gullì D. (2008) - *La necropoli di contrada Tranchina di Sciacca*, Agrigento.

Guzzardi L. (1996) - *L'area degli iblei tra l'età del bronzo e la prima età del ferro*, in Guzzardi L. (a cura di), Civiltà Indigene e Città Greche nella Regione Iblea, Ragusa, pp. 9-42.

Hoskin M. (1998) – *Studies in iberian archaeoastronomy: (5) orientations of the megalithic tombs of northern and western Iberia*, JHA, XXIX, pp. S39-S87.

Hoskin M., Allan E. and Gralewski R. (1994) – *Studies in iberian archaeoastronomy: (1) orientation of the megalithic sepulchres of Almeria, Granada and Malaga*, JHA, XXV, pp. S55-S82.

Hoskin M. and Zedda M. (1997) – *Orientations of Sardinian Dolmens*, JHA, XXVIII, pp. S1-S16.

Hoskin M. and Palomo y Perez (1998) – *Studies in iberian archaeoastronomy: (4)the orientation of the megalithic tombs of eastern Catalunya*, JHA, XXIX, pp. 63-79.

La Rosa V. (2004) (ed.) - *Le presenze micenee nel territorio siracusano*, Atti I simposio siracusano di preistoria siciliana, Padova.

Orlando A. (2015) - *Studio archeoastronomico della 'spirale megalitica' di Balze Soprane (Bronte, CT): analisi preliminare*, Nicolosi, Parco dell'Etna, pp. 86-89.

Orlando A. (2016) - *Archaeoastronomy In Sicily: Megaliths And Rocky Sites*, Silva F., Malville K., Lomsdalen T. and Ventura F. (eds.), The Materiality of the Sky, Sophia Centre Press, pp. 301-316.

Orlando A. (2017) (eds.) - *The Light, The Stones and The Sacred*, Proceedings of SIA2015, Springer.

Orlando A. and Veca C. (in press) - *Gli orientamenti delle tombe a pozzetto della necropoli protostorica di Thapsos (Siracusa): analisi preliminare*, XVI Convegno della Società Italiana di Archeoastronomia (SIA) - "Quis dubitet hominem coniungere cielo?" (Politecnico di Milano, 3-4 Novembre 2016).

Orsi P. (1892) - *La necropoli sicula di Castelluccio (Siracusa)*, Bullettino di Paletnologia Italiana, s. 2 VIII, (XVIII), pp. 1-34, 67-84.

Orsi P. (1895) - *Thapsos*, Monumenti Antichi dei Lincei, Vol. VI, coll. 89-150.

Procelli E. (1981) - *Il complesso tombale di contrada Paolina ed il problema dei rapporti tra Sicilia e Malta nella prima età del Bronzo*, Bollettino d'Arte, vol. 9, pp. 83-110.

Tanasi D. (2011) - *Living and dying in a foreign country. Maltese immigrants in Middle Bronze Age Sicily*, in Tanasi D., Vella N.C. (eds.), *Site, Artefacts, Landscape: Prehistoric Borġ in-Nadur, Malta*, Praehistorica Mediterranea, vol. 3, Monza: Polimetrica, pp. 283-337.

Tanasi D. (2015) - *Borġ in-Nadur pottery abroad: a report from the Sicilian necropoleis of Thapsos and Matrensa*, Tanasi D., Vella N.C. (eds.), The late prehistory of Malta: essays on Borg in-Nadur and other sites, Archaeopress Archaeology Series, pp. 173-184.

Tomasello F. (1995-96) - *Le tombe a tholos della Sicilia centromeridionale*, Cronache di Archeologia, 34-35, Catania.

Tusa S. (1990) - *La preistoria nel territorio di Trapani*, Palermo.

Tusa S. and Foderà Serio G. (2001) - *Rapporti tra morfologia ed orientamento nelle architetture rituali siciliane dal IV al II millennio a.C.*, Atti dei convegni lincei 171, pp. 297-323.

Veca C. (2014) - *Contenitori "per i vivi" e contenitori "per i morti" a Thapsos (Siracusa): un approccio tecnologico a un problema interpretativo*, Rivista di Scienze Preistoriche, LXIV, pp. 203-225.

Veca C. (2016a) - *Nuovi dati per la definizione delle spade "tipo Thapsos" dai contesti siciliani del Bronzo Medio*, IpoTESI di Preistoria, vol. 8, pp. 27-36.

Veca C. (2016b) - *Archeologia funeraria. Architettura, riti e liturgie nella Sicilia sudorientale del Bronzo Medio (1450-1250 a.C.)*, Lecce 2016, ISBN 9788892643925

Veca C. (2017) - *Pugnali, daghe e spade. Nuovi elementi per l'inquadramento delle armi da taglio e da punta nei contesti funerari del medio Bronzo pugliese e siciliano*, XLVII Riunione Scientifica dell'Istituto Italiano di Preistoria e Protostoria - Preistoria e Protostoria della Puglia (Ostuni 9-13 ottobre 2012), Firenze 2017, pp. 993-998.

Voza G. (1972) - *Thapsos, primi risultati della più recenti ricerche*, in Atti XIV RS IIPP, Firenze, pp. 175-205.

Voza G. (1973) - *Thapsos: resoconto sulle campagne di scavo del 1970-71*, Atti XV RS IIPP, Firenze, pp. 133-157.

Voza G. (1980-1981) - *L'attività della Soprintendenza alle Antichità della Sicilia Orientale. Thapsos*, Kokalos 26-27, II,1, pp. 675-680.

Voza G. (1984-1985) - *Attività nel territorio della Soprintendenza alle antichità di Siracusa nel quadriennio 1980-1984. Thapsos*, Kokalos 30-31, pp. 666-668.

Voza G. (1985) - *I contatti precoloniali col mondo greco*, in Pugliese Caratelli G. (a cura di), Sikanie. Storia e civiltà della Sicilia greca, Milano, pp. 543-562.

17

IBERIAN APPROACHES TO ASTRONOMY DURING THE SIXTEENTH CENTURY

Walmir Thomazi Cardoso[1*] and Roberto de Andrade Martins[2]

[1] Pontifícia Universidade Católica de São Paulo; Grupo de História, Teoria e Ensino de Ciências, Universidade de São Paulo
[2] Universidade Federal de São Paulo; Fundação Municipal de Ensino Superior de Bragança Paulista; Grupo de História, Teoria e Ensino de Ciências, Universidade de São Paulo

**Corresponding author: Walmir Thomazi Cardoso (walmir.astronomia@gmail.com)*

ABSTRACT

The aim of this work is to describe different kinds of astronomical and astrological works written in the sixteenth century, with special emphasis on Portuguese and Spanish texts. This period is especially relevant in the development of astronomical culture in Portugal and Spain, since this was the age of the great overseas explorations, when astronomical navigation became of fundamental importance. The astronomical culture of that time involved ancient and new concepts, and old revisited narratives about the cosmos and its influence in many facets of the sublunary world. The present research describes the influence and transformation of the classical and medieval astronomical culture in the specific context of sixteenth century Iberian texts. There were different publics involved in the production and study of those works, such as navigators, priests, physicians, farmers, astrologers and scholars. Those different aspects of astronomical and astrological knowledge were not distinct, there was a significant overlap between them; and the educated public required some broad acquaintance with those several approaches. Some popular treatises, called *Chronographia*, or *Reportorio dos tempos*, provided astronomical and astrological information required by the general public. This overview of those sources can contribute to a better and more comprehensive understanding of the astronomical culture in that period.

KEYWORDS: Portuguese Astronomy, Spanish Astronomy, Sixteenth Century

1. INTRODUCTION

Although the study of Iberian science during the Renaissance has been overlooked by the majority of historians, its relevance is recognized nowadays (Goodman, 1992; Cañizares-Esguerra, 2004). There are several historical studies of science and, more specifically, astronomy and astrology in Portugal and Spain, during this period (Martins, 2004; Costa and Leitão, 2008; Fabiani, 1995). All over Europe, astronomy and astrology were both respected and deeply related fields, up to the beginning of the seventeenth century (Allen, 1966). Therefore, no attempt will be made here to disentangle those two fields.

The aim of this paper is neither to provide a full account of this subject not to discuss particular works. Its purpose is calling the attention to the different publics interested in the production and study of astronomy and astrology in the Iberian Peninsula at that time.

Figure 1. A representation of some of the uses of astronomy in the sixteenth century (Fra Mauro, 1537, frontispiece).

In Portugal and Spain, during the sixteenth century, there was a variety of interests concerning astronomical and astrological knowledge – as it usually happens in any culture (Macedo, 1975). Different publics were involved with diverse contents and uses of astronomy. Among the several cultural aspects of Iberian astronomy in this period (Martins, 2004), we may include the astronomical prospects of:

- priests and other people concerning with religion;
- pilots, navigators and cartographers;
- physicians, surgeons and other persons concerned with health;
- farmers, hunters and fishermen;
- astrologers;
- scholars, that is, university students and teachers;
- other groups.

During the sixteenth century, astronomy was regarded as an important and practical science, useful for several purposes (Figure 1). Although there was an overlap of those different snippets of astronomy, each of them made a specific use of selected portions of astronomy and astrology, producing distinct texts.

2. PUBLICS AND INTERESTS

The religious dimensions of astronomy, at that time, included the biblical and theological view of the cosmos; the relation between the heavenly spheres and the different orders of angels and saints; the locality of God and of demons; and technical calendar problems such as ascertaining the Easter days.

The astronomical profile of pilots, navigators and cartographers embraced the study of celestial circles and corresponding terrestrial circles; the annual motion of the Sun (from a geocentric point of view) and the motion of the Moon; the influence of the Moon and the Sun on the tides; the constellations and the celestial poles; the astronomical determination of time; the study of the terrestrial zones and climes; ascertaining the main geographical directions; and determination of latitude (by the stars and the Sun) and longitude (especially by lunar eclipses).

Physicians and surgeons of that time required a knowledge of the accepted influences of the celestial bodies (especially the Moon) on health and diseases; the connection between the bodily humours and the sky; the relation between drugs and heavenly influences; the correlations between parts of the human body and the planets and constellations; the adequate astronomical settings for specific treatments (such as bleeding); the relation between celestial phenomena and epidemics; and the precise relation between the individual horoscope and health problems.

The astronomical profiles of farmers, hunters and fishermen involved the study of constellations and their associations with the seasons; the seasonal influences on animals and plants; the association between the phases of the Moon and biological phenomena; and the relation between celestial occurrences and the weather.

Although judicial astrology was banned for theological religious reasons in Spain and Portugal (Baldini, 2001), it was practiced; and the study of collective (not individual) astrological influences was permitted. Both required the detailed study of the celestial bodies, including the computation of the motion of the planets, and knowledge of their influences on sublunary events. Of course, not every astrologer was able to compute the motions of the celestial bodies, most of them used tables produced by a few experts.

The astronomical contingent of students and teachers of liberal arts at the universities and colleges embraced the study of philosophical views of the universe; the structure of the cosmos; discussions about the form and position of the Earth; a general study of the heavenly motions; the understanding of astronomical phenomena; the philosophical views of astronomy; and astronomical knowledge discussed or cited in classical literature. This general basis was also the basis of the study of Medicine, but medical students required a more detailed study of the motion of the planets and of astrology.

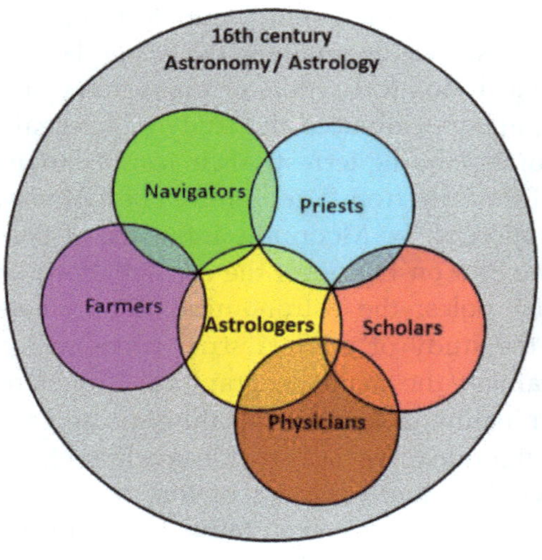

Figure 2. A schematic representation of several interest groups that were concerned with astronomy and astrology in the Iberian Peninsula, during the sixteenth century. No attempt was made to show all intersections or to provide a quantitative estimate of each approach.

The general public did not require a deep or technical acquaintance with astronomy and astrology, but the broad astronomical knowledge of the time was an important part of their culture and practical life.

None of those groups was interested in the whole astronomical and astrological knowledge of that time (Figure 2). Of course, the different cultural profiles of astronomy had some overlap; and any specific person could be uniquely interested in a smaller or larger set of astronomical and astrological issues, not restricted by his or her professional involvement.

Our classification of interest groups does not include "astronomers". During that period, astronomical observations were rarely made in Portugal or Spain; and theoretical (mathematical) astronomy was a subject that was not restricted to a single group. It was part of the university teaching and it was practiced mainly among astrologers and a few physicians.

3. ASTRONOMICAL PUBLICATIONS

The sources used and produced by the several cultural profiles of astronomy varied; sixteenth-century literature produced in the Iberian Peninsula reflected the astronomical concerns of several different interest groups.

There was a very strong development of astronomy and astrology in the Spanish-speaking world, since the reign of Alfonso X de Castilla, "el Sabio" (1221–1284). The Iberian Peninsula was one of the places where the classical astronomical and astrological knowledge was brought back to Europe, accompanied by its recent Islamic developments (Hasse, 2006; Procter, 1945). From that time to the sixteenth century, the study of Greek, Roman and Arabic works on the subject became gradually widespread, with the help of the recently created European universities and, later, of the invention of the printing press. Important Latin Medieval treatises were also produced, such as Johannes de Sacrobosco's *Tractatus de Sphaera*, and the more technical works called *Theorica planetarum* by Gerard of Cremona, Campanus de Novarra and, later, by Georg von Peuerbach.

Beginning in the late fifteenth century, a second wave of astronomical and astrological interest developed in the Iberian Peninsula (Chabá, 2006), stimulated by the beginning of the great oversea explorations developed by Portugal and Spain, when astronomical navigation became of fundamental importance (Bensaúde, 1912; Costa, 1934). The wealth brought back by the conquerors produced a stream of new literary, philosophical, medical and scientific developments, with the production of some relevant original production related to astronomy (Navarro-Brotons, 1992). The poet Luís Vaz de Camões (1525–1580) wrote the epic *Os Lusíadas* (first printed in 1572), describing the Portuguese conquests. In this work we can find a profusion of astronomical allusions (Silva, 1943).

The Portuguese and Spanish navigators produced new geographic, cosmographic and cartographic works (Mosley, 2009), describing and mapping the lands they visited and the new ship routes they followed (Cortesão, 1935). The production and use of the maps required technical astronomical knowledge and the use of instruments (such as the nautical astrolabe) to find out the geographical latitude of a place. In clear nights, the latitude can be found from the position of the stars close to the celestial North Pole, in the northern hemisphere; but during the day, or after crossing the Equator, the best method is measuring the altitude of the Sun and using tables of the solar motion to establish the latitude. This method and the required tables were produced by the Spanish mathematician Abraham ben Samuel Zacuto (c.1450–c.1522), who produced in the decade of 1470 the first version of his *Almanach perpetuum* (Goldstein, 1965), in Hebrew (*ha-hibbur ha-gadol*). This work was translated into Spanish and afterwards in Latin by his disciple, Mestre José Vizinho, and published in Leiria (Zacuto, 1496). It was one of the first books published in Portugal with a movable type printing press and it became the basis of the tables used by navigators.

All over Europe, Sacrobosco's *Tractatus de Sphaera* was the main astronomical textbook, and the universities of Spain and Portugal followed the main current (Navarro–Brotons, 2006). Even in the case of Iberian works directly based on former treatises (such as Sacrobosco's *Sphaera*) we find a mixture of tradition - with plenty of quotations of ancient thinkers and natural philosophers as Aristotle, Pliny the Elder or Ptolemy, for example - and innovation - such as the new geographical knowledge that corrected and complemented the old tradition (Martins, 2004). Pedro Nunes (1502–1578), professor of the University of Coimbra, published in 1537 a Portuguese translation of Sacrobosco's *Tractatus de sphaera*, to which he added not only several comments (as was usual), but also other tracts on the motion of the Sun and the Moon, the translation of the first book of Ptolemy's geography, and two treatises on nautical maps and astronomical navigation (Nunes, 1537). Jerónimo de Chaves (1523–1574) published an erudite Spanish translation of and commentary on Sacrobosco's work, transforming the short original into a treatise of over 200 pages, with the addition of tables and practical elucidations (Chaves, 1545).

Towards the end of the century, the Jesuit priests of the "Colégio das Artes" (School of Arts) of Coimbra produced a series of treatises on Aristotle's philosophy (Casalini, 2017), including a detailed commentary of his book *De coelo* (Colégio das Artes de Coimbra, 1608), first printed in Lisbon, in 1593. This work, containing the Greek and Latin versions of the original work, with lots of additional material, attempted to update the Aristotelian cosmology, introducing much new information. The treatises produced by the *Conimbricenses* had a strong influence all over Europe.

Astrological medicine was taught in the Iberian peninsula, as in most other European countries (Navarro & Tayra, 2006); and a book by Jerónimo de Miranda (1562) highly recommends its study by physicians. The production of specific works associated to the several cultural aspects of Iberian astronomy in this period was a rare occurrence. Only a few of the astronomical/astrological works produced in this period were highly specific, such as Enrique de Cuellar's treatise on astrological medicine (Cuellar, 1543). This book was written for medicine students and physicians and it is likely that it did not come to the attention of other people. Most sixteenth-century Iberian astronomical works, however, were addressed to a wide public.

A representative work of that time is a manuscript treatise on the sphere written in the form of a dialogue between a teacher and a student, written around 1530 (Castro, 1940). It is usually attributed to Dom João de Castro, Vice-Roy of Portuguese India, although its authorship is not easily ascertained (Cardoso, 2004). It contains many elements and astronomical concepts in common with other texts of that period that were used to teach navigators or to serve as reference to an extensive public interested in themes related with astronomical subjects in that time (Cardoso and Martins, 2001). Another relevant unpublished work of that time was Duarte Pacheco Pereira's *Esmeraldo de Situ Orbis* (Pereira, 1991), a book primarily containing geographical information but also exhibiting a remarkable acquaintance with the astronomical knowledge of the period (Carvalho, 1983).

4. DISTINCTIVE IBERIAN ASTRONOMICAL WORKS

Much of the astronomical and astrological production of sixteenth-century Portuguese and Spanish authors exhibit no Iberian peculiarity and they could have been written in any other European country - such as Pedro de Espinosa's commentary on Sacrobosco's *Sphaera* (Espinosa, 1550). There were, however, two special types of publications that were distinctively Iberian: those related to cosmography and navigation; and the astronomical / astrological compilations called *Reportorio de los tiempos*.

4.1. Cosmography and navigation

Astronomical navigation required a basic knowledge of spherical astronomy that was uncommon among pilots of that day. In Spain, the *Casa de la Contratación* provided the necessary training of navigators (Martínez, 2010). Around 1510 two Portuguese seafaring guides were published, including a translation of Sacrobosco's *Tractatus de sphaera* to provide the fundamental concepts required for understanding the nautical methods (Albuquerque, 1965). They were the first examples of a new genre of Iberian astronomical literature, the so-called "nautical guides", that contained a description of basic astronomical concepts; explanations of the methods for determining the latitude of a place using the Sun, the Pole Star and the Southern Cross; the method of ascertaining the time, at night, from observations of the Ursa Minor and the Southern Cross; tables of the daily declination of the Sun; and other relevant geographic and astronomical information. There were also several attempts to evaluate longitudes, during the sixteenth century (Randles, 1985). This kind of practical astronomical literature grew in Portugal and Spain from small manuals to huge detailed treatises that also described astronomical instruments and their use (González, 2006). Some of those treatises had a strong influence all over Europe, during that period (Waters, 1970). They combined traditional astronomical knowledge with new methods, tables, instruments and practical wisdom (Lisboa, 1903; Faleiro, 1535).

Pedro de Medina (1493–1567) was one of the most important Iberian cosmographers during the sixteenth century (Lamb, 1972). The title of his most famous work was *Arte de navegar* ("The art of navigation"), published for the first time in 1545, in Valladolid. This very successful book was published three times in Spanish, six times in French, once in Italian, four times in German, three times in Dutch and twice in English, during the sixteenth century - besides other editions in the next century. It's awesome if you consider that time! He also published other works, including a cosmographical treatise (Medina, 1980).

4.2. *Reportorio de los tiempos*

In addition to the nautical guides, there were other peculiar Iberian astronomical and astrological works produced during the sixteenth century. There were, for instance, many popular treatises called *Chronographia,* or *Reportorio de los tiempos,* which provided the astronomical and astrological information required by the general public and by some special groups. The oldest example of this kind of literature was Andrés de Li's *Reportorio de los tiempos*, published in Zaragoza, in 1495 (Li, 1999). This small work was published several times in Spain and in Portugal, where it was translated and improved by Valentim Fernandes (1552).

Li's book inspired other larger works with similar title and content, the most famous being Jerónimo de Chaves' *Chronographia o reportorio de los tiempos* (Chaves, 1550), republished nine times in Spain and once in Portugal. A similar work was produced in Portugal by André do Avelar (1546–c.1625): *Reportorio dos tempos, o mais copioso que até agora sahio a luz* (Avelar, 1585; Costa, 2007). It underwent four editions (from 1585 to 1602).

The highly popular works of the genre *Reportorio de los tiempos* are a good example of the amalgamation of a variety of approaches and interests (Costa, 2007). Its chronological part included information required by priests to calculate the religious calendar and the correlation between the hours of each weekday and the planets, according to astrologers; it described the division of the day and night in natural and artificial hours; the heavenly spheres, and the astrological influence of each planet; the Zodiac is described both from the point of view of its influence on the sublunary world and regarding the details of the motion of the Sun along the ecliptic, and the duration of the days and nights. The work includes a fairly detailed description of the universe and its parts, of the heavenly circles and their terrestrial counterparts, introducing the zones and climes. The religious calendar, with the names of the catholic saints associated to each day, is combined with recommendations concerning what should be done or avoided in each lunar phase, for each month. A considerable part of the book is devoted to astrological medicine, explaining the four humours and temperaments, and the astronomical choice of treatments according to the temperament and the configuration of the heavens. Eclipses are described both as purely astronomical phenomena, providing tables of their occurrences; and as astrological causes of storms and other wars. The treatise deals with the influence of the Moon on the weather and on crops; and its final part presents information concerning navigation and tides.

The encyclopaedic character of works of this kind shows that it was not aimed at a particular public with peculiar interests - indeed, any person of that time might profit from reading some part of the book. This circumstance probably explains the wide circulation and popularity of those works.

5. FINAL COMMENTS

When studying the history of astronomy in the past, it is important to note desirable to notice that this field of knowledge was not the domain of a specific class of "astronomers"; it was a subject studied from different points of view by different publics. Instead of being a single discipline, astronomy (together with astrology, in the period studied here) can be regarded as a set of subfields, each of them contemplating specific interests.

It is impossible, of course, to present in a single paper the whole variety of Iberian astronomical works of the sixteenth century. We hope, however, that his overview can contribute to a better and more comprehensive understanding of the astronomical culture in that period.

ACKNOWLEDGEMENTS

One of the authors (R. A. Martins) is grateful for support received from the Brazilian National Council of Scientific and Technological Development (CNPq).

REFERENCES

Albuquerque, L. M. (1965) *Os Guias Náuticos de Munique e de Évora*, Lisboa, Junta de Investigações do Ultramar.

Allen, D. C. (1966). *The Star-Crossed Renaissance*, London, Fank Cass.

Avelar, A (1585) *Reportorio dos Tempos.* O Mais Copioso que Ate Agora Sahio a Luz, Conforme à Noua Reformação do Sancto Padre Greg. XIII. Anno 1582, Lisboa, Manoel de Lyra.

Baldini, U. (2001) The Roman Inquisition's condemnation of astrology: antecedents, reasons and consequences. *Church, Censorship and Culture in Early Modern Italy*, G. Fragnito (ed.), Cambridge, Cambridge University Press, pp. 79–110.

Bensaúde, J. (1912) *L'Astronomie Nautique au Portugal à l'Epoque des Grandes Découvertes*, Bern, Akademische Buchhandlung von Max Drechel.

Cañizares-Esguerra, J. (2004) Iberian science in the Renaissance: Ignored how much longer?, *Perspectives on Science*, Vol. 12, No. 1, pp. 86–124.

Cardoso, W. T. (2004) *Conceitos e Fontes do Tratado da Esfera em Forma de Diálogo Atribuído a João de Castro.* São Paulo, Educ.

Cardoso, W. T. and Martins, R. A. (2001) Conceitos e fontes do "Tratado da esfera" atribuído a João de Castro. *Actas do 1°. Congresso Luso-Brasileiro de História da Ciência e da Técnica*, Évora, Centro de Estudos de História e Filosofia da Ciência da Universidade de Évora, pp. 95–106.

Casalini, C. (2017) *Aristotle in Coimbra: The Cursus Conimbricensis and the Education at the College of Arts*, translated by Luana Salvarani, London, Routledge.

Castro, J. (1940) *Tratado da Sphaera, da Geografia, Notação Famosa, Informação sobre Maluco*, ed. A. Fontoura da Costa, Lisboa, Agência Geral das Colônias.

Chabá, J. (2006) The university of Salamanca and the renaissance of astronomy during the second half of the 15th century. *Universities and Science in the Early Modern Period,* M. Feingold and V. Navarro-Brotons (ed.), Dordrecht, Springer, pp. 29-36.

Chaves, J. (1545) *Tractado de la Sphera que Compuso el Doctor Ioannes de Sacrobusto: con Muchas Additiones*, Sevilla, En casa de Juan de Leon.

Chaves, J. (1550) *Chronographia o Reportorio de los tiempos*: el mas copioso y preciso q[ue] hasta agora ha salido a luz, Sevilla, Christoval Alvarez.

Colégio das Artes de Coimbra (1608) *Commentarii Collegii Conimbricensis Societatis Iesv, in Quatvor Libros de Cœlo, Meteorologicos, Parua Naturalia, & Ethica Aristotelis Stagiritæ*, Lugduni, Ioannem Pillehotte.

Cortesão, A. (1935) *Cartografia e Cartógrafos Portugueses dos Séculos XV e XVI*, Lisboa, Seara Nova, 2 vols.

Costa, A. B. (2007) *O Reportório dos Tempos de André do Avelar e a Astrologia em Portugal no Século XVI*, Rio de Janeiro, Booklink.

Costa, A. F. (1934) *A Marinharia dos Descobrimentos*, Lisboa, Agência Geral das Colónias.

Costa, P. F. and Leitão, H. (2008) Portuguese Imperial science, 1450–1800: a historiographical review. *Science in the Spanish and Portuguese empires, 1500–1800,* D. Bleichmar, P. De Vos, K. Huffine, and K. Sheehan, K. (ed.), Stanford, Stanford University Press, pp. 35–53.

Cuellar, E. (1543) *Opus Insigne ad Libros Tres Predictionum Hippocr.*, Conimbrie, Ioha[n]nis Aluari & Ioha[n]nis Barrerii.

Espinosa, P. (1550) *Sphera Ioannis de Sacro Busto.* Cum Commentariis Petri a Spinosa Artium Magistri, Celeberrimiq[ue] Praeceptoris Salmanticensis Gymnasij Aeditis, Salmanticae, Ioannes Iunta.

Fabiani, V. L. S. (1995). La cultura científica y técnica en la España de los siglos XVI y XVII, *Bulletin Hispanique*, Vol. 97, No. 1, pp. 233–259.

Faleiro, F. (1535) *Tratado del Sphera y del Arte del Marear*, Sevilla, Juan Croberger.

Fernandes, V. (1552) *Reportorio dos Tempos em Lingoagem Portugues.* Com as estrellas dos signos. E com as condições do que for nascido em cada signo. E o crecer & mingoar do dia & da noite. E das quatro compreições. E a declinaçam do sol com seu regimento. E ho regimento da estrella do norte. Com outras muytas cousas acrecentadas de nouo, Lixboa, Germão Galharde.

Fra Mauro (1537) *Sphera Volgare.* Novamente tradotta con molte notande additioni, Venetia, B. Zanetti.

Goldstein, B. R. (1965). The medieval Hebrew tradition in astronomy, *Journal of the American Oriental Society*, Vol. 85, No. 2, pp. 145–148.

González, F. J. G. (2006) Del "Arte de marear" a la navegación astronómica: Técnicas e instrumentos de navegación en la España de la Edad Moderna, *Cuadernos de Historia Moderna,* Vol. 31, No. 5, pp. 135–166.

Goodman, D. C. (1992) The scientific revolution in Spain and Portugal. *The Scientific Revolution in National Context*, R. Porter and M. Teich (ed.), Cambridge, Cambridge University Press.

Hasse, D. N. (2006) The social conditions of the Arabic- (Hebrew-) Latin translation movements in medieval Spain and in the Renaissance. *Wissen über Grenzen: Arabisches Wissen und lateinisches Mittelalter*, A. Speer, and L. Wegener (ed.), Berlin, de Gruyter, pp. 68–86.

Lamb, U. (1972) *A Navigator's Universe - The Libro de Cosmographía of 1538 by Pedro de Medina*, Chicago, Chicago University Press.

Li, A. (1999) *Reportorio de los Tiempos*, edited by Laura Delbrugge, London, Tamesis.

Lisboa, J. (1903) *Livro de Marinharia. Tratado da Agulha de Marear,* ed. J. I. B. Rebello, Lisboa, Libânio da Silva.

Macedo, J. B. (1975) Livros impressos em Portugal no século XVI. Interesses e formas de mentalidade, *Arquivos do Centro Cultural Português*, Vol. 9, pp. 183–221.

Martínez, A. S. (2010) Los artífices del Plus Ultra: pilotos, cartógrafos y cosmógrafos en la Casa de la Contratación de Sevilla durante el siglo XVI, *Hispania*, Vol. 70, No. 236, pp. 607–632.

Martins, L. A. (2004) *O Livro Científico dos Séculos XV e XVI: Ciências Físico-Matemáticas na Biblioteca Nacional.* Lisboa, Biblioteca Nacional Portugal.

Martins, R. A. (2004) La herencia de Sacrobosco en España: intercambio entre estudios universitarios y la práctica de navegación durante el siglo XVI. *Epistemología e Historia de la Ciencia. Selección de Trabajos de las XIV Jornadas*, P. García, and P. Morey (ed.), Córdoba, Universidad Nacional de Córdoba, pp. 371–379.

Medina, P. (1980) *Svma de Cosmographía*, ed. J. Fernández Jiménez, Valencia, Albatros.

Miranda, J. (1562) *Dialogo da Perfeyçam e Partes que sam Necessarias ao Bom Medico*, Lixboa, Ioam Aluarez.

Mosley, A. (2009) The cosmographer's role in the sixteenth century: a preliminary study, *Archives Internationales d'Histoire des Sciences*, Vol. 59, No. 163, pp. 423–439.

Navarro, L. and Tayra, M. C. (2006) Medical astrology in Spain during the seventeenth century, *Cronos*, Vol. 9, No. 1, pp. 59–83.

Navarro-Brotons, V. (1992) La actividad astronómica en la España del siglo XVI: perspectivas historiográficas, *Arbor*, Vol. 117, pp. 185–216.

Navarro-Brotons, V. (2006) The cultivation of astronomy in Spanish universities in the latter half of the 16th century. *Universities and Science in the Early Modern Period,* M. Feingold and V. Navarro-Brotons (ed.), Dordrecht, Springer, pp. 83–98.

Nunes, P. (1537) *Tratado da Sphera com a Theorica do Sol & da Lua. E ho Primeiro Liuro da Geographia de Claudio Ptolomeo Alexãdrino*, Lixboa, Germão Galharde.

Pereira, D. P. (1991) *Esmeraldo de Situ Orbis*, ed. J. B. Carvalho, Lisboa, Fundação Calouste Gulbenkian.

Procter, E. S. (1945) The scientific works of the court of Alfonso X of Castille: the king and his collaborators, *The Modern Language Review*, Vol. 40, No. 1, pp. 12–29.

Randles, W. G. (1985) Portuguese and Spanish attempts to measure longitude in the 16th century, *Vistas in Astronomy*, Vol. 28, pp. 235–241.

Silva, L. P. (1943) *A Astronomia dos Lusíadas*, Lisboa, Agência Geral das Colónias.

Simoes, A., Carneiro, A. and Diogo, M. P. (2008) Perspectives on contemporary history of science in Portugal, *Nuncius*, Vol. 23, No. 2, pp. 237–263.

Waters, D. W. (1970) *The Iberian bases of the English art of navigation in the sixteenth century*, Coimbra, Universidade de Coimbra.

Zacuto, A. S. (1496) *Almanach Perpetuum Celestium Motuum Astronomi Zacuti: cui[us] radix est 1473*, translated by Joseph Vecinho, Leiria, Abraham d'Ortas.

18

SKYSCAPE ARCHAEOLOGY AS A ROAD TO CULTURAL INSIGHT

Andrew M. Munro* **and Steven R. Gullberg**

The University of Oklahoma, College of Professional and Continuing Studies, 1700 Asp Avenue, Norman, OK 73072, USA

**Corresponding author: Andrew.M.Munro@ou.edu*

ABSTRACT

Archaeoastronomers have made great strides in development of research methodologies, yet there is limited curriculum available to train new practitioners. If we seek results that address current archaeological research questions, then our work must necessarily be pertinent to such questions and grounded in rigorous archaeoastronomy fieldwork and analytical methods. Furthermore, the inferences we create should be supported by the points of intersection between archaeoastronomical data and archaeological theory (Iwaniszewski, 2015). To achieve these objectives, practitioners must be aware of current archaeological research questions, trained in archaeoastronomy methods, and aware of the intersections between archaeoastronomical data and archaeological theory. Historical and ethnohistorical information from a wide variety of cultures demonstrate that visual astronomy may be interconnected with cosmovision, politics, ritual, religion, and economics in variable and unique ways. Research must be iterative and interdisciplinary. To illustrate variations in interdisciplinary sources, we briefly present two case studies. These case studies underscore the variability of sources supporting archaeoastronomy research; therefore, a curriculum and supporting instructional materials to train practitioners must by definition be interdisciplinary (see, e.g., Magli, 2016). Such a curriculum is under development for the University of Oklahoma's College of Professional and Continuing Studies (OU PACS). One key charter of OU PACS is interdisciplinary study. The OU PACS Archaeoastronomy program will integrate astronomy, anthropology, archaeology, history of science, and history of religion. The program is initially planned to include five (5) graduate courses offered as a graduate certificate. The program prominently features North American archaeoastronomy and will include a field methods practicum.

KEYWORDS: Archaeoastronomy, research epistemology, interdisciplinary studies, archaeoastronomy curriculum, online education, archaeoastronomy graduate certificate

1. INTRODUCTION

This paper discusses specific interdisciplinary challenges in archaeoastronomy research, and their implications for curriculum design. Progress continues in the development of standard methods for archaeoastronomy fieldwork, analysis, and interpretation. Notwithstanding, significant methodological variation continues to be manifested in published archaeoastronomy work (see, e.g., Aveni, 2001; Belmonte, 2015; Prendergast, 2015; Ruggles 1996, 2015). The lack of methodological consensus is well illustrated by varied perspectives regarding the evidence required to demonstrate intentionality or cultural pertinence for proposed lunar standstill alignments (González-García, 2016; Malville, 2016; Sims, 2016). It seems reasonable to suppose that variations in epistemology are driven, in part, by the varied disciplinary backgrounds and training and assumptions of archaeoastronomy researchers, as discussed by McCluskey (2015).

Unquestionably, continuing debate on the fundamentals of archaeoastronomy research methods complicates curriculum design. Furthermore, no two fieldwork-based archaeoastronomy research designs can be grounded in entirely consistent published peer-reviewed research due to the variations in research previously conducted for site(s) and culture(s). To illustrate this point, we briefly review two case studies. These demonstrate the benefit of ensuring that an archaeoastronomy curriculum includes explicit treatment of the varied epistemologies and taxonomies applied by related fields. Subsequently, we present an archaeoastronomy curriculum currently under development at the University of Oklahoma's College of Professional and Continuing Studies (OU PACS).

2. CASE STUDIES

To illustrate the variability of interdisciplinary sources supporting archaeoastronomy research, we briefly review two case studies, one at Chaco Canyon, New Mexico, and one at Inca sites in Peru.

Research at Chaco Canyon, New Mexico has demonstrated that 10th through 12th century CE monumental structures exhibit four orientation and placement traditions. Temporal analysis of these architectural traditions and photographically confirmed astronomical alignments provide evidence for shifting multi-cultural collaboration over time at Chaco, and highlight elevated cultural importance for solstices after 1100 CE (Munro, 2012; Munro et al., 2016).

The Inca Empire in the Peruvian Andes is replete with examples of intentional light and shadow effects, as well as orientations to key solar positions on the horizon. When such research data is taken in cultural context with reference to works such as those of Zuidema (1981), Bauer (1998), and Urton (1981) it emphasizes a society tightly interwoven with astronomy. Precise archaeoastronomical research methods and interpretation are key (Gullberg, 2009; Gullberg and Malville, 2016).

2.1. *Chaco Canyon, New Mexico*

Over the past decade, Munro and colleagues have conducted ongoing fieldwork at Chaco Canyon focused on validation of previously published archaeoastronomy findings, as well as on completion of an encompassing survey of all monumental "Great Houses" in the canyon. This work has relied on published and archived archaeological research for sites, theodolite survey, integration with ephemerides, and photo confirmation of predicted events.

This research program has yielded some negative results; it has not identified convincing evidence in support of lunar standstill hypotheses previously published for Chaco. In addition, published equinox alignment claims for the site known as Pueblo Bonito have been determined to be non-functional due to horizon elevations. Survey validation and photo confirmation of the claimed equinox alignments at Pueblo Bonito was attempted and failed (Munro, 2012, pp. 132–137, 216–220). In light of the number of previously claimed alignments at Chaco that are demonstrably non-functional, we have come to the view that photo-confirmation of alignment claims is an important and underutilized tool, as discussed by Malville and Ninnemann (2017).

The work has expanded upon and reinforced a convincing pattern involving four architectural traditions linked to cosmology. Three of these are orientation traditions among Ancestral Pueblo monumental structures, to the cardinal directions, to the east-southeast, and to the south-southeast. The fourth is a consistent pattern of construction of monumental "Great Houses" at observation points for solstitial horizon foresights, especially for structures built after 1110 CE, as shown in Figure 1. There is well-constrained dating of the architecture at Chaco. Temporal analysis of the astronomical alignment findings provide support for archaeological theories of multi-cultural collaboration at Chaco, as well as highlighting significant changes in cultural intent during the period after 1100 CE (Munro, 2012; Munro et al., 2017).

Figure 1 June solstice sunrise interacting with a horizon feature at Rabbit Ruin

2.2. *Solar Associations with Inca Huacas*

Gullberg has performed multiple seasons of fieldwork-based archaeoastronomy research at Inca sites in Peru. In contrast to studies at Chaco, where centuries elapsed between construction of monumental architecture and historical documentation of the sites, fewer than 100 years elapsed between the rise of Pachacuti and his empire, and its eventual conquest by Spanish conquistadores. There is a rich history of archaeoastronomy, however, and it is both to the horizon, similar to Chaco, and as well internal to many huacas (shrines).

One example can be found at a huaca called Lacco. Lacco is a large granite outcropping that includes three caves with solar orientations. One of these is a cave with its opening on the southwest face that includes a light-tube aimed for a carved altar. The light-tube is oriented to receive direct sunlight at the time of the zenith sun, as shown in Figure 2.

Figure 2 Lacco altar at the time of the zenith sun

An example regarding horizon events is found at a small, obscure huaca that we refer to as the Huaca of Solar Events. Two circles, one larger than the other, were sculpted from the granite in such a manner that they can serve as a visual guide to the positions on the horizon for each of the principle annual sunrises and sunsets. Tangential lines drawn between them point to the four solstice sunrises and sunsets and a line drawn across them does likewise for the two equinoxes. There additionally are the remains of carved seats positioned to view each principle event.

The Incas also constructed solar pillars on the horizons of Cusco for calendrical purposes to mark dates such as those for planting, harvesting, and religious festivals. Extant examples are above the palace of Qwespiwanka, which thrived on a site near the present-day village of Urubamba in the Sacred Valley. They indicate the June solstice sunrise when viewed from a large granite boulder near the center of the palace courtyard.

Research shows that there are many huacas throughout the regions extending away from Cusco that exhibit astronomical orientations. The most common are for the June solstice sunrise, followed by those for sunrise at the December solstice. There are other orientations for June solstice sunset and December solstice sunset, to a lesser extent the equinox sunrises and sunsets, and as well a few for sunrise or sunset at the time of the zenith sun and the anti-zenith sun.

Astronomical alignments in carved rock huacas were found to be oriented toward the June solstice sunrise 63% of the time at the sites studied. Some sites had more than one orientation. The frequency of alignments toward the December solstice sunrise was found to be 56%; the June and December solstice sunset alignments were 38% and 25%, respectively (Gullberg, 2009; Gullberg and Malville, 2016).

2.3. *"Knowing the Literature" and Implications for Curriculum Design*

We do not believe a strict processual model for archaeoastronomy research is achievable or desirable. However, we do see a structural approach founded on "knowing the literature" for each research question to be a useful framework to support iterative interdisciplinary research, including fieldwork-based archaeoastronomy. With the benefit of hindsight, we see that a significant weakness of some published work that has been subsequently nullified by reassessment (for example with respect to equinox alignment claims at Chaco's Pueblo Bonito) is that they were inadequately grounded in peer-reviewed published work from varied disciplines pertinent to the culture or site(s) of interest. This can be as simple as a failure to consider well-demonstrated dating of archaeological sites, and as complex as over-stressing the importance of aesthetics (as provided by well-developed art history) without due consideration of less subjective factors from other disciplines, such as engineering. We suggest that it is critically important to develop familiarity with a broad sample of the literature for any given site or culture of interest across disciplines.

Figure 3 and Figure 4 provide a simple comparative illustration, based upon our two case studies, of the variable nature of source material available for interdisciplinary assessment. As depicted, the research at Chaco described in section 2.1 has been conducted in the context of an extraordinarily robust body of published archaeology (see, e.g., Lekson, 2006; Mathien, 2005). That body of literature is supplemented by a variety of published work from other disciplines, with special focus on anthropology, the history of religion, and past interdisciplinary archaeoastronomy work. In contrast, while Inca sites are well studied archaeologically, there is a much broader range of published historic period literature to include anthropology of descendant peoples in Peru, as well as historical accounts of Inca culture from the perspective of conquering 16th century Spaniards (see, e.g. Bauer, 1998; Urton, 1981; Zuidema, 1981).

For any given site or culture, the availability of peer-reviewed literature to assess is driven by temporal factors, available physical evidence, and the history of academic research. Publication of work that does not incorporate a broad scope of pertinent peer- reviewed research from associated disciplines is fraught with risk, and undermines the potential for archaeoastronomy findings to be taken seriously by colleagues working in those disciplines.

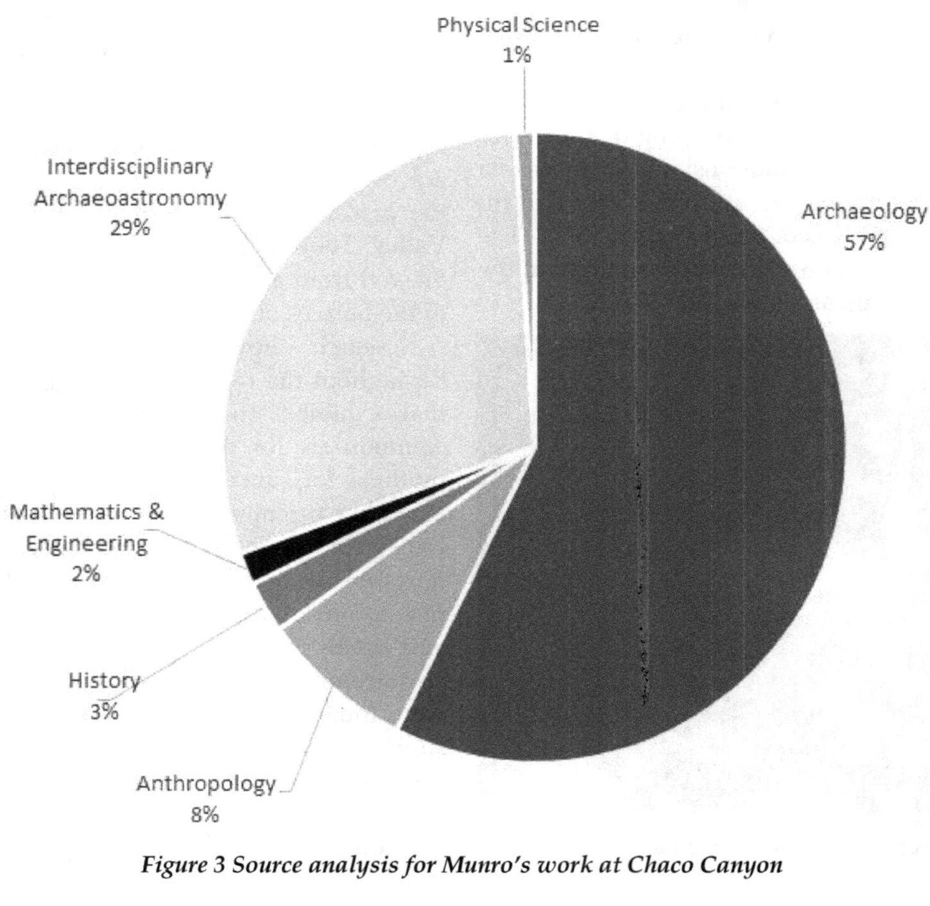

Figure 3 Source analysis for Munro's work at Chaco Canyon

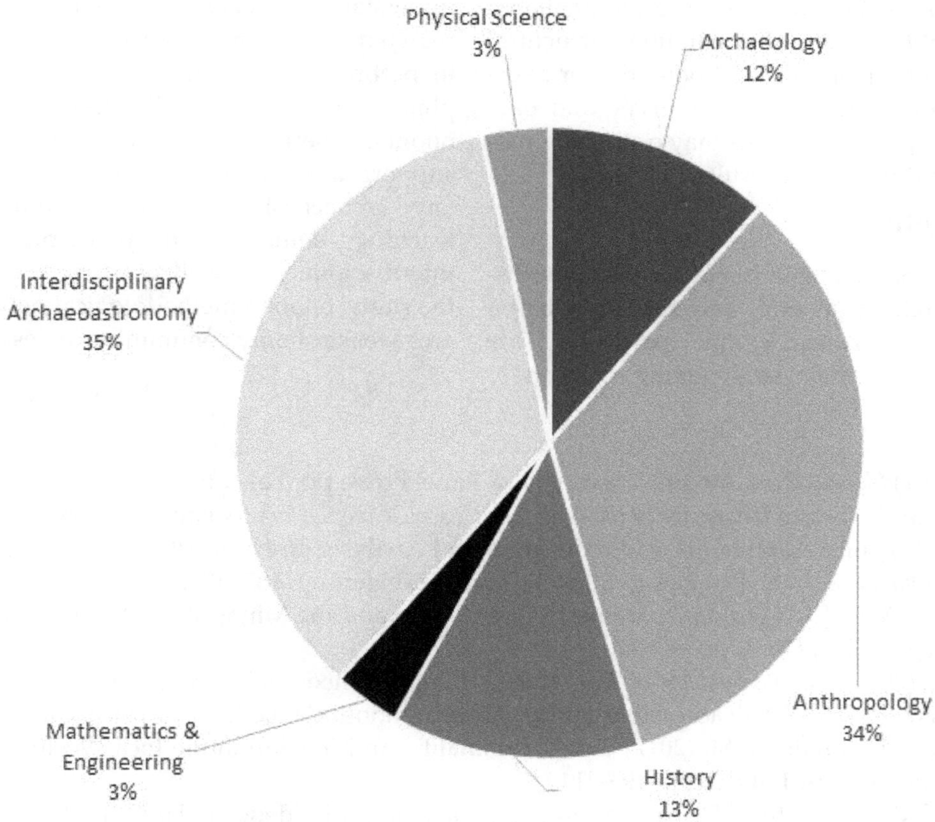

Figure 4 Source analysis for Gullberg's work on the astronomy of Inca huacas

3. THE PLANNED OU PACS ARCHAEOASTRONOMY CURRICULUM

The authors are currently developing a five-course archaeoastronomy curriculum to be offered as a primarily online program through OU PACS. These courses will be offered as graduate certificate program. The certificate will provide individuals who have earned degrees in other fields (e.g. astronomy, archaeology or anthropology) with the training necessary to apply archaeoastronomy tools and techniques. Given that OU PACS emphasizes holistic learning, our objectives are to use archaeoastronomy research as a training ground for interdisciplinary research methods, as well as to train archaeoastronomy practitioners.

The five courses under development include *Archaeoastronomy and Methods, Archaeoastronomy of Chaco Canyon and Cahokia, Latin American Archaeoastronomy, World Archaeoastronomy,* and *Field Work in Archaeoastronomy*. The courses include reviews of the evidence for astronomical associations among a wide variety of selected sites and cultures from around the world. The courses will include visual motions of celestial objects in the day and night sky, and archaeoastronomy field survey and interpretation methods. Common themes will include the use of astronomical principles in monumental design and city planning, using astronomical stories to encode socially important cultural reference information, linkage to ritual practices including pilgrimage, and functional uses of sky knowledge to include navigation and the development of agricultural and ritual calendars. We also intend to explore the human use of astronomical observation and prediction in diverse ways, highlighting visual astronomy's role in the development of religion, social hierarchy, and science. Selected archaeoastronomy "false positives" will also be included in the curriculum to assist students in development of advanced analytical and critical thinking skills.

The Field Work course will require an onsite practicum. It will consist of three components: online contextual and site literature review, a one-week onsite field survey with faculty that includes preliminary collaborative data interpretation, and optional post field-school research publication. We will stress the benefits of framing research questions in collaboration with archaeologists who are active and reputable researchers for the culture and site(s) in question. We intend to provide students with hands-on survey, data integration, interpretation, and publication experience at sites within North America. During the fieldwork course students will develop horizon sketches, conduct compass and clinometer surveys, collect GPS and photographic data, and conduct theodolite surveys of architecture and horizon features. Data reduction will be performed collaboratively to identify structural orientations, structural alignments, and horizon foresight potential for celestial object interactions. After field survey and data reduction is completed, students may optionally participate in the publication of results.

4. CONCLUSION

The issues to be managed in conducting interdisciplinary research in general, and fieldwork-based archaeoastronomy research in particular are complex, and will remain so. Training students to understand the challenges inherent in framing research questions and research designs with reference to pertinent source material from all related disciplines is just a start. Explicit treatment of varied taxonomies, methods, and standards of evidence from supporting disciplines should be core elements of any archaeoastronomy curriculum. In addition, teaching demonstrated best practices for the interdiscipline, as well as directly communicating the status of open methodological debates within the archaeoastronomy community, are essential.

REFERENCES

Aveni, A. F. (2001) *Skywatchers*. Austin, University of Texas Press, pp. 124–126.

Bauer, B. S. (1998) *The Sacred Landscape of the Inca: The Cusco Ceque System*. Austin, University of Texas Press.

Belmonte, J. (2015) Solar Alignments – Identification and Analysis. In *Handbook of Archaeoastronomy and Ethnoastronomy*, C. L. N. Ruggles (ed.), New York, Springer, pp. 483–492.

González-García, A. C. (2016) Lunar Extremes, Lunar Cycles and the Minor Standstill, *Journal of Skyscape Archaeology* Vol. 2, No. 1, pp. 77–84.

Gullberg, S. R. (2009) *The Cosmology of Inca Huacas*. Unpublished Ph.D. Thesis, James Cook University, Townsville, Aus. (Web accessible at http://researchonline.jcu.edu.au/10346//).

Gullberg, S. R., and Malville, J. M. (2017) Caves, Liminality, and the Sun in the Inca World, *Culture and Cosmos* Vol. 21, Nos. 1 and 2, pp. 193-214.

Iwaniszewski, S. (2015) Cultural Interpretation of Archaeological Evidence Relating to Astronomy. In *Handbook of Archaeoastronomy and Ethnoastronomy*, C. L. N. Ruggles (ed.), New York, Heidelberg, Dordrecht, London, Springer, pp. 316–324.

Lekson, S. H. (ed.), (2006) *The Archaeology of Chaco Canyon: An Eleventh-Century Pueblo Regional Center*. Santa Fe, School of American Research Press.

Magli, G. (2016) *Archaeoastronomy, Introduction to the Science of Stars and Stones*. New York, Springer.

Malville. J. M. (2016) The Enigma of Minor Standstills, *Journal of Skyscape Archaeology* Vol. 2, No. 1, pp. 85–94.

Malville, J. M. and Ninnemann, J. L. (2017) Archaeoastronomy and Cultural Astronomy as Scientific Disciplines: Falsifiability and Photo Documentation, *The Marriage of Astronomy and Culture*, a special issue of *Culture and Cosmos*, Vol. 21, Nos. 1 and 2, pp. 173–192.

Mathien, F. J. (ed.) (2005) *Culture and Ecology of Chaco Canyon and the San Juan Basin*. Santa Fe, National Park Service.

McCluskey, S. C. (2015). Disciplinary Perspectives on Archaeoastronomy. In *Handbook of Archaeoastronomy and Ethnoastronomy*, C. L. N. Ruggles (ed.), New York, Heidelberg, Dordrecht, London, Springer, pp. 227–237.

Munro, A. M. (2012) *The Astronomical Context of the Archaeology and Architecture of the Chacoan Culture*. Unpublished Ph.D. Thesis, James Cook University, Townsville, Aus. (Web accessible at http://researchonline.jcu.edu.au/40277/).

Munro, A. M., Hull, T., Malville, J. M., Mathien, F. J., and Morrow, C. (2017) Investigation of Solstice Horizon Interactions at Chacoan Monumental Architecture, *The Marriage of Astronomy and Culture*, a special issue of *Culture and Cosmos*, Vol. 21, nos. 1 and 2, pp. 151–171.

Prendergast, F. (2015) Techniques of Field Survey. In *Handbook of Archaeoastronomy and Ethnoastronomy*, C. L. N. Ruggles (ed.), New York, Heidelberg, Dordrecht, London, Springer, pp. 389–409.

Ruggles, C. L. N. (1996) Summary of the RAS Specialist Discussion Meeting on Current Issues in Archaeoastronomy. *The Observatory*, Vol. 116, pp. 278–285.

Ruggles, C. L. N. (2011) 'Interpretive Archaeoastronomy' Thirty Years On. In *Archaeoastronomy and Ethnoastronomy: Building Bridges between Cultures*, C. L. N. Ruggles, (ed.), Cambridge, Cambridge University Press, pp. 255–265.

Ruggles, C. L. N. (2015) Analyzing Orientations. In *Handbook of Archaeoastronomy and Ethnoastronomy*, C. L. N. Ruggles (ed.), New York, Heidelberg, Dordrecht, London, Springer, pp. 411–425.

Sims, L. (2016) What is the Minor Standstill of the Moon?, *Journal of Skyscape Archaeology*, Vol. 2, No. 1, pp 67–76.
Urton, G. (1981) *At the Crossroads of Earth and Sky: An Andean Cosmology*. Austin, University of Texas Press.
Zuidema, R. T. (1981) Inca Observations of the Solar and Lunar Passages Through Zenith and Anti-Zenith at Cuzco. In *Archaeoastronomy in the Americas*, R. A. Williamson, (ed.), Los Altos, Ballena Press, pp. 319–342.

19

THE GENESIS OF HIPPARCHUS' CELESTIAL GLOBE

Susanne M. Hoffmann

Independent Scholar, formerly: Excellence Cluster TOPOI
Hannoversche Str. 6, 10115 Berlin, Germany
(akademeia@exopla.net)

ABSTRACT

The globe of Hipparchus is not preserved. For that reason, it has been a source of much speculation and scientific inquiry during the last few centuries. This study presents a new analysis of the data given in the commentary on Aratus' poem by Hipparchus, in comparison with other contemporary Babylonian and Greek astronomical data, as well as their predecessors in the first millennium and their successors up to Ptolemy. The result of all these studies are the following: i) although the data of Ptolemy and Hipparchus are undoubtedly correlated, it is certainly also wrong to accuse Ptolemy having simply copied and transformed it without correct citation; ii) although Hipparchus presumably observed most of his star catalogue with his own instruments, we cannot neglect Babylonian influences. Hipparchus was educated in Greek astronomy but, in his time, there are traces of Babylonian influences since at least two centuries. Since we are unable to definitely prove that Hipparchus used Babylonian data, we are not sure if there are direct Babylonian influences in his time or as a consequence of his education only. Finally, we present a virtual 3D–image showing what the globe of Hipparchus might have looked like.

KEYWORDS: Hipparchus, Hipparchus' globe, globes, Babylonian mathematical astronomy, Ptolemy, Greek mathematical astronomy.

1. INTRODUCTION

This presentation is a summary of some of the results of the book on Hipparchus' celestial globe (Hoffmann, 2017) which appeared recently, but due to university laws had to be written in German. In this article, I would like to provide access to the book's information for the international community. The book contains four chapters: The first, of course, is a summary of the state of the art before my research, the second and third chapter describe the reconstruction of Hipparchus' star catalogue and analysis of this data with the aim of the virtual reconstruction of the globe. The fourth chapter presents numerous texts of different genres from both Babylonian and Greek astronomy including comparisons of their accuracy, style and data format with the equivalent in Hipparchus' text. The fifth chapter is a summary telling the (hi)story with the facts derived from the above research. In this article, I focus on the second chapter's reconstruction and the results of its analysis.

2. QUESTIONS AND AIMS

For nearly 400 years it has been known that Ptolemy's ecliptical longitudes have a systematic error. For at least 200 years, there are accusations that Ptolemy stole data from Hipparchus and transformed it with inaccurate precession. Although Ptolemy refers to Hipparchus' globe and cites data and constellation details from it, the Hipparchian globe cannot be completely reconstructed. There is a lack of information for some details and any reconstruction will only show a subset of the astronomical knowledge of Hipparchus.

However, the desire to build a model of the Hipparchian globe cannot be accomplished by simply using the Almagest star catalogue without any changes because Ptolemy in some issues reports deviations, e.g. "The star over the head, which Hipparchus [calls] 'the one on the muzzle'".[1] In some cases, he mentions deviations, but in other cases, he describes an asterism differently without mentioning any changes, e.g. the constellation of the crater has two handles and one star in the foot for Ptolemy, while for Hipparchus it does not have any handle and the foot contains four stars. Thus, Ptolemy's star catalogue is not a simple copy of Hipparchus' catalogue and we need an independent reconstruction of Hipparchus's catalogue to trace its roots and further development.

For the central question on the genesis of Hipparchus's globe we, therefore, have to pursue two general directions. First, how it is possible for us to reconstruct the historical (but missing) object, and second, how the historical object had been made. Did Hipparchus really observe everything by himself (which is generally assumed) or did he - like Ptolemy - also use Babylonian data in addition?

2.1. *Historical Questions*

The questions of the historian are related to the process of transfer and transformation of data, measuring and computing methods, as well as concepts (like frames of reference, metrical units, worldview, etc.) of uranometry in ancient science. It is known that Ptolemy used independent lists of longitudes (meridians) and lists of latitudes (*klimata*) to combine them to geographical coordinates as (longitude, latitude) pairs. Consequently, one of the historical questions is whether Hipparchus did something similar in astronomy combining earlier information. According to Pliny, Hipparchus was the first person to compile a star catalogue, but from what we know about the development of knowledge and scientific standards and conventions, they all have predecessors. What are the predecessors of Hipparchus? Did he use only earlier Greek data (which is attested to in the *Almagest*, Alm. VII, 1) or did he also use foreign sources of data - for instance, Babylonian sources?

There are numerous questions about Hipparchus' own observations: What kind of instrumentation did he use? Did he apply Babylonian methods to achieve a similar accuracy or did he use the instruments Ptolemy describes in the Almagest? Or does his scheme and method differ from both?

2.2. *Archaeoastronomical Aims*

The archaeological part of the research project is the virtual reconstruction of the missing globe. We know that the globe of Hipparchus existed because it is mentioned in the *Almagest*. Ptolemy cites the idea of an exact globe from Hipparchus but in some lines Ptolemy has a view which differs from that of Hipparchus, e.g. concerning the interpretation of constellation depictions (quoted above). We conclude from this that the globe of Hipparchus might have looked similar but not identical to Ptolemy's globe. Thus, we will use Ptolemy's exact descriptions as a more developed (later epoch) version of Hipparchus' object. Using only the mentioned differences in the *Almagest* in addition to Hipparchus's own descriptions and comparing the descriptions of Ptolemy with the appropriate parts of Hipparchus's one leads to a proper reconstruction of the two globes and the history in between them.

---

[1] Toomer and Ptolemaios, 1984, p. 361.

3. METHOD

The method of this research is quickly summarised: First, we reconstruct the star catalogue of Hipparchus by evaluating the data from the preserved two sources. Second, we evaluate the resulting database with the common and suggested algorithms of computational astronomy.

The preserved sources do not compile orthogonal coordinates of stars: Instead, the data they include are given in relative positions and events at the horizon. The subset of alignment-stars in the *Almagest* (Alm. VII, 1) are constructed to form lines of three stars. Thus, the astrometrical information in this text is only the relative position of three or four stars to each other. Interestingly, some of those stars quoted from Hipparchus are not in Ptolemy's star catalogue. This does not only suggest that Ptolemy's catalogue is not a simple copy of Hipparchus's one but also suggests an external source for Hipparchus's alignments.

The one and only preserved document penned by Hipparchus, the commentary on Aratus' poem on the *Phaenomena*, gives a corrected and much more systematic, even though not poetic, version of risings and settings of constellations. Hence, the data in this text is in the best cases a right ascension or declination, but in most cases it is a very broad range at the eastern or western parts of the horizon. If, and only if, a star happens to be mentioned two or three times, it is possible to compute its coordinates from this given risings, settings and culminations. The algorithms had been previously applied e.g. by (Vogt, 1925) and (Graßhoff, 1990) but repeated and improved by us.

In the text, the smallest unit Hipparchus distinguishes is a half degree in longitude. The accuracy is, therefore, a quarter of a degree and in declination only a half degree. However, Hippachus's text does not list all stars in his catalogue or a random subset of them but a carefully selected subset of rising, setting or culminating stars with well-defined constrains. That is why the average deviation of the selected stars from their real position (measured by modern measurements) may differ from this (Figure 1).

Although any reconstruction will be a subset and only a part of Hipparchus' full star catalogue, we discussed the error bars in different parts of the texts and compared the deviations of Hipparchus' data from modern data with the deviation of Ptolemy's and Babylonian data. The results are visualised in 2D diagrams and 3D plots as well as a 3D virtual reconstruction of Hipparchus's globe.

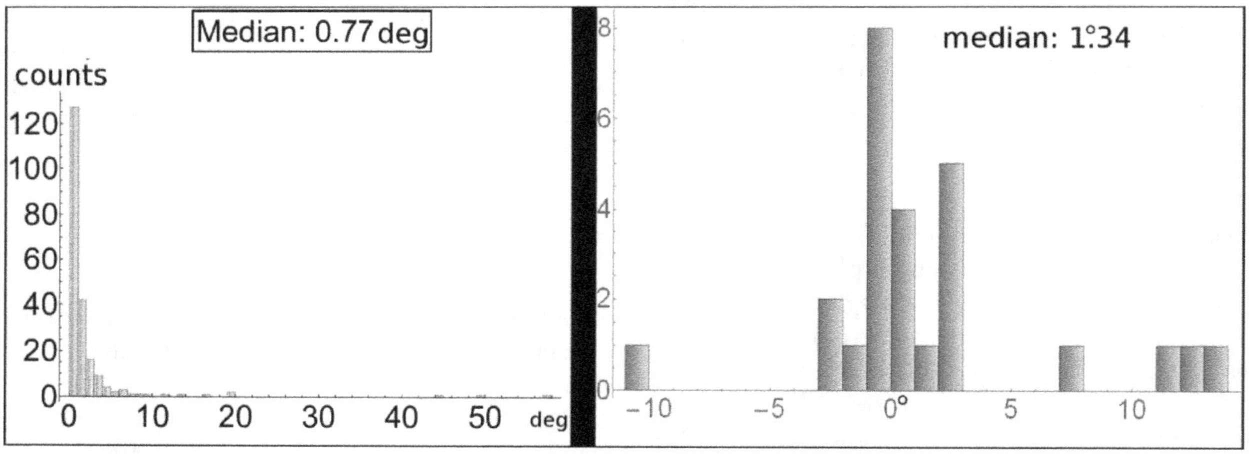

Figure 1. Deviations of Hipparchus' right ascension from those in the HR catalogue: The data (left) in Hipparchus' second part with his own version of risings and settings has a smaller median error than the data (right) in the first part, where he discusses Aratus' verses. Does this indicate different sources?

4. RESULTS

Testing the deviations of the reconstructed Hipparchian coordinates from the modern values we achieve different results in the direction of longitude and latitude and also different results for the different parts of his text. In the first part, where he lists his critique on Aratus' data Hipparchus tells us some declinations. Those declinations have a median deviation from the modern measurements by only 0°.088, while the right ascensions given in this part deviate by 1°.34 (median).[2] This average deviation for right ascensions is astonishingly big - larger than in the median deviation in the second part, where he achieved an average deviation of 0°.77 considering only the culminations or 0°.92 for all stars given.[3]

[2] See Figure 1 and for further information see (Hoffmann, 2017), pp. 194-211, especially p. 201 and p. 203.

[3] See Figure 1 and (Hoffmann, 2017), p. 118 (culminations) and p. 104 for all stars in the second part.

This is not a very good result but not surprising because his scale of his instrument allows him to distinguish only half degrees. Hence, the smallest possible average deviation is 0°.25λ on the ecliptic which transforms into different parts of °RA on the equator. However, with much bigger uncertainties in the observational data, this theoretically possible accuracy was maybe not achievable.

Concerning the huge differences between the first and second part of Hipparchus' text, we can suspect different sources. Could it be that the more accurate data has been observed directly, while the latter is read from a globe? This appears plausible, because Hipparchus might have denoted and published his own observations in a very systematic way, while the discussion of a poem verse-by-verse he probably did not during the night but on the desk. On the other hand, the declinations he gives in the first part have a surprisingly tiny median deviation (roughly 0°.1); is this only because of the small subset and the criteria of selection?

In the third part, Hipparchus defines 46 hour-stars by their right ascensions. Those right ascensions are also very accurate having a median deviation of 0°.16 from the modern measurement.[4] This might either also indicate an alternative source or be a result based upon much more observations than the other stars due to their importance or is caused by Hipparchus' selection. Choosing the median to average the data gives the most likely estimate for the errors of measurement because the biggest errors in our reconstructed data are probably not errors by Hipparchus but from two thousand years of copying tradition.

These results concerning the errors and deviations of Hipparchian stellar positions derived from the only preserved text make the accusation of Ptolemy being a fraud very suspicious: The globe of Hipparchus obviously had a scale with one or a half degree as the smallest unit while Ptolemy's star catalogue distinguishes 5 arcminutes as smallest unit. Additionally, we guess that Hipparchus' catalogue had about 800 stars out of which we can reconstruct 62 to 67 (depending on identification) full coordinates of declination and right ascension plus roughly 230 right ascensions (without declination) from culminating stars.

The original coordinate system of Hipparchus had very likely been the equatorial system:[5] Since the declinations are much more precise than the right ascensions, he probably used a different instrument and method to determine them: Since right ascensions are easily measurable by clocks (via siderial time) and ancient clocks had accuracies up to 20 minutes[6] this measurement might have been less accurate than a (probably more difficult) measurement of declinations with an angular instrument.

Measuring declinations and right ascensions while writing declinations and (apparent) ecliptical longitudes, Hipparchus used a different frame of reference for his observations and for his writing, and both systems are not the frame of reference of the *Almagest* (longitude, latitude). The latter is an orthogonal coordinate system with the ecliptic as baseline and Ptolemy in *Alm.* V, 1 even describes the construction of an instrument (a sort of spherical astrolabe) to measure those coordinates directly. However, Hipparchus at an earlier state of the art observed declinations and right ascensions separately with different methods and instruments.

4.1. *Archaeoastronomical Result*

With the data of the reconstructed part of a star catalogue, we can easily model a globe. The model presented here follows the description in the Almagest. Ptolemy instructs to build a (wooden) sphere, paint it dark and set the star-dots according to the description (coordinates, brightness, and colour) in the star catalogue, but generally in light colours on the dark background.

Concerning the other elements described in the building instructions of the Almagest, we should carefully select which of them we are sure about for Hipparchus' globe. Hipparchus' globe must have had a horizon because he reads risings and settings for his commentary. It must have had a meridian because he reads culminations and this wooden ring serves great for a detailed scale of declinations. He must have been able to easily figure out a 24th part of the equator to read the hour stars so the globe probably had a pattern of 24 right ascension-hours. Writing the commentary Hipparchus was not aware of precession.[7] Therefore, his globe had probably only one axis to turn around and not two axes like Ptolemy's globe. Nonetheless, we assume that Hipparchus placed the stars as exactly as described in the Almagest and afterwards somehow summarized the stars of the same constellation with a decent drawing or dashed borderline like it has been the astronomical tradition of globe makers until Joh. E. Bode in the 19th century.

The result of our reconstruction is displayed in

Figure 2. The drawn declination lines are not sure and the polygons summarizing the constellations are the convex hulls of the reconstructed coordinates of the stars which are surely preserved: Since this is

[4] Hoffmann, 2017, p. 122-135, especially p. 125.

[5] Duke, 2002, and Hoffmann, 2017, p. 612-643.

[6] Steele, 2000.

[7] Hoffmann, 2017, record 2.3.6 on p. 181.

only a subset of Hipparchus' catalogue, certainly the constellation polygons are too small but this is the only data we have at the moment.

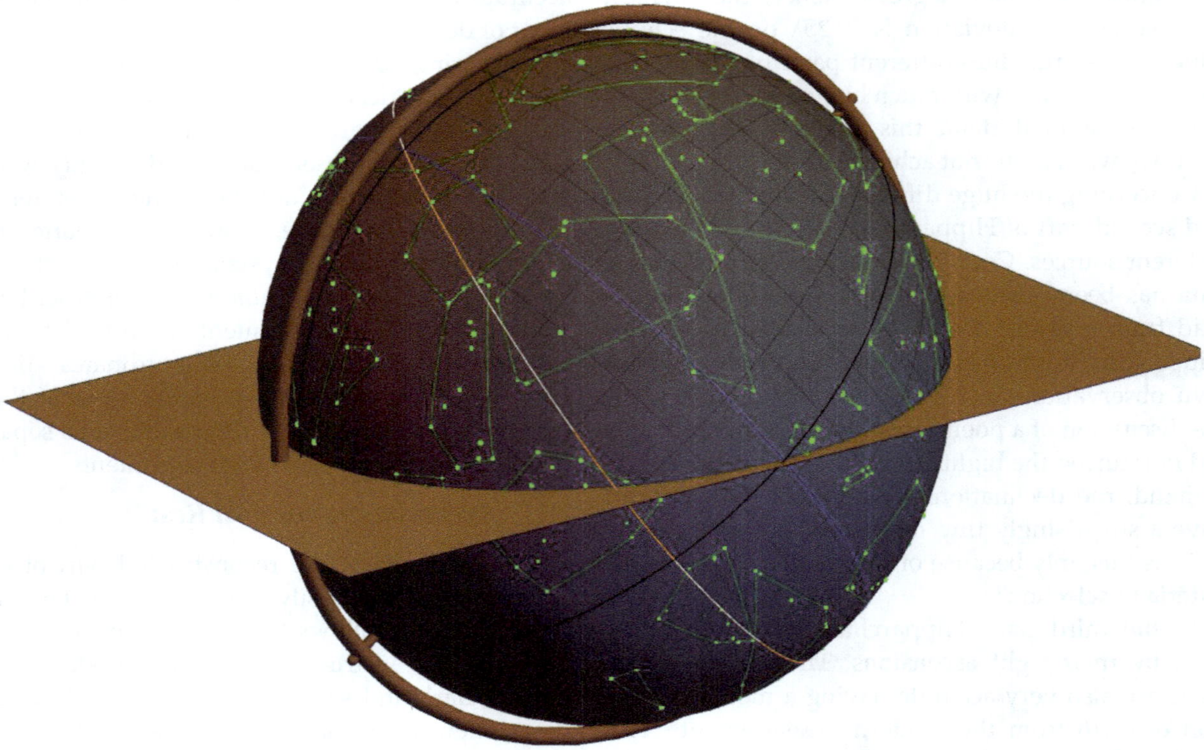

Figure 2. This is how Hipparchus' globe roughly looked like: It had a horizontal plane and a meridian circle, the "longitude"-lines were 24 right ascension hours and the scale in declination has been either none or equally broad but there was a very detailed declination scale on the wooden meridian ring. The ecliptic - either drawn on the globe or modelled by another wooden ring - was divided into 12 signs times 30 parts, 360 in total. Whether or not the celestial equator was marked is not preserved.

4.2. *Historical Results*

Aiming to answer the historical questions on the relation of Hipparchus's data to the Almagest and to Babylonian predecessors, we analysed the metrical units in which the data was given as well as the errors of the coordinates. Since, for example, the star pi Hydrae already had been a key witness (Graßhoff, 1990) for a correlation of the catalogues of Hipparchus and Ptolemy because it has the same error in both of them, we have been surprised by some important differences. The fact that the two catalogues differ and nevertheless are correlated suggests that Hipparchus's original works had been corrected and transformed before Ptolemy: It might have been one of the many sources but not the only one.

5. RELATION TO BABYLONIAN MATHEMATICAL ASTRONOMY

Babylonian astronomy in Hellenistic time considered the zodiac only and used an ecliptical coordinate system there. We do not know if the zones outside the zodiac played any role in that Late Babylonian astrometrical culture. There is no Babylonian star catalogue preserved. The list in MUL.APIN (which is far too old for our considerations here) which is often called "star catalogue" is not a catalogue but belongs to the text genre of commentaries. Hence, it is almost useless for any comparison with Hipparchus's style and contemporary Greek and Babylonian style.

Since Babylonians in Hellenistic epochs used the so-called normal stars[8] to give positions of planets as distances to those stars, the positions of normal stars must have been written down somewhere. Unfortunately, only thirteen lines of such a list are preserved in the fragments BM 36609+ and BM 46083: The list which is reconstructed by (Roughton, Steele, and Walker, 2004) using both tablets, gives only ecliptical longitudes and no latitudes of normal stars. The accuracy is 0°.25 and therefore higher than in Hipparchus' text but what we estimated as the possible maximum accuracy of Hipparchus' data source. Therefore, it was worth questioning possible correlations, but the correlation plots do not lead to any link between this Babylonian and the later Greek data.[9]

---

[8] "Normal" derives from the Latin word "norma" for "the measure".

[9] See Figure 3 and analysis in (Hoffmann, 2017), pp. 449-460, especially diagrams on p. 459.

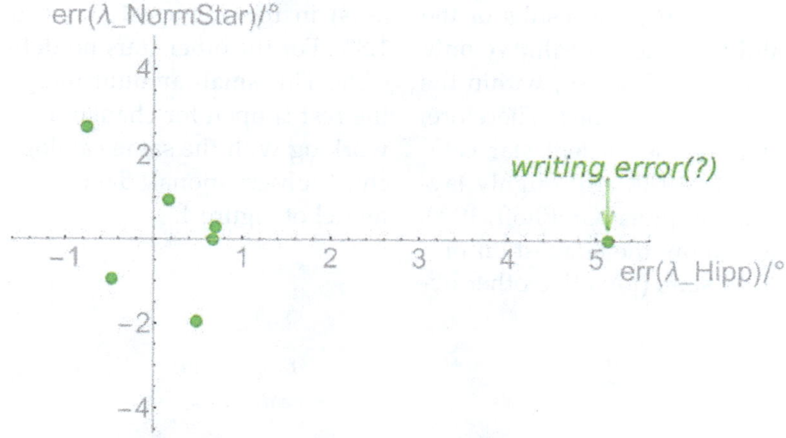

Figure 3. Correlation plot for longitudes in degree of stars in Hipparchus' text and the Babylonian normal star lists.

In Hipparchus' commentary, he does not use ecliptical coordinates. He thinks in equatorial coordinates giving declinations directly, and when he uses terms like "Leo 3°" this is a right ascension named after the degree of the ecliptic it passes.[10] That is why the hypothesis that Hipparchus might have directly used Babylonian data appears unlikely.

However, Hipparchus is not free from Babylonian influences. His circle is divided sexagesimally into 360 parts, he uses the Babylonian signs to divide the ecliptic and in the *Almagest* citations he uses the units "finger" and "cubit" for the tiny deviations from the lines which seems to be similar to the Babylonian units SI and $KUŠ_3$ respectively. The comparison of the real deviations and the values given by Hipparchus does not allow us to decide whether or not this hypothesis is true (50% chance). But in fact, this data leaves some room for interpretation.

However, the 2nd century BCE is known for some Greek innovations in mathematical astronomy. Maybe, Hipparchus learned from only Greek teachers who had been influenced by Babylonian astronomy directly or through the generations of teachers before. This hypothesis is very likely because Hipparchus cites an ecliptic divided in the same way as Eudoxus two centuries before him. That makes the Babylonian influence for Hipparchus only secondary and confirms the hypothesis of his own observations and developments in Greek mathematical astronomy. This result is especially interesting because Hipparchus' famous contemporary, Hypsikles, witnesses Babylonian influence on mathematical astronomy. Hence, we now found different development for computational and observational astronomy in the second century BCE.

[10] I tested both possibilities but this one gives the better result (Hoffmann, 2017) and this is what e.g. (Duke, 2002) already found before me.

6. RELATION TO PTOLEMY

Even the first plot of all Hipparchian stars on the same map as Ptolemy's stars shows that there are some stars mentioned by Hipparchus (at least in the alignments, but also in the commentary) that are missing in the *Almagest* star catalogue. So, if Ptolemy had stolen any data, why would he leave out some ten to twenty stars?

Considering the biggest errors in the reconstruction of Hipparchus' catalogue we find different positions than in Ptolemy's catalogue.[11] This is an allusion to a correction of Hipparchus' biggest errors before the writing of the *Almagest*. Considering that there are almost three hundred years of scientific progress between those two historical figures, we are allowed to assume some re-work of data in between. We suggest that "the ancient star catalogue" has not only been touched by Hipparchus himself and the school of Ptolemy writing the *Almagest*. The more likely model is a continuous usage and re-working of the data which we presume to have been originally observed by Hipparchus.

Ptolemy additionally used some Babylonian data, for instance, solar and lunar eclipse data. Since we concluded that Hipparchus likely did not have direct access to Babylonian data, this leads to the conclusion of new Babylonian influences in the time of Ptolemy or shortly before him.

A very likely model of the development is displayed in Figure 4.

The commentary had been written without considering precession. Later, Hipparchus discovered precession. Further generations of astronomers maybe did not observe the whole catalogue but checked and re-worked some apparent errors. It is still plausible that small stars without any practical meaning (like pi Hydrae) were never re-observed and, therefore, the error remained - even in the *Al-*

[11] Hoffmann, 2017, p. 618.

magest star catalogue - and, thus, the results of the dependency are still valid (Graßhoff, 1990). We only admit that most of the data is, of course, within the margin of error of ancient observations. Therefore, only roughly twenty stars of the ancient star catalogue analysed in (Graßhoff, 1990) are roughly laying on a line in the correlation plots (Graßhoff, 1990, pages 192-197) and, thus, allow the suggestion of a correlation. Out of the 1024 stars (plus five other objects) in the Almagest star catalogue, this is 1.7 to 1.8%. For the other stars no definite statement is possible. This small amount indicates a correlation and the rest is open for change due to a long tradition of working with the same catalogue rooting in Hipparchus' observational data and again supports our model of Figure 4.

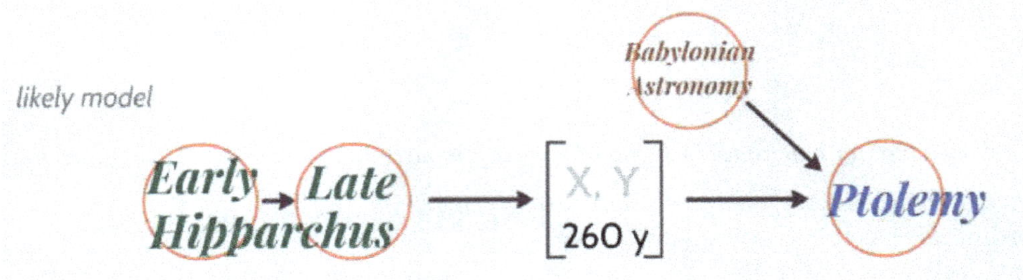

Figure 4. A model for the most likely development of data from Hipparchus's commentary to Ptolemy's Almagest.

ACKNOWLEDGEMENTS

We thank the anonymous reviewers for their constructive comments. This work was a part of the Excellence Cluster TOPOI (Berlin), funded by the German Research Foundation (DFG). Special thanks to Gerd Graßhoff as a very helpful consultant.

REFERENCES

Aratus, and Kidd, D. (1997) *Phaenomena - edited with introduction, translation and commentary*. Cambridge: Cambridge University Press.

Bennet, G. (1982) The calculation of astronomical refraction in marine navigation. *Journal of Navigation*, pp. 255–259.

Duke, D. (2002) The Measurement Method of the Almagest Stars. *The International Journal of Scientific History*, 12, pp. 35–51.

Graßhoff, G. (1990) *The history of Ptolemy's star catalogue*. Springer.

Hipparchus, Macfarlane, and Mills. (unpublished) English translation of Hipparchus' commentary on Aratus.

Hoffmann, S. M. (2017) *Hipparchs Himmelsglobus*. Heidelberg, New York: Springer.

Hunger, H. (1989) *MUL.APIN - An Astronomical Compendium in Cuneiform*. Horn: Ferdinand Berger und Söhne Gesellschaft m.b.H.

Meeus, J. (1988) *Astronomical Algorithms*. Richmond: Willmann-Bell.

Pingree, D., and C. Walker (1988) *A Babylonian Star Catalogue*: BM 78161. Philadelphia.

Roughton, N. A., Steele, J. M., and Walker, C. B. (58 2004) A Late Babylonian Normal and Ziqpu Star Text. *Arch. Hist. Exact Sci.*, pp. 537–572.

Steele, J. (2000) *Observations and Predictions of Eclipse Times by Early Astronomers*. Dordrecht, Boston, London.

Toomer, G., and Ptolemaios. (1984) *Ptolemy's Almagest*.

Vogt, H. (1925) Ein Versuch der Wiederherstellung von Hipparchs Fixsternverzeichnis. *Astronomische Nachrichten*, Bd. 224, Nr. 5354–55, pp. 17–54.

20

ORGANISTRUM IN SANTIAGO DE COMPOSTELA: SYMPHONIA COELESTIS

Giuseppe Severini*[1] and Andrea Orlando[2]

[1]*Casa della Musica di Randazzo, Italy*
Istituto di Archeoastronomia Siciliana, Italy
[2]*Institute of Sicilian Archaeoastronomy, Italy*
Fondazione Floresta Longo, Italy

**Corresponding author: Giuseppe Severini (info@liuteriaseverini.it)*

ABSTRACT

The famous Portico of the Glory in the Cathedral of Santiago de Compostela (Galicia), dedicated to St. James, recently also examined in an interesting archaeoastronomical perspective (Vilas Estevéz and Gonzalez-Garcia, 2016), has been built following an iconographic project inspired by St. John's book of Apocalypse (Moralejo, 1988). At the top of the main arch, in the middle of the 24 venerable men's row around Christ's throne, there is a peculiar musical instrument - equipped with a wheel - played by two of them. The particular position, the extremely accurate details and the peculiarity of the instrument underline the importance of an object whose origins, symbolism and actual musical role (Lopez-Calo, 1988) have never been completely explained.

The most intriguing features are: a) 12 interval division of the octave; b) wheel used to produce sound; c) general shape and decorations. In this article we introduce a possible interpretation of Symphonia (the so-called Organistrum) as a sampler of cosmological and astronomical knowledge typical of European culture from the ninth to the twelth century (Eastwood, 1987; McCluskey, 1998). Close relationship between astronomy and music in platonic-pythagorean doctrines is confirmed through detailed analysis of astronomical texts in manuscripts copied in Benedictine scriptoria from Carolingian Renaissance onwards (Eastwood, 2007).

Beyond general reference to Plato's Timaeus, through Macrobius and Calcidius commentaries (Martello, 2011), the main suggestions for our study come from interesting diagrams (Eastwood and Grasshoff, 2004) found in these manuscripts (e.g.: clm 14436, f.61r, Munich Bayerische Staatsbibliothek).

KEYWORDS: Symphonia Organistrum, Cathedral of Santiago de Compostela, musical instrument, cosmology

1. INTRODUCTION

Probably at the end of the 11th century a wheel and a keyboard were added to a large-size bowed Lyre, like those painted in San Millàn de la Cogolla Beatus manuscript (Figure 1). The playing position had to be changed, in order to allow two musicians to operate properly on the same instrument (Figure 2). In some sculptures no keyboard is actually visible (e.g. Ahedo de Butron, Soria, Spain).

Figure 1. Beatus de la Cogolla, chapter XV, 2-3.

Figure 2. Upper: Capital from Boscherville Abbey, France; Lower: Ms. Hunter 229 f. 21v (Glasgow University Library).

Since this object does appear only in sacred contexts, many authors conclude that it was devoted to accompany early polyphony such as *Organum parallelum*. Unfortunately, there are reasons to reject this idea:

1. *Organum parallelum* had been described since the ninth century in texts as *Musica enchiriadis*, *Scholica enchiriadis* (late ninth-century), *Micrologus* (eleventh-century), *Winchester Troper* (eleventh-century). No instrumental accompaniment was needed of course, since natural consonances as Octave, Fifth and Fourth are easily tuned by ear.
2. Whether you want to accompany simple polyphonies like those, you don't need such a complex instrument as the "two men lyre". A large-size bowed Lyre would work better. The instrument painted in the quoted manu-

script from San Millan had two long strings passing over the fingerboard (tuned a fifth or a fourth apart, we guess), while the middle one ends under the fingerboard, acting as a drone (possibly an octave above the bass string).

Much more likely the "two men Lyre" had been invented in order to play and accompany the *two voices* polyphonies becoming popular in monasteries during the 11th and 12th centuries, such as *organum per motum contrarium, Organum floridum or melismaticum* and free organum to be found in several manuscripts as *Ad organum faciendum (about* 1100), *S. Martial de Limosges music collection, Codex Calixtinus (Santiago de Compostela), Magnus liber organi (Paris).*

After that, we note that the instrument was quickly dropped out, due both to the increasing complexity of thirteenth-century polyphonic music and to the developing of Organs (Ferrari Barassi, 1979; Ferrari Barassi, 1983). Then simpler instruments derived from it, played by one musician, were adopted.

2. SANTIAGO DE COMPOSTELA GATE OF GLORY

The Gate of Glory is the main door of Saint James Cathedral in Santiago de Compostela, has been sculpted by an artist known as *Magister Mateus* and opened in 1188 (Figure 3). The instrument at the top of the gate is famous for excellent workmanship and beautiful decorations.

Figure 3. Full representation of the Gate of Glory.

The shape of the sound box is an "8" made of two identical circles connected by two little lobes. The outline between the sound box and the keyboard is a broken line. The two little instruments sculpted at both sides of it have the same shape but different proportions. The musicians are holding the instrument on their knees, looking at each other (Figure 4): the first is turning the crank with his right hand and his left hand is on the middle of the sound box. The second musician is handling the keys with both hands, thumbs outward. No one knows which was the mechanism of the keyboard. A widely adopted solution is to make keys that you can lift up and drop down. Some authors suggest that the keys were turned, but all attempts to make instruments with revolving keys have been rather unsuccessful.

Figure 4. The two musicians with the "double instrument".

3. SYMBOLISM

The Gate of Glory is crowded by statues bearing different kinds of stringed instruments (Moraeljo, 1988). At the very top of the gate two of them are playing a beautifully carved and decorated instrument, the so-called *Organistrum.* This curious object has been largely studied and several reconstructions have been proposed and carried out by scholars and luthiers during the past 40 years, nevertheless some aspects are still uncertain, such as its symbolic significance and its peculiar musical role (Luengo, 1988).

Its special location is the first sign of its importance. On the opposite side of Saint James shrine, illuminated magically by an architectural device (Vilas Estévéz and González-García, 2016), the statues of the extrados of the Gate of Glory represent the 24 elders of the Apocalypse. They can be symbols of Time (24), Music (musical instruments) and Divine symphony (they are heavenly creatures) (Lopez-Calo, 1988).

The two performers at the top are very close to Christ's head, a position usually occupied by angels (Figure 5). Compared to the other musicians they look to be the only two actually playing their instrument, the most complex of all. Closer observation of its features sheds more light on its deep symbolic significance. A turning object, the wheel, creates the sound from strings stretched over a sound box made of two equal circles, to which an unusual keyboard is added, equipped with 11 keys, so that the octave is divided into 12 parts. One needs to refer to theological, philosophical, mathematical and musical theory of the age to give a reasonable interpretation of all these unusual features. Medieval Latin Platonism, for example, mainly related to the Chartres cathedral school, and well-connected with Paris and Santiago, brings to us interesting suggestions (Albertazzi, 2010).

Figure 5. Christ's head and the two central Elders.

4. COSMOS ACCORDING TO PLATO

In *Timaeus* Plato says that, at the beginning of creation, making the World Soul, the Demiurge mixed three abstract elements: Being, Sameness and Difference. Having divided this compound into the right proportions of a Pythagorean diatonic musical scale he finally gave the World Soul the shape of two circles one inside the other and put them into the material World, thus creating both the Sphere of fixed stars and the plane of the planetary orbits. Circular motions of planets and stars were thus connected in perfect harmony, like the strings of a well-tuned Lyre or the pipes of a well-cut Pan flute (Martello, 2011).

5. THE INSTRUMENT: STRUCTURE AND DECORATIONS

The first circular part of the sound box, the Circle of Sameness, contains the wheel, symbol of planetary motion, the Circle of Difference; this is the origin: the invisible World Soul. Then the other circle is added, the visible world, symbol of Cosmos, with its charming beauty. Finally we find the keyboard, with 11 keys dividing the octave in parts, the symbol of human mind, which divides and measures in order to understand natural phenomena (Figure 6).

Figure 6. Symphonia coelestis.

Therefore, the widespread philosophical principle "*Veritas est adaequatio intellectus et rei*" ("Truth is an adjustment between intellect and things") is well described here, as we observe that the vibrating string-length (rectilinear thought) is equal to the circumference of each circle of the sound box (nature and the ideal world). Further suggestions come from the analysis of instrument decorations. A long line of *dots* runs all around the outline of it. They are symbols of the stars. From left to right: first circle, the idea of the stars in the abstract world; second circle, the visible stars; keyboard, our knowledge of the stars (straightened).

The large *rosette* in the middle represents the Visible World, a circular figure divided in four equal sections by two perpendicular axes. Each section is occupied by five lobes of leaves and twigs (these leaves are similar to those sculpted all around the 24 Elders). The division into four sections may refer to the four elements of the material world.

This figure recalls the diagrams of the seasons frequently drawn in glosses of 9th–12th century astronomical manuscripts. The main goal of any scholar was to calculate accurately the length of each season in order to obtain an exact and affordable calendar to establish with great accuracy both the dates of Christian feasts and the right hours for prayer in monasteries (McCluskey, 1998). Finally, *keyboard lid* decoration, unique among all other decorations sculpted in the Gate, shows a pattern that had been popular in some areas in northern Italy (Figure 7), Spain and Ireland between 9th and 12th centuries.

Figure 7. Decoration from S. Maria foris portam church in Lucca (Italy).

6. PLANETARY DIAGRAMS

Each knot of the interlace corresponds precisely to one of the 11 keys protruding out of the rectangular box. This division in fact is the most puzzling feature of all, as nobody could justify the introduction of a chromatic scale in 12th century, when only Guidonian hexachords were allowed. But we discovered that, curiously enough, both the unusual decoration and the division of the octave in 12 semitones could be related to diagrams of planetary latitudes, as well as musical scales, in contemporary astronomical manuscripts (Figure 8).

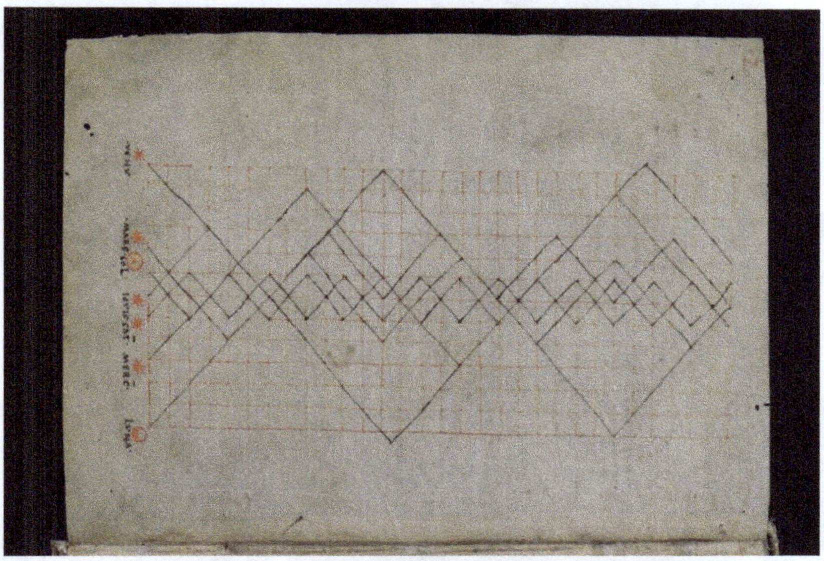

Figure 8. Planetary latitudes diagram from Cod. Sang. 250 (St. Gallen, Stiftsbibliothek).

Such diagrams were intended to describe planetary latitudes across the zodiacal band, divided vertically in 12 degrees. As usual the horizontal line of these grids was divided in 30 parts, but since the eleventh century we often find the number 12 signifying the 12 zodiacal signs. The Sun and Saturn were given a serpentine path within the two middle degrees of the zodiac, Jupiter had three degrees, Mars four, Mercury eight, the Moon covered the zodiacal band with its 12 degrees of latitude. Venus was assigned latitude of 14 degrees, one degree beyond the zodiac on each side (Eastwood, 1997).

Then, in a manuscript by Abbo de Fleury (about 940–1004) we find a horizontal column list of the Plinian intervals between planetary orbits attached to planetary latitudes grid (Figure 9) (Eastwood and Grasshoff, 2004). Whether 14 semitones can be counted in total, the musical octave spanning from Moon to Saturn is divided into 12 semitones (semitone being the unit of measurement of the scale, clearly enough). A keyboard with keys ranging about two full chromatic octaves demonstrates that this arrangement, despite the diatonic stiffness exhibited in sacred music, fits quite well into twelfth-century two voice polyphonic music, in original settings, in *mutationes* and *permutationes*. The sixth key indicates the Tritonus, in the middle of the chromatic scale: musician's right hand is dangerously operating just aside.

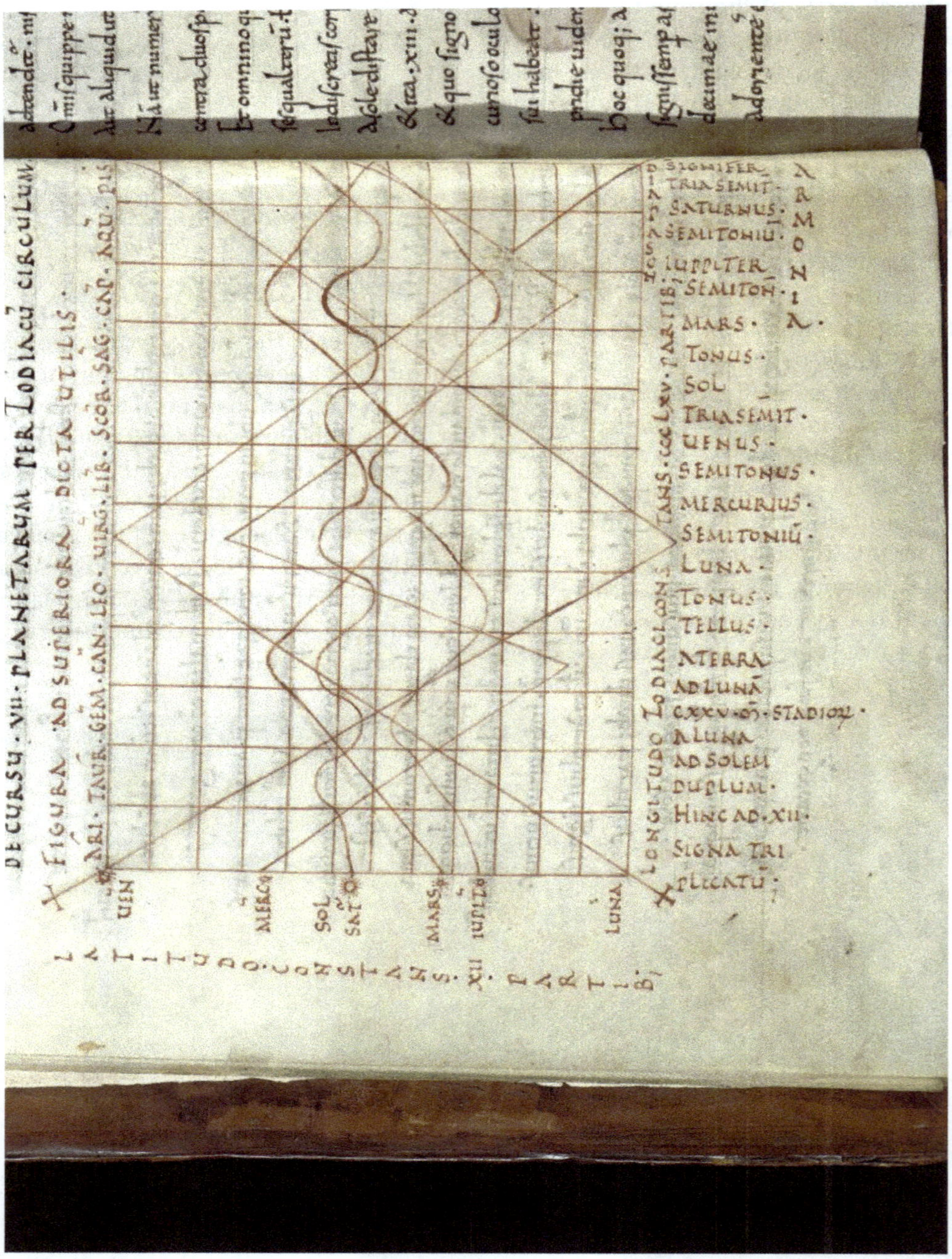

Figure 9. Planetary latitudes diagram and Plinian intervals from ms. R. 15. 32 (James 945), f. 4v (Trinity College, Cambridge, UK).

7. SYMPHONIA COELESTIS: A POLYPHONIC INSTRUMENT

By tuning the strings either (A–d–a) or (A–e–a) we get an overall extension of two chromatic octaves. Before lifting each key, the performer can choose which string he is going to touch, simply by turning the key to the proper position: the first one allows him to act on the bass string, the second on the middle one, the third on the higher string (Fig. 10). Thus it is possible to play two different melodic lines simultaneously. In the Santiago sculpture the hands of the musician on the right are on the third and on the fifth key. This means he is playing (c–g) rather than (d–f) stops on the bass and middle strings, or (g–c′) rather than (f–d′) stops on the middle and higher strings (first tuning) (Table 1).

In position 4 the key is provided with two tangents in order to act on the second and third strings (tuned either in Fifth or Fourth) at the same time to perform *Organum parallelum*, leaving the bass string as a drone. On the other side, the musician who is in charge of turning the crank, by a smooth, even movement of his right hand, can lift each string from the edge of the wheel in order to stop it vibrating with his left hand. This way you can either avoid conflicts between the voices or stop undesired drone effects.

Figure 10. Detail of polyphonic keyboard.

Table 1. According to Helmholtz's notation; e.g. # = sharp.

Head 1	Head 2		Head 3	Head 3	Head 4	Sum
First string	A	a#	b	c'	c'#	d'
Middle string	D	d#	e	f	f#	g
Bass string	A	A#	B	c	c#	d
Keys	0	1	2	3	4	5

8. CONCLUSIONS

Organistrum looks like a summary of the main concepts of *musica mundana* (Boethius, *De Institutione Musica*), a retelling of the ancient myth of Music of the Spheres (Godwin, 1993; Lindberg, 1992). Thus *Symphonia Coelestis* could be its most meaningful name. It is not the first time in history that a musical instrument was assigned the role of representation of Cosmos (for example, the Arabic Lute in its Persian origin and tradition) (Severini, 2015). Its representation in Santiago de Compostela could be an exclusive message for a restricted group of scholars capable of understanding the cosmological and musical implications of a non-traditional object that had been invented by scholars trained in the liberal arts of *Quadrivium*.

REFERENCES

Albertazzi, M. (ed.) (2010) *Philosophia. Lavìs.* La Finestra editrice.

Eastwood, B. (1987) Plinian astronomical diagrams in the early Middle Ages. *Mathematics and its Applications to Science and Natural Philosophy in the Middle Ages.* Edited by Edward Grant and John E. Murdoch. Cambridge: Cambridge University Press.

Eastwood, B. (2007) *Ordering the Heavens. Roman Astronomy and Cosmology in the Carolingian Renaissance.* Brill.

Eastwood, B. and G. Grasshoff (2004) Planetary Diagrams for Roman Astronomy in Medieval Europe, ca. 800–1500, *Transactions of the American Philosophical Society*, Vol. 94, No. 3.

Ferrari Barassi, E. (1979) *Strumenti musicali e testimonianze teoriche nel Medio evo.* Cremona.

Ferrari Barassi, E. (1983) *Testimonianze organologiche nelle fonti teoriche dei secoli X–XIV.* Cremona.

Godwin, J. (ed.) (1993) The Harmony of the Spheres: A Sourcebook of the Pythagorean Tradition. *Music, Inner Traditions International*, Rochester, Vermont.

Lindberg, D. C. (1992) The Beginnings of Western Science: the European Scientific Tradition. *Philosophical, Religious and Institutional Context, Prehistory to A.D. 1450.* Chicago, University of Chicago Press.

Lopez-Calo, J. (1988) La musica en la Catedral de Santiago, A.D. 1188. *El Portico de la Gloria. Musica, Arte y pensamento. "Cuadernos de Musica en Compostela II"*, Santiago de Compostela.

Luengo, F. (1988) Los instrumentos del portico. *El portico de la Gloria. Musica, Arte y pensamento. Cuadernos de Mu-sica en Compostela II*, Santiago de Compostela.

Martello, C. (2011) *Platone a Chartres.* Palermo, Officina di Studi Medievali.

McCluskey, S. C. (1998) *Astronomies and Cultures in Early Medieval Europe.* Cambridge University Press.

Moraeljo, S. (1988) Marco historico y contexto liturgico en la obra del Portico de la Gloria. *El Portico de la Gloria.Musica, Arte y pensamento, "Cuadernos de Musica en Compostela II"*, Santiago de Compostela.

Severini, G. (2015) La reconstitution des Rebabs d'après les peintures du XII siècle de la Chapelle Palatine à Palerme. *L'instrumentarium du Moyen Age. La restitution du son.* Paris, l'Harmattan.

Vilas Estévéz, B. and González-García, C. (2016) Illuminating effects at the cathedral of Saint James (Galicia): first results. *Mediterranean Archaeology and Archaeometry*, Vol. 16, No. 4, pp. 465–471.

21

E LUCEVAN LE STELLE: ENGAGING THE PUBLIC OF ROME IN A CULTURAL REPOSSESSION OF THE URBAN SKY

Stefano Giovanardi, Gabriele Catanzaro, Giangiacomo Gandolfi and Gianluca Masi

Planetarium and Astronomical Museum of Rome

**Corresponding author: Stefano Giovanardi (stefano_giovanardi@yahoo.it)*

ABSTRACT

During the Summer of 2016, the Planetarium of Rome organized a successful and innovative public observing event: E Lucevan le Stelle. Intended to revamp the attention for the local Planetarium and Astronomical Museum - closed since 4 years for renovation works - the event was a call to citizens and amateurs to join the Planetarium astronomers at 8 different locations in the city with their own telescopes for free stargazing sessions, like a diffuse urban star party. Each place was selected in relation to an astronomical theme, with "guided tours" to the evening sky narrated by the Planetarium astronomers, illustrating the cultural relevance of each location with reference to the history of astronomy in Rome.

By connecting the squares and the parks used as observing spots, the circuit of E Lucevan le Stelle invited the public to create a new constellation over Rome and name it. Inspired by the historic effort of Pope Sixtus V, who traced the streets around the Basilica of S. Maria Maggiore in the shape of a star, and by other astronomical readings of the network of churches and landmarks, the goal of this event was to bring to light the ancient connections between Rome and the stars, in the spirit of the European Year of Cultural Heritage. By switching off the main lights at each location, the project aimed at encouraging a direct participation to a collective repossession of the urban sky. The overall attendance to the event was estimated to be approximately 7000 people.

KEYWORDS: Astronomy, Communication of science, Planetarium, Rome, Star Party, Public Outreach, Cultural Heritage.

1. INTRODUCTION

The Planetarium of Rome is one of the oldest Planetaria in the world: opened in 1928, it was the first planetarium outside of the German-speaking region (Firebrace, 2017). Its history of almost 90 years, however, was quite a troubled one: the Planetarium was shut down several times, including an impressive 23 year continuous stop (1981–2004) when Rome remained the only European capital city without a planetarium.

As history often repeats itself, a new and unexpected shut down occurred in 2014, less than 10 years after the reopening of the Planetarium and the new Astronomical Museum at a different location – the Museum of Roman Civilization in the Eur district; the reason for closing being renovation works needed primarily by the neighboring Museum. Yet those works have begun only in 2017, and will continue at least for a couple of years before the site will finally be reopened.

In order to revamp the attention of the public for the Planetarium, whose activity is suffering from an undeserved oblivion, in 2016 the Counsellor for Cultural Growth of the Municipality of Rome asked the Planetarium staff to create an event to be held during the summer, inviting people in Rome to explore the night sky and observe the stars. This event would be part of the cultural offer of the Roman Summer, and was intended as a sign of attention for the Planetarium by the Municipality, a first move toward its reopening.

Working on this remarkable input, we chose to combine entertainment with culture and the history of Rome, adding value to the event in terms of Cultural Heritage promotion and fruition, as described by the guidelines of the European Year of Cultural Heritage 2018[1]. The purpose was to aim for "culture beyond the event", as reviewed in Romano et al. (2009), for other major events organized in Rome. The innovative format for the event was conceived with no educational goal, however it was based on the outcomes of previous visitor studies on the mixed public visiting the Planetarium of Rome in the summertime (Giovanardi et al., 2010).

This was the genesis of E Lucevan le Stelle[2], a series of 8 public astronomical observing sessions dislocated in different areas of the city over the months of August and September 2016, in order to reach out to different layers of the public: Roman citizens living in different parts of Rome, tourists from Italy and abroad, amateur astronomers, students, children and their families (Figure 1).

Figure 1. The promotion banner of the event E Lucevan le Stelle.

2. ADAPTING THE STAR PARTY FORMAT TO A CULTURAL HERITAGE PROJECT

Named after a famous aria from Puccini's Tosca, E Lucevan le Stelle was conceived in the format of a diffuse urban star party. A choice that is coherent with the previous activity of the Planetarium of Rome: since 2009 it revisited the popular concept of the star party, typically a gathering of telescopes and amateur astronomers taking place as far away as possible from city lights, by bringing it back into the city regardless of the limits imposed by the sumptuous light pollution of Rome. The underlying idea is to encourage people to make eye contact again with the starry sky even inside the city, because it still holds a cultural value for everybody, although it is not so starry any longer.

Thus for 6 years (2010-2016, with a break in 2014) the Planetarium hosted the Rome Star Party, ironically named "For a fistful of stars", an event that used to attract dozens of telescopes and several hundred participants: a clear proof of how appropriate was the intuition of not discarding the city sky as a target of observation just because it is devastated by the glare of urban lights. Instead, the Rome Star Party became an ideal occasion to raise and spread an awareness about the problem of light pollution, and invite people to stand up against it in concrete ways. In fact, a correct use of artificial illumination would be crucial to enjoy several historical and archaeological landmarks of Rome, featured in the UNESCO World Heritage List[3], together with the night sky, that is also protected by UNESCO as astronomical heritage[4].

---

[1] http://europa.eu/cultural-heritage/about_en

[2] www.planetarioroma.it/mostre_ed_eventi/programma_degli_spettacoli/e_lucevan_le_stelle

[3] https://whc.unesco.org/en/list/91

[4] https://www3.astronomicalheritage.net/index.php

3. EIGHT BRIDGES TOWARDS THE SKY

A natural evolution of our approach to involving the public in stargazing (described in Gandolfi et al., 2005), seeking to valorize the depleted sky of Rome, was then the creation of a itinerant star party, as if the astronomers - and their telescopes - were going on tour, touching different parks and squares of Rome to gather with anyone interested to join in and have a look at the stars. Thus we selected eight separate locations scattered across the urban area, from the very center of Rome to farther suburban places. We scheduled the calendar of the events over the months of August and September 2016, and associated to each date an astronomical theme, chosen according to the historical references given by the location itself, or by the occurrence of celestial events such as the visibility of certain stars or planets and ultimately by the celebration of international events like the European Researchers Night.

The full calendar of E Lucevan le Stelle resulted as follows (date, place and title are indicated):

1. August 8 – Piazza del Campidoglio
   Roma Caput Coeli: the historical astronomical observatories of Rome
2. August 9 – Janiculum Terrace
   Galileo in Rome: from the observations of the moons of Jupiter to the Juno mission
3. August 10 – Pincio Terrace
   The tears of St. Lawrence: when the stars are falling down
4. August 22 – Piazza Navona
   Water on Earth and on other worlds: from the Rivers fountain to extrasolar planets
5. September 10 – Villa Torlonia
   Stars and constellations, from the Planetarium to the sky
6. September 14 – Villa Carpegna
   The Moon, 47 years after Apollo 11
7. September 23 – Villa Lais
   On the birthday of Augustus, the sky of the Fall equinox
8. September 30 – Schuster Park
   Modern astrophysics in the European Researchers Night

Each of these dates and themes suggests an interesting reason to look at the sky; the astronomers of the Planetarium gave short lectures under the sky – better defined as "guided tours of the night sky" – using laser pointers and tablets equipped with a sky map application to illustrate the constellations and the main stars visible by the naked eye and through the telescopes, in connection with the chosen themes. The goal of those guided tours was to frame the contents of the observations within a broader perception of their cultural value (i.e.: Jupiter with its moons from the Janiculum, where Galileo observed it himself in 1611 and while the Juno probe is exploring the planet; or the pale light of the stars 51 Pegasi and HD189733 as the suns of extrasolar planets where water has been found, from Piazza Navona, next to the Rivers fountain and not far from the statue of Giordano Bruno). Following the definition by Griswold (2013) of culture and society as reciprocal mirrors within the cultural diamond, these are all attempts to operate as producers of a cultural object - the act of an astronomical observation through a telescope, shared by a large group of receivers – that goes beyond pure aesthetics, yet it exploits the enchantment of the moment to attribute a specific meaning to the act itself, connected with the promotion of Rome's historical and cultural heritage. Ultimately, this responds to a fundamental task of the Planetarium's mission: sharing astronomy with the public not as a notion but as a live experience reconnected to popular culture.

4. A NEW CONSTELLATION FOR ROME

The whole scheme of the eight observing sessions was in turn framed within the underlying concept of E Lucevan le Stelle: focusing the attention of the public to the relationships between Rome and the sky, from ancient history to current research. Perhaps the most outstanding historic reference is found in the effort by Pope Sixtus V to re-shape the street system around the Basilica of S. Maria Maggiore - one of the four major churches in Rome - in the shape of a star ("in forma sideris"), as shown in Figure 2, as it is well documented by Bordini (1588).

Figure 2. The street plan around the Basilica of St. Maria Maggiore "in forma sideris" by G. F. Bordini (1588).

Further astronomical descriptions of the network of churches and other landmarks of Rome have been proposed, although their authenticity is questionable. It can notably be mentioned a pattern created by seven churches resembling the Big Dipper, including St. Peter's Basilica, S. Maria del Popolo, Trinità dei Monti, S. Maria Maggiore, S. Giovanni in Laterano, S. Croce in Gerusalemme and S. Lorenzo al Verano (Chirri et al., 2016).

Whether that is real or not, we played with the idea of using the eight locations of E Lucevan le Stelle as "stars" drawn on the map of Rome, to engage the public in a social game linked to the event: create a new constellation over Rome and give it a name. This game had the purpose of spreading the voice about E Lucevan le Stelle on the social networks, notably Twitter in association with the hashtag #elucevanlestelle. Everyone was invited to post the drawings of their constellations online, and a selection of the most interesting ones were published on the Planetarium newsletter (Figure 3).

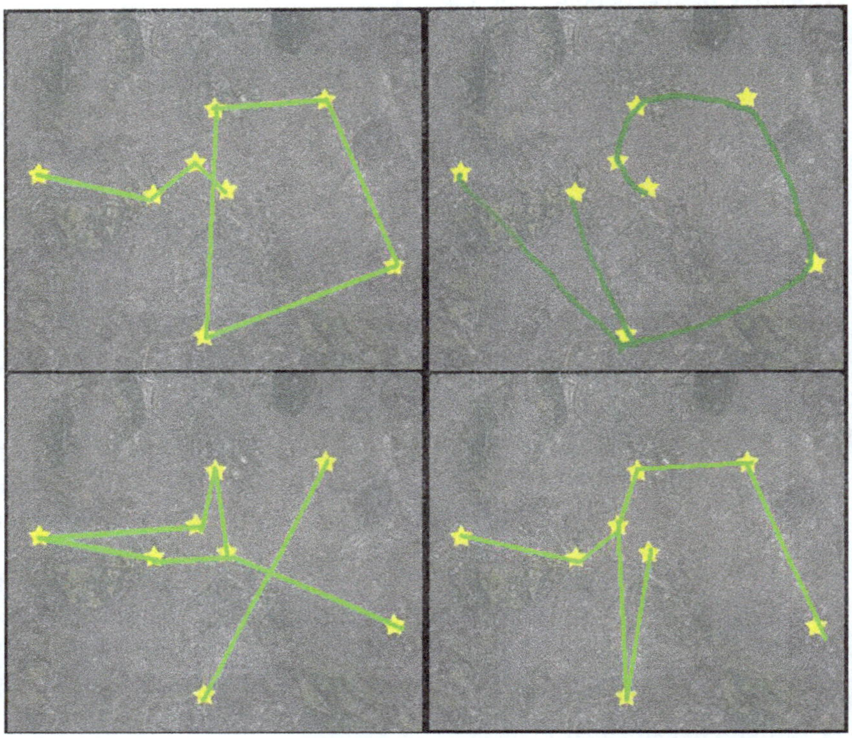

Figure 3. Some constellations proposed by the participants to E Lucevan le Stelle, based on the locations of the event: the Kite, the Snail, the Sparrow and the Elephant.

The overall attendance to E Lucevan le Stelle was approximately 7000 people. Since the events were free, there were no ticket counts available to know the exact number of participants to each session, thus the figures are rough estimates. Nevertheless, there has been a remarkable spread in the number of participants between different dates, partly due to the centrality of the location, partly to the weather conditions and partly to the appeal of the theme. In fact, the most popular date was by far August 10 (Figure 4), known in Italy as the Night of St. Lawrence and strongly associated with the shower of the Perseid meteors (dubbed "the tears of St. Lawrence"): the turnout for that evening alone was estimated about 3000 participants. On the opposite side, the observations scheduled for September 10 at Villa Torlonia had to be cancelled because of bad weather. In general, the four dates in August registered a larger attendance because they were located in very central squares of Rome and benefited from warmer temperatures; the remaining four September dates were based inside urban parks but attracted a lower number of people, due to colder temperatures and regardless of the fact that they offered somewhat better observing conditions. Considering the huge success of E Lucevan le Stelle, the Municipality asked for a supplementary date to be held at one more suburban location (Tor Bella Monaca): it was planned for 8th October 2016, but was canceled because of the cloudy sky.

Figure 4. A guided tour of the night sky during the observing session of August 10, 2016 at the Pincio Terrace. Note the public street lamps switched off for the event. (Photo by F. Barbati).

5. "OPEN SPACE" PLANETARIUM

E Lucevan le Stelle did not mark the first time that the activity of the Planetarium of Rome was organized outside of the dome. Over the years there were several important occasions for astronomical public events to be set up in external locations, that acquired a specific meaning precisely because of their connection to the site.

Besides the already mentioned Rome Star Party, that usually took place in the large square before the entrance to the Planetarium at the Museum of Roman Civilization and was still connected to simultaneous planetarium shows, it is relevant to note that the exportation of astronomical events from the planetarium to other venues (theaters, schools, auditoriums etc.) had started early since the opening of the Planetarium in 2004: it is a natural consequence of the staff approach to communicating science, constantly seeking contaminations with other cultural values, a well defined feature of our productions that goes under the metaphoric label "cupola convessa", or the convex dome.

Among other events organized by the Planetarium astronomers over the years taking place under the city sky, it is especially poignant to remind Infiniti Soli, Innumerabili Mondi (Infinite suns, countless worlds), dedicated to the discovery of extrasolar planets: it was staged in 2009 for the International Year of Astronomy in Campo dei Fiori below the statue of Giordano Bruno. The event included a real time observation of the transit of an exoplanet (TrES-3b) in front of its host star, using a telescope set up next to the statue, to share the resulting lightcurve with about 1000 people who attended the event.

Again in 2011 we presented, in partnership with the American Academy in Rome, Una Filosofica Ragunata (A philosophical gathering) to celebrate the 400th anniversary of Galileo's visit to Rome to present his telescope. The event was located on the Janiculum hill, on the very same grounds where Galileo made his presentation (now within the premises of the Academy): with a replica of the same instrument we showed visitors the Moon and a few objects that shone also in Galileo's sky, 400 years before.

Now temporarily without a dome, the Planetarium needs to push in this direction to maintain its contact with the public by looking for other spaces: the format of E Lucevan le Stelle represents an inversion in the relationship between the (closed) Planetarium and the city, modulating its cultural offer in a diffuse, or "open space", fashion. On this respect, E Lucevan le Stelle has been the most remarkable experiment in migrating outside of the dome to meet the public in the open air.

6. ASTRONOMICAL STORYTELLING IN THE OPEN AIR

There are a few interesting consequences of delivering a message about astronomy in an open environment: it is so different from the protective dome of a planetarium that it radically changes the nature of the contact with the public. Besides the obvious limits imposed by light pollution, surrounding noises and obstruction by buildings and trees, performing astronomical communication in the open air provides stimulating perspectives for scientific storytelling. In the first place, the speaker does not have any control over the aspect of the sky, or on the flow of time as inside a planetarium, but needs to rely on the actual sky and what it shows at the present time.

This strongly influences the narration of what is happening in the heavens, as it forces to keep up with the actual pace of celestial events, like the rising of the Moon or the setting of a planet - including the "special effects", like the unexpected flash of a real shooting star, or a predictable bright pass of the International Space Station (during the dates of E Lucevan le Stelle the ISS made multiple shiny passes overhead and became a treat for viewers). Hence a special attention is required when composing the observing plan - going side by side with the narrative plot -, in order to choose certain visible objects that allow to create a link to the selected theme of the session[5].

Instead of considering this a restriction of the "superpowers" of an astronomer working in a planetarium (i.e. the freedom to disregard any space and time constraints), working under the real sky as in the eight sessions of E Luceavan le Stelle opens up the way to a new approach to astronomical storytelling, that due to a tight connection with the surrounding environment resumes the cultural value of the urban sky and becomes very effective in engaging the public in a direct and personal repossession of the sky of Rome - or of any other place.

The huge turnout of participants - with and without their own telescopes - to the public observing nights of E Lucevan le Stelle clearly shows that citizens from all backgrounds responded with genuine enthusiasm to the invitation of the Planetarium of Rome, and despite the high levels of light pollution the old sky of Rome could be regarded again as something worth sharing with a large local community.

May that be a good omen toward the reopening of the Planetarium dome, possibly sometime next year, where all the stories of the sky will once again find their own home.

[5] One of the authors (S. Giovanardi) had been experimenting for years with the open air communication style, by conducting a series of astronomical "starwalks" through Rome since 2011: in partnership with a touristic guide he drew dozens of thematic itineraries touching historical sites all over the city, often reminiscent of the history of astronomy. In the "starwalk" format, the commented walk leads to an observing spot - located in the darkest nearby area available - where participants continue their exploration by watching the stars through a telescope, guided by an astronomer. The project "Starwalks in Rome" aims at creating a network of links between the city and the sky, in the spirit of casting cultural bridges that allow the public to reconnect with the superior as well as with the inferior half of the panorama.

7. CONCLUSIONS

E Lucevan le Stelle was conceived without any educational or pedagogical goal. It simply was intended as a public outreach project, gauged with a specific attention toward going beyond pure scientific entertainment, in order to promote the historical and cultural heritage of Rome. As such, this event taught us a few interesting lessons worth sharing with the community of science communicators, concerning the cultural approach to communicating astronomy with the public.

The format of a diffuse urban star party adapts so well to the complex reality of a city like Rome that it might be replicated every year; on the other hand the connections between the archaeology and the history of Rome with astronomy are so abundant and diverse - from ancient times to the present - that they would easily offer valuable material, just waiting to be brought to light in terms of promotion of Cultural Heritage.

It would be possible to create a different calendar of itinerant observations every year, developing different astronomical themes and connecting each time new locations among the many squares, parks, panoramic points and historic buildings of the city.

The formula of E Lucevan le Stelle is quite flexible and offers the possibility of realizing a wide public event at very low costs: they are basically reduced to transporting the telescopes to the observing spots and to temporarily switching off the surrounding public lights.

Ultimately, a periodic public event like E Lucevan le Stelle would qualify as a new tourist attraction for Rome, helping the Eternal City to enter - in connection with the network of museums and archaeological sites and despite the level of light pollution - among the destinations of astro-tourism, an emerging share of touristic activities worldwide (Fayos-Solá et al., 2014).

Besides its Roman location, the format of E Lucevan le Stelle could be exported to any other place, especially big cities with a relevant astronomical heritage although suffering from light pollution. Thus we encourage other planetariums, astronomical observatories, universities and research institutes active in public outreach to use this format, with all the suitable adjustments to their local reality, to create popular events engaging communities everywhere in a cultural repossession of the urban sky.

ACKNOWLEDGEMENTS

We thank the Municipality of Rome and in particular the Counsellor for Cultural Growth Luca Bergamo for inviting us to create the event E Lucevan le Stelle, as well as the personnel of Zetema who supported the logistics of the event and the amateur astronomers who spontaneously contributed to the observations with their own telescopes. The authors would especially like to thank the approximately 7000 Roman citizens for participating with enthusiasm and curiosity to the event.

REFERENCES

Bordini, G. F. (1588) De rebus praeclare gestis a Sixto V Pon. Max., pp. 50–52.

Chirri, M., Lombardi, L., Ruggiero, L. and Fabrizi, L. (2016) Rome is star shaped: S.T.A.R. the first Italian app for the spread of cultural astronomy. *Mediterranean Archaeology amd Archaeometry*, Vol. 16, pp. 525–531.

Fayos-Solá, E., Marín, C. and Jafari, J. (2014) Astrotourism: no requiem for meaningful travel. *Pasos. Revista de Turismo y Patrimonio Cultural,* Vol. 12, pp. 663–671.

Firebrace, W. (2017) Star theatre. The story of the planetarium. *Reaktion Books,* p. 209

Gandolfi, G., Catanzaro, G., Giovanardi, S., Masi, G. and Vomero, V. (2005) New perspectives in planetarium lectures: how to tell science under the dome while preserving the enchantment. *Proceedings of the Conference Communicating Astronomy with the Public 2005*, pp. 108–117.

Giovanardi, S., Catanzaro, G., Gandolfi, G., Masi, G. and Vomero, V. (2010) Astrosummer, ovvero raccontare il cielo d'estate: un visitor study sul pubblico del Planetario di Roma. *Museologia Scientifica Memorie* n. 6/2010, pp. 252–255.

Griswold, W. (2013) Cultures and societies in a changing world. *Sage Publications.*

Romano, M. and Melorio, F. (2009) La cultura oltre l'evento. Nuove professionalità tra il concept e la realizzazione di un grande evento. Una ricerca dell'Istituto Luigi Sturzo. *Luca Sossella Editore.*

22

STARS AND THEATRE: FROM RENAISSANCE STAGE ASTROLOGERS TO ASTRONOMY-FLAVORED SCIENCE PLAYS

Giangiacomo Gandolfi

Planetario e Museo Astronomico di Roma Capitale

**Corresponding author: Giangiacomo Gandolfi (giangiacomo.gandolfi@gmail.com)*

ABSTRACT

Theatre has a longstanding and surprising tradition of familiarity with the starry night and its investigators but alas, since Ariosto's *Negromante,* Della Porta's *Lo Astrologo* and the many *comedias* of the Golden Age of Spanish Theatre the adepts of the stars are almost invariably portrayed as tricksters, buffoons and greedy cheaters pretending to be experts of astrology and magic. Comedy is everywhere in modern Europe the only genre associated to such characters and to the study of the universe, at least until the 19th century, when farcical dramaturgy is complemented by some minor tragedies (e.g. Nievo's pioneering *Galilei,* Andreev's *To the Stars,* Gsantner's *Tolternicus* and Ogilvie's *Hypatia*) that slowly pave the way to 20th century dramas centered on science (the prototype is Brecht's *Life of Galileo*). I propose a voyage through the history of theatrical astrologers/astronomers tracing the evolution of the complex relationship between stars and stage and at the same time analyzing the ascent of the contemporary science-play format where the dramaturgy either inflates becoming verbose, philosophical and sometimes ironical or tends to dissolve in a multisensory experience of cosmos, history and society called Postdramatic Theatre.

KEYWORDS: Astronomy, Astrology, Modern Theatre, Post-dramatic Theatre.

---

1. INTRODUCTION

Is there a connection between astronomy and astrology and modern Theatre? And if the connection does exist, is it possible to trace a history, an evolution of such a relationship, beyond the usual anecdotal examples?

The Science of the Stars and its practitioners are often portrayed since the beginnings of Renaissance Theatre, but almost invariably in the category of the Comic and in the genre of Comedy. It is no coincidence: the first fiction about astronomers is a joke, the tale of Thales falling in a pit while observing the heavens, transmitted by Aesop, Plato and Diogenes Laertius, propagated along the centuries through stories like the one contained in the Italian medieval collection "Il Novellino" or through the 17th century fables of La Fontaine, up to the cinema of Georges Méliès in the first years of the 1900s (Gandolfi & Sandrelli, 2009). This demonstrates the stereotypical character of fictional observers and diviners, their funny presumption of being above human worries and pettiness but also the risk of going astray, violating laws of nature or even morality.

In the present essay I will try to explore the presence and the role of astronomers/astrologers concealed in modern farces and dramas, an operation rarely attempted on a large scale involving European stages at all latitudes and epochs. I do not pretend to reach completeness in this survey, nonetheless I hope to isolate relevant themes and tendencies, bringing to light significant episodes (often forgotten) and important moments of transformations in the perception of the world of the stars and of their adepts.

2. STAGE ASTROLOGERS

At the beginnings of European Modern Theatre, when medieval plays were leaving space to more sophisticated representations aware of the classical Greek and Latin heritage, science and wisdom were often present on the stage through the recognizable figures of physicians or astronomers, but also through magicians and necromancers. Such figures clearly overlap at the time in the popular imagination, despite the well known social differences between the academic culture, the esoteric tradition of wizards and occultist and the world of charlatans and street healers.

2.1. *From Bernardo Dovizi to Ariosto and Gnaphaeus*

La Calandria (1513) by Bernardo Dovizi from Bibbiena - a Tuscan erudite cardinal - is usually considered the first modern theatrical representation, drawing themes from the ancient roman comedy of Plautus, but also introducing vernacular and prose, in a modern context and with a modern staging (D'Amico, 1991). This essential humanistic play was produced in Renaissance Italy, during the pontificate of Leo X. It presents the character of Ruffo, the Astrologer/Necromancer, a real archetype for many future astrologers–charlatans on the stage, even if the figure has only a limited space in the comedy.

Il Negromante is a real Ferrarese masterpiece by Ludovico Ariosto that develops with fine insight the character of the astrologer Mastro Iachelino as a prince of the fraudsters, both in the first version of 1520 and in that of 1528 (Coluccia, 2001). The description made by Nebbio, his assistant, is revealing:

Per certo, questa è pur gran confidenzia
che mastro Iachelino ha in sé medesimo,
che mal sapendo leggere e mal scrivere
faccia professione di filosofo,
d'alchimista, di medico, di astrologo,
di mago, e di scongiurator di spiriti;
e sa di queste e de l'altre scienzie
che sa l'asino e 'l bue di sonar gli organi;
benché si faccia nominar lo astrologo
per eccellenzia, sì come Virgilio
il Poeta, e Aristotele il Filosofo;
ma con un viso più che marmo immobile,
ciance, menzogne, e con altra industria
aggira et avviluppa il capo a gli uomini;
e gode, e fa godere a me (aiutandoci
la sciocchezza, che al mondo è in abbondanzia) l'altrui ricchezze.
(*Il Negromante* II, 473)

Mastro Iachelino is particularly interesting because a passage of the text shows hints of his possible connection with a real astronomer: Celio Calcagnini or even Copernicus himself, identified by a reference to the movement of the Earth[1]. But the key role of the astrologer in this and other plays, beyond the identification of relevant historical figures, is certainly revealing of a large diffusion of similar practices in the Italian society of the Renaissance. The strong cultural impact of Astrology is in fact confirmed at the same time and in a nearby territory (Padua and Mantova) by the presence of a consistent number of funny practitioners described in the satirical texts of many Macarronee, vernacular poems based on an Italian–Latin pastiche, for example in

---

[1] Describing the feats of Mastro Iachelino to the servant Themolo the young Cinthio says *"He maketh the Earth move if he wishes so" (Il Negromante II, Second Act, First scene).*

the *Macaronea* (1488–89) by Tifi Odasi and in the *Baldus* (1517) by Teofilo Folengo.

Another important play related to the Science of the Stars in the first half of the 16th century is *Morosophus* (1541), *The Learned Moron*, an allegorical comedy in Latin by the Dutch Protestant Gulielmus Gnaphaeus (Demoed, 2008). The protagonist of the text, conceived for a university audience, may allude to Copernicus (in a mocking mood): the play is a clear exception in the "canon" of comical representations with its religious and metaphysical subtext (the fool Morophosus that pretends respect when mimicking the knowledge of astronomers, converts in the end to Divina Sophia, the spiritual wisdom) and differentiates itself from the tradition of medieval comedies and drama. Its modernity is inspired by the Erasmian *Praise of Folly* and by the *Ship of Fools* created by Sebastian Brant half a century before.

2.2. Ariosto, Della Porta and their Legacy

Lo Astrologo (first version in 1591 and second one in 1606) by the Neapolitan Giovambattista della Porta is another important prototype for the figure of the astrologer-scoundrel, soon imitated and cited everywhere in Europe (Rodda, 2016). Albumazar, the main character, is clearly based on Mastro Iachelino, but his technical jargon and his name are even more telling about the intentional parody of the world of professional astrologers, paving the way for many charlatan-heros on the stage, with funny names that range from Eclipticus and Solsticius to Stargaze and Giropolice.

Occasionally this same astrological knowledge, technical and specific, emerges in other plays as well but in general these comedies are dominated by a widespread and gross contamination of necromancy, magic and other illicit activities, together with an omnipresence of the debate about Astral Fate and Human Free Will.

The influence of Della Porta's play was very broad and enduring, especially in Spain, where Calderon de la Barca proposes, complicating the plot, a similar swindler in the celebrated *El Astrologo Fingido* of 1624 (an Aporta is explicitly cited in the comedy as an astrologer) and in England, where Thomas Tomkis draws heavily on the Italian work for inspiration when creating his successful *Albumazzar* (1615). This last comedy is notable for its introduction on the stage of a Galilean "perspicill", cited together with an "otacousticon," a telescope for hearing sounds from far away consisting in a simple couple of ass's ears.

Many scholars wrongly inferred from such *clichés* a general refusal of Astrology, while they just indicate the usual ambiguity about a dangerous topic among clerics and humanists, and the general consensus even among astrologers themselves, that charlatans dominated a complex and very difficult field of research. On the other hand, as I noted earlier, parody implies a pervasive presence of the science of the stars in society and the relevance (deserved or not) of its practitioners, even during the Renaissance, all along the 17th century.

The legacy of *Il Negromante* and *Lo Astrologo* includes many Italian comedies, ever more contaminated by amorous misunderstandings in the vein of Calderon de la Barca, from *Lo Astrologo Impazzito* (1607) by Giovanni Ralli, to *Astrologia Amorosa* (1610) by Domenico Ravicio and *L'Astrologo non Astrologo e gli Amori Sturbati* (1665) by Ottone Lazzaro Scacco. The same invasion of stage astrologers happens in England, where the work of della Porta, reverberated in Tomkis' *Albumazzar* and in *El Astrologo Fingido*, generates *An Evening Love or the Mock Astrologer* (1668) by John Dryden and *The Astrologer* (1744) by James Ralph, and even in France where Thomas Corneille, brother of the more famous poet Pierre, publishes *Le Feint Astrologue* (1651).

2.3. Astrologers in the Commedia dell'Arte

The tradition of buffoon astrologers propagates into the Commedia dell'Arte, creating a fashion for astral voyages and celestial divinations performed on the stage by the classical Italian masks. The famous actor Flaminio Scala (alias Flavio Comico) published a collection of 50 canovaccios, *Il Teatro delle Favole Rappresentative* (1611), where no less than three scenarios are dedicated to the stars: *Il Finto Negromante*, *Flavio Negromante* and *Isabella Astrologa*.

The celebrated Evaristo Gherardi, chief of the *Comediens italiens du Roi* who relocated to Paris the Italian tradition, produced at the end of 17th century another very important collection, *Le Theatre Italien* (1694) where among many texts of anonymous authors at least two comedies, *Arlequin Mercure Galant* and *Arlequin Empereur dans la Lune*, present astronomical and astrological elements and themes. The *Theatre Italien* terminated its activity at the end of the century but the Commedia dell'Arte and the astrologers remained on the stage, proposed by French artists that exploited that form of Theatre and gave birth to the Opera Comique. Still in 1785 we find on the scenes jokes and quips by *Cassandre Astrologue* (1780) by Pierre-Antoine-Augustine de Piis, a funny comédie-parade, and even in the 19th century *Le Dernier Jour d'un Astrologue* (1864) by Albert Laporte and Ernest Rigodon which is, notwithstanding its dramatic title, a comédie-vaudeville dedicated to the same old character of the cheating astrologer.

3. STARS IN EUROPEAN THEATRE

3.1. *Astrology in Spanish and French Theatre*

Astrologers and the Science of the Stars are very important in Spanish Theatre too. Frederick de Armas studied extensively the Siglo de Oro with an eye for the stars and their role on the stage. He not only found the usual charlatans in comedies, but also a fascinating zodiacal symbolism and a complex astrological subtext, even in serious, dramatic plays. In general, as even Robert Lima (1995) affirms: "There are countless plays in which astrology and other forms of divination are basic motifs and frequent elements".

Pedro Calderon de la Barca represents a wonderful example for both the register of Comedy (the above mentioned *El Astrologo Fingido*) and Drama (*La Vida es Sueno,* 1635), where the deterministic Astrological Fate is the evil force that generates the tragedy), but the Science of the Stars is everywhere in his operas, for example in *El Laurel de Apolo* (1658), in *Las Cadenas del Demonio* (1635), in *Los tres Afectos del Amor* (1648) and in *Fieras afeminas amor* (1670).

Other important Spanish authors dedicating works and characters to Astrology include Lope de Vega, with his *La desdichada Estefanìa* (1604) and *El Vellocino de Oro* (1622), Bances Candamo's *El astrólogo tunante* (1687), Andrés de Claramonte's *La Estrella de Sevilla* (1623) and the Count of Villamediana with *La Gloria de Niquea* (1622).

French stages during the 17th and 18th centuries were equally filled with astrologers and astronomers: I have cited *Cassandre Astrologue*, but the theme was approached even by first–rate men of letters as Bernard Le Bovier de Fontenelle, who wrote in 1681 his funny comedy *La Comète* making fun of superstitions about long–haired stars. Very light comedies in the same vein did not disappear after the Revolution: *L'Eclipse Totale* (1782) by Auguste Etienne Xavier Poisson de La Chabeaussiere and *L'Eclipse de Lune* (1796), a revision of the same work by the same author, are two perfectly equivalents comédies–vaudeville, before and after the storming of the Bastille.

3.2. *Elizabethan Stars and Beyond*

In English Theatre Elizabethan and Jacobean playwrights display frequent interest for stars and planets, from the classic Ptolemaic voyage to the stars of Marlowe's necromancer Faust to Tomkit's charlatan Astrologer I have cited before, to Ben Jonson's explicit Galilean Masque about the telescope and the Moon–world (*News on the New World discovered on the Moon*, 1620), to the reprise of *El Astrologo Fingido* by Dryden: *An Evening Love or the Mock Astrologer*. Sometimes the celestial bodies descend directly on the stage, as in Robert Greene's *Planetomachia* (1585) and in John Lyly's *The Woman in the Moon* (1595). On the other hand, the Commedia dell'Arte farce of *Arlequin Empereur de la Lune* is transformed by Aphra Behn in a masterpiece comedy that pays again a direct homage to della Porta and Ariosto: *The Emperor of the Moon* (1688).

Many scholars, for example Rowan Robinson (2015) noticed an abundance of astral references even in Shakespeare's works and inferred a special interest in astronomy and astrology on his part. Apart from some wild speculations about a possible Shakesperean knowledge of a pre–Galilean telescope (Usher, 2006), many clues of a proximity to British heliocentrists and more in general to the debate about world–systems are offered in his plays, as summed up in a recent New Scientist article by Dan Falk (2014). Hamlet especially, "the king of infinite space" as he defines himself, is very suggestive of the Copernican debate, with an adversary named Claudius (Ptolemy?), the citation of two families of relatives of Tycho Brahe (Rosencrans and Guilderstern) and the appearance as a ghost heralded by a westward star in the sky of Elsinore, strongly recalling the epochal supernova of 1572, again described and studied by Tycho. In this perspective astronomers highlight the probable connection of the playwright with Leonard Digges, the erudite son of Thomas, a leading Copernican in England, suggested by the fact that both lived nearby in North London and that a verse by Leonard contributed to the first published collection of Shakespeare plays.

3.3. *Drammi Giocosi and Melodrammi*

Carlo Goldoni, the prince of Italian Theatre in the eighteenth–century, revived the Comedy tradition and dedicated his *Il Mondo della Luna* to the old buffoon–astrologer mask. The musical component was expanded through the collaboration of maestros of the calibre of Galluppi, Haydn and Paisiello, all working on the same libretto, and in the end this and other similar operations gave birth to the new format of Dramma Giocoso. This theatrical genre was one of the antecedent of the classic form of 19th century melodrama, where the usual astrologers were again in the spotlight. Examples of practitioners of the Science of the Stars for music lovers are *Gli Astrologi o Amor non Può Celarsi* (1791) by Luigi Crippa and Francesco Marconi, *La Casa dell'Astrologo* (1811) by Luigi Romanelli and *Etra l'Astrologa* (1857) by Francesco Cortesi.

4. NINETEENTH AND TWENTIETH CENTURIES

During the nineteenth-century the Astrologer becomes finally a professional Astronomer and the dramatic implications of his job and life condition enter the stage with a Romantic aura. History and ideology become dominant: dramas are filled with strong emotions and prejudices (racism, sexism, classism, etc). Hero-scientists usually suffer in the wake of Inquisition and the Great Wheel of Progress promises them a bright, inextinguishable memory in the heart of their descendants. Tragedy is the main key, while the comic register of Comedy tends to attenuate.

4.1. *Astronomer-heros and Anti-heros*

An interesting example of the new trend is *Hypatia* (1893) by Stuart Glencairn Ogilvie, based on the novel by Charles Kingsley, a typical anti-Catholic text full of metaphysical and religious concerns. The Victorian era produces some literary texts about this exotic alexandrine astronomer and proto-scientist, who since the age of the Enlightenment assumes many dramatic and instrumental roles: last of the Hellenes, martyr of science, embryonic feminist and symbol of Christian obscurantism. Ogilvie's stage adaptation of the tragedy is particularly complex and full of prejudices.

Two late pre-modernist views of astronomers on the stage may be also cited: *To the Stars* (1905) by Leonid Andreev and *Tolternicus Astronomer of Thorn* (1915) by Otto Carl Gsantner. Andreev play portrays the typical scientist conceived as a "wise man" that doesn't mix with politics and human tragedies, looking at the stars from his isolated observatory, while the semi-unknown author Gsantner describes the last days of a fictional Copernicus (perhaps the distorted name intends to justify the countless artistic licenses) tortured and poisoned by the Inquisition (sic!).

In 1879, instead, an American comic opera parody stands out for being the first representation starring real scientists on the stage. Harvard astronomer Winslow Upton rewrote in fact Gilbert and Sullivan's *pièce H.M.S. Pinafore* of 1878 a year after and in 1929 the staff (with Harlow Shapley, Cecilia Payne and Bart Bok among others) performed it with the title *The Observatory Pinafore*.

This musical comedy recalls the legendary rewriting of *Faust* performed by quantum physicists at Niels Bohr Institute in 1932 and in a light vein represents the daily activities of director, observers and computer women in their natural settings: dome, library and computer room.

4.2. *The Galileo Paradigm*

One of the historical figures of astronomy most exploited by modern theatre is without doubt that of Galileo Galilei, who has become a real symbol of Science in his conflicts with Religion and Society. Galileo on stage constitutes in a sense a perfect mirror for the perception of Astronomy, a paradigm of its practitioners.

Since the Enlightenment this character has received a continuous attention, generating some indisputable masterpieces and an abundance of interpretations, sometimes divergent, from the philosophical draft of *Galilée* (1843) by Pierre-Joseph Proudhon, to the positivist, Mazzinian *Gli Ultimi Anni di Galileo Galilei* (1854) by Ippolito Nievo and to the trivial *Galilée* (1867) by François Ponsard, dominated by the romantic liaison of his daughter that prepares his final *abiura*.

After many anticlerical Galileos, invariable champions of free thought, the twentieth-century brings in more nuances and the exceedingly famous *Life of Galileo* by Bertolt Brecht in his three celebrated versions (1938, 1947, 1955). While in the initial play the Pisan astronomer represents optimistically the classical hero of Truth, in the following versions a strong criticism to the responsibility of Science creeps in, due to the new post-Hiroshima scenario. In the same years, to use the words of Shepherd-Barr (2006), "the air was thick with Galileo plays": Barrie Stavis wrote *Lamp at Midnight* (1947), in Germany Arthur Trebitsch published *Galilei, Ein Trauerspiel in funf Akten* (1920) and Frank Zwillingern *Galileo Galilei. Schauspiel* (1953). But the process of convergence towards the complexities of the historical situation conducts to a really significant episode with *Galileo* (1970) by Tom Stoppard, a crucial play unfortunately never performed and only recently published for the first time. The new drama wittingly juxtaposes Brecht's vision, proposing an anti-ideologic humanization of the character: the scientist is no more an abstract symbol but a real person in flash and blood, immersed in a wide, multilevel storytelling. The aversion to the historical inaccuracies of Brecht's text finds his apotheosis (but also a similar ideological partiality) in an unpublished theatrical dialogue by Lewis Wolpert: *Good Evening, Galileo*, where the author interviews Galileo after assisting to a production of *Life of Galileo*.

5. THE POST-DRAMATIC SCENE

Between the end of 19th and the beginning of 20th century a deep crisis dominated the scenes of western drama. According to Peter Szondi (1956) it was a phase of transition from pure classical drama, absolute and out of time[2], to the epic theatre championed by Bertolt Brecht, historical and well steeped in past and future.

But Szondi didn't perceive a more profound change, questioning the relevance of the text and opening to a radical redefinition of stage conventions, formats and languages: beyond Ibsenian drama, Brecht Epos and the Theatre of Absurd, stands the Post–dramatic Theatre, a Theatre no more *fabula*-centred, for the Postmodern Age (Lehmann, 2006).

Astrology evaporates from the scene, but Science, Astronomy and the world of stars become frequent inspirations for plays, installations and happenings. Celestial bodies and their investigators shine among dumbshows, ballets, music, digital and video performances, even circus, in addition to classical text-driven dramaturgy, while real and fictional scientists enter the stage in their everyday activities and not only in epochal paradigm shifts (as in the premonitory *Observatory Pinafore*).

5.1. *Theatrical Biographies*

Dramaturgy, however, survives in this new scenario up to the present 21st century, and playwrights abound, concentrating on topics, themes and points of view rarely considered in depth in modern Theatre. Among them Science is especially appreciated and explored mainly through theatrical biographies. English language stages are particularly rich of historical characters whose life is condensed and exemplified in a few acts and scenes, where philosophy dominates verbose scenarios and postmodernism seeds irony and citations. One may cite the classical confrontation of Johannes Kepler and Tycho Brahe, represented by *A Short History of Night* (1990) by John Mighton, *The Stargazers* (1998) by Joseph Cowley and *Reading the Mind of God* (1994) by Patrick Gabridge, or the relationship between Halley and Flamsteed, finely told by Gavin Hood in his *The Astronomer's Garden* (1991), or still the unlucky story of Guillaume Le Gentil in Maureen Hunter's *The Transit of Venus* (1992). A fascinating feminine figure like that of Caroline Herschel has been portrayed at least twice, by Chiori Miyagawa's *Comet Hunter* (2003) and by Siobhan Nicholas in her *Stella: A Play about Women, their Men and Astronomy* (2013), but even contemporary astrophysical superstars have their multiple place on stage: it is the case of Stephen Hawking, honored by the couple of ironical plays *Hawking's Dream* (1995) by Erwin Riess and *God and Stephen Hawking* (2000) by Robin Hawdon.

Celebrated astronomers and science–writers joined the company of playwrights with *Eccentric Planet* (1980–2013) by John Barrow, an exploration of George Joachim Rheticus science and homosexuality, and *And the Sun Stood Still* (2011) by Dava Sobel, a delicate reenactment of Copernicus' last years.

The fashion of theatrical biographies of scientists has produced even a best–selling author, a real queen of Science Plays: Lauren Gunderson, an American award–winning writer who was rated USA's most produced living playwright in 2017. Gunderson, whose interests oscillate between the role of women and the impact of science on society, wrote at least four astronomy–related *pièces*: *Background* (2001), in memory of Ralph Alpher and his contribution to the discovery of the Cosmic Background Radiation, *Leap* (2004), a comedy about Isaac Newton and the law of Gravitation, *Emilie, la Marquise du Chatelet Defends Herself Tonight* (2008), a memorable feminine portrait in the Age of Enlightenment, and *Silent Sky* (2011), the life of Henrietta Leavitt, pioneer of the measurement of cosmological distances during the difficult era of the computer women of nineteenth–century.

5.2. *Post–dramatic Stars*

But Post–dramatic Theatre has a more experimental side, where narration dilutes and leaves space to other sensorial experiences. An exceptional example of such tendency is the show *Infinities* (2002) by the director Luca Ronconi and the astrophysicist John Barrow, whose first representation was staged in the Spazio Bovisa in Milan, a highly unconventional space. Barrow's mini–essays and theatrical suggestions about infinity were read and represented with high impact visuals during a promenade through this vast, post industrial structure, resulting in a seminal "manifesto" for post-dramatic science theatre.

Sette Variazioni sul Cielo was another interesting post–dramatic play in Italy, a country that became a real epicentre for such a type of productions in the first decade of the new millennium. The work explored the fascination for astronomy and cosmology on the main Italian stages in 2005, again with videos and images. The late popular astrophysicist Margherita Hack wrote the script and appeared as an actor in the *pièce* together with a dancer and some musicians.

Again in Italy a fascinating treatment of the old unforgotten character of Galileo should be cited: *I.T.I.S. Galileo* (2010) by Marco Paolini. Paolini is a

---

[2] In the words of Peter Hoyng (2009): *"derived from Aristotelian poetics in its opposition to epic elements, Szondi's absolute Drama consists of the following three requisites: a) the dramatis personae represent autonomous individuals who are motivated b) to resolve their interpersonal conflicts by speaking to one another. The dialogues unfold c) in a linear time sequence in front of the spectator who watches the actors".*

pioneer of *Teatro di Narrazione*, a style of representation based on non-fiction monologues about social issues, akin to *Documentary Theatre*, and in this case the monologue revolves around the figure of the Pisan and its modernity, his first telescopic observations, the impact of his scientific endeavours on society. But the expansion of scopes and means of Post-dramatic Theatre is testified in Galileo's test case by productions like the opera *Galileo Galilei* (2002) by Philip Glass and Mary Zimmermann and the acrobatic scenario of *Galileo. An Astronomical Story* (2016) by the French company Deus Ex-Machina Cirque.

In general sciences as astronomy and astrophysics are rapidly becoming on international stages the focus of many vital metaphors, conditioning sometimes the entire structure of the plays, while mathematics mixes with psychology and emotions and different genres collide to produce innovative experiences, even in text-driven, literary dramaturgy. Among these works stand out dramas like *Constellations* (2012) by Nick Payne, with its vertiginous cascade of possible quantum universes, *Observatory* (1999) by Daragh Carville, a strange hybrid of science, gothic tale and political spy-story located in Dublin between 1799 and 1999, and productions like *Birth of Stars* (2014) by Michael Chemers and Joan Raspo, exploring the inspiration of astronomy and its impact on the life of two very different individuals, or *The Blue Ball* (1995) by Paul Godfrey, a witty enquiry about the magic and awe of space exploration.

5.3. *A bright future for Storytelling*

Finally, it is worth mentioning a recent British production, innovative and very significant. Space and human exploration of Mars are the main topics of *Pioneer* (2014), a gripping work by Curious Directive, a multi-awards winning Theatre Company that indicates new and fresh paths for exploring the contemporary world of Science and in this particular case of the stars. Its promising technique of multi-layered and intricate narrative brings again with decision the text at the centre of theatrical experience, beyond conventional biopics and with a twist of digital age complexity and Netflix storytelling. *Pioneer* adopts global thinking between Europe, Russia and Mars and presents a number of interconnected themes and mythologies that weave together politics, history and the possible space future of our race, while exploiting simple but stunning visuals.

6. CONCLUSIONS

As I have shown, a long and noble history of connections between stars and Theatre may be traced back from the Renaissance, but until recently almost all plays were centered on the people that study and investigate the celestial bodies more than on the cosmos itself (with an obvious penchant for the most direct of the disciplines interested in their influences on humans: astral divination). Since the Enlightenment Astrology faded away and the dialogue between astronomy and dramaturgy flourished, especially in the era of Positivism and Neopositivism, a dialogue which is still alive and well in this beginning of the XXI century, in the context of what has been termed Science Theatre.

Science in general and astronomy in particular are also often associated with Post-dramatic Theatre, with its productions dominated by directors and actors rather than by the traditional playwright. Such theatrical productions (and more in general the so called "Science Plays", including biographies on stage) have known a "peak" of popularity in the first decade of the new century, but they continue to be highly appreciated in recent years even if staged a bit less frequently.

There is still a lack of detailed comparative studies on style and content of this new works, but with a few notable exceptions: Kirsten Shepherd-Barr (2006), Eva-Sabine Zehelein (2009) and Liliane Campos (2012), all three analyze Science on stage, hence even astronomy and cosmology. Their efforts, together with those of a new inchoate generation of scholars, will help to draw a map of the vast and complex connections between Stars and Theatre in the third millennium.

REFERENCES

Armas, F. (2006) El Rey Astrologo en Lope de Vega y Calderon de la Barca, in Lorenço, L. G., *El Teatro Clasico Espanol a travès de sus Monarcas*. Madrid: Editorial Fundamentos.

Armas, F. (2011) Zodiacal Plays: Astrology and the Comedia, in *A Confluence of Words: Studies in Honor of Robert Lima*. Eds. W. H. Finke and B. J. Luby. Newark, DE: Juan de la Cuesta.

Campos, L. (2012) *Sciences en Scène dans le Theatre Britannique Contemporain*. Rennes: Presses Universitaire de Rennes.

Coluccia, G. (2001) *L'Esperienza Teatrale di Ludovico Ariosto*. Lecce: Manni.

D'Amico, J. (1991) Drama and the Court in La Calandria. *Theatre Journal*, Vol. 43, No. 1.

Davenport, R. (2014) Pioneer Review. *https://www.timeout.com/london/theatre/pioneer-review.*

Demoed, V. E. M. (2008). Stultitia on stage: Gnaphaeus's 'Foolish scientist' and the 'Praise of folly' of Erasmus. In J. Bloemendal, and P. Ford (Eds.), *Neo–Latin drama: forms, functions, receptions*, Hildesheim: Olms.

Falk, D. (2014) Shakespeare: Did Radical Astronomy inspire Hamlet? http://www.newscientist.com/article/mg22229654.800-shakespeare-did-radical-astronomy-inspire-hamlet.html.

Gandolfi, G., and Sandrelli, S. (2009) Dal Pozzo alla Luna. *Piccolo Atlante Celeste. Racconti di Astronomia*, Torino: Einaudi.

Hoyng, P. (2009) Peter Szondi's *Theorie des modernen Dramas* (1956/63): From Absolute Drama to Absolute Theory, in *Monatshefte*. Volume 101, N. 3.

Lanuza–Navarro, T. (2012) The Representation of Astrology in Golden Age Spanish Theatre: Different Points of View in Calderon's *El astrologo fingido*, eds. Nicholas Campion and Rolf Sinclair, *Culture and Cosmos*, Vol. 16 nos. 1 and 2.

Lehmann, H.–T. (2006) *Postdramatic Theatre*, Abingdon: Routledge.

Lima, R. (1995) *Dark Prisms: Occultism in Hispanic Drama*. Lexington: The University Press of Kentucky.

Perkowitz, S. (2011) Galileo Through a Lens: Views of His Life and Work on Stage and Screen, in Corsini E.M. *The Inspiration of Astronomical Phenomena VI. Proceedings of a conference held October 18–23, 2009 in Venezia, Italy*. ASP Conference Series, Vol. 441. San Francisco: Astronomical Society of the Pacific.

Reeves, E. (2014) Astrology and Literature, in *A Companion to Astrology in the Renaissance*. Ed. B. Dooley. Leiden: Brill.

Rodda, G. (2016) Liquore di pianeti e rugiade di stelle fisse: commentando l'Astrologo dellaportiano, in Baldassarri, G. (ed.) *I cantieri dell'italianistica. Ricerca, didattica e organizzazione agli inizi del XXI secolo*. Roma: Adi Editore.

Rowan–Robinson, M. (2015) Shakespeare's Astronomy. *http://astro.ic.ac.uk/public/mrr/Shakespeare.astron.pdf.*

Shepherd–Barr, K. (2006) *Science on Stage*. Princeton: Princeton University Press.

Szondi, P. (1956) *Theorie des Modernen Dramas*. Frankfurt: Suhrkamp Verlag.

Usher P. (2006) *Hamlet's Universe*. San Diego: Aventine Press.

Zehelein, E. S. (2009) *Science: Dramatic. Science Plays in America and Great Britain, 1990–2007. American Studies*, Volume 180, Heidelberg: Universitatsverlag Winter.

23

THE ORIENTATION OF THE MITHRAEA IN OSTIA ANTICA

Silvia Sclavi[1], Marzia Monaco[1], Flavio Carnevale[1], Marcello Ranieri[2], Silvia Gaudenzi[1,2], Francesco Vito Polcaro[2,3] and Claudia Scatigno[1]

[1]*La Sapienza University of Rome*
[2]*INAF-IAPS Rome*
[3]*ACHe, University of Ferrara*

Corresponding author: Marzia Monaco (marzia.monaco@uniroma1.it)

ABSTRACT

We conducted an investigation on the orientations and geometrical content of the Mithraea of Ostia Antica. A geometrical CAD study based on the most reliable plans from the Archives of the *Sovrintendenza Speciale per il Colosseo, il MNR e l'Area Archeologica di Roma – Ostia Antica,* aside with the identification of geometrical and numerical schemes and associated length units, allowed us to identify the axes of the Mithraea. Then the orientations of the axes were measured on field by means of a professional compass with ± 0.5° uncertainty. The distribution of these measured azimuths follows the topography of the city. On the contrary, the *Decumanus,* the main street of the town, presents an indubitable orientation toward the Winter Solstice Sunset. This was ascertained by measuring the coordinates of four significant points along the street with the help of a palmar Trimble GPS which averages 100 measures per point.

The exceptional density of Mithraea in Ostia led us to suppose that a symbolic cosmic-solar value is to be searched in the orientation as a whole of the town itself, founded half a millennium before the first presence of Mithraism in Rome. The *Decumanus* could be considered the very holder of this symbolic value with its orientation toward the Winter Solstice Sunset, possibly making of Ostia a very special place for Mithraicism.

KEYWORDS: Ostia Antica, Mithraicism, Winter Solstice, Decumanus, Orientation of Ancient Shrines, Ancient Geometry, Triads of Integers, Roman Length units.

1. INTRODUCTION

Mithra is the god of a religion of salvation and of a mystery cult widely spread throughout the Roman Empire from the end of the 1st Century AD until the destruction of Mithraea after the establishment of Christianity as the state religion of the Roman Empire in the second half of the 4th century AD.

A cosmological cult of a god Mithra is also attested from the 2nd millennium BC in India, Iran and the Near East. Many historians of religion perceives a continuous line from the God of the Vedas and the Indo-European area of the 2nd millennium BC to the Roman Mithra (see, e.g. Turcan, 2000; Ries 2004; Sauer, 2012 and references therein). However, Beck (1998; 2006), introducing a new, hermeneutical method to study the mysteries of Mithra, warns against exaggerating the importance of the Iranian inheritance on the Roman Mithraism.

Whatever was its origin, the mystery cult of Mithra, in the form in which it has developed in the Roman Empire, is a cosmic religion. It aim to join the Earth and the Sky with an astrological doctrine that explains the universe with the devotees ascending to the divine world and win immortality with a series of ritual practices (Merkelbach 1984). It is therefore not surprising that this cult is full of astronomical symbolism. Many are the astrological references in the decorations of Mithraea. In the Mithraeum of the *Sette Sfere* (Seven Spheres) in Ostia a representation of a cosmic geography has been recognized, where the seven planets (as in the Mithraeum of *Felicissimo*) and the twelve signs of the zodiac are represented (Gordon, 1976). It has been suggested that each of the seven spheres drawn on the mosaic floor corresponds to a step of the hierarchy that characterized the initiatory path but also to a step of the ladder that through the seven celestial spheres leads to Heaven (Gordon, 1976).

This strong influence of astrology is certainly related to the influence of the neo-Pythagorean and Platonic doctrines, very popular in the Roman Empire in the first centuries of our era (Turcan, 1975; Ries, 2004). Following Beck (2006), the Mithraeum is a model of the Graeco-Roman universe as it was conceived in Platonic philosophy: the Mithraeum allowed the initiate to experience once again the descent of his soul from the heavens and to experience its exit back to the heavens.

The same Tauroctony, focal point of all Mithraea in the Roman Empire, has a cosmological content: the sacrifice of the bull is the saving act that generates the cosmos and all forms of life (Ries, 2004 and references therein). The Sun and the Moon appear at the edges of the representation of the Tauroctony. Sometimes, the signs of the zodiac are above the scene.

In a rich epigraphic documentation, we find also the dedicatory formula "*Dei Soli invicto Mithrae*" and surely Mithraism was advantaged by the popularity of solar cults in imperial times (Ries, 2004).

Ostia Antica has the highest known concentration of Mithraea. Because of this reason, as part of a broader research on the architectural planning of the Mithraea in Ostia Antica (Gaudenzi et al., 2012), we have performed on field measurement of their orientation, searching for a possible astronomical symbolism.

2. EXPERIMENTAL METHOD AND DATA ANALYSIS

At first we performed a geometrical CAD study on the most reliable plans from the Archives of the *Sovrintendenza Speciale per il Colosseo, il MNR e l'Area Archeologica di Roma - Ostia Antica* for the 16 presently known Mithraea of Ostia (Figure 1). For each of these Mithraea we identified the geometrical and numerical scheme, the associated length units and the axis.

Then, with the collaboration of the *Soprintendenza Speciale per il Colosseo, il MNR e l'Area Archeologica di Roma - Ostia Antica,* we measured in Ostia Antica the orientation of 15 Mithraea (being not accessible the 16th, the *Aldobrandini* one) and of the *Decumanus,* that is the leading directional element of the Roman street system (*Figure 4*).

We successively performed a DBSCAN Cluster Analysis on the measured azimuth data of the Mithraea.

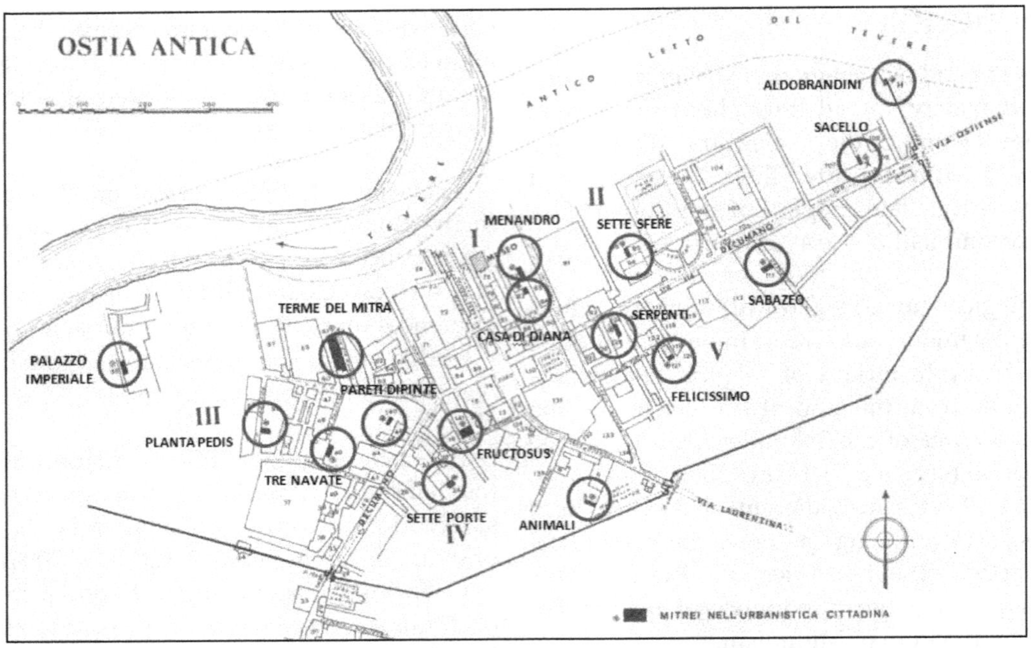

Figure 1. The positions of the 16 Mithraea in Ostia Antica

3. GEOMETRICAL ANALYSES

This type of analysis has already demonstrated a valid and suited tool to arrive at the geometrical features of ancient architectural structures: the intended geometric forms, the length-units used at the ancient building yard and the associated integer numbers describing the lengths. In particular, the squaring of rectangular parts is attributed to those combinations of small integer numbers that are suited to determine a right angle (that is, the Pythagorean Triads of integers) or an "almost right" angle, that is, sufficient for building purposes (Quasi-Pythagorean Triads). A complete explanation of the method and its background can be found in Ranieri (1997). A list of these combinations is reported in table I.

Table I. Pythagorean and Quasi-Pythagorean Triads of integers

Symbol	TRIAD (A-B-C)			r (B/A)	α (° ')	
Q	5	5	7	1.0	1	9
Q	7	7	10	1.0	1	10
Q	12	12	17	1.0	0	12
Q	17	17	24	1.0	0	12
V	20	21	29	1.05	0	0
VA	15	16	22	1.0667	0	21
VB	13	14	19	1.0769	0	38
VC1	10	11	15	1.1	1	2
VC2	18	20	27	1.1111	0	24
S	8	9	12	1.125	0	24
SA	13	15	20	1.1538	0	53
SB	19	22	29	1.1579	0	16
SC	11	13	17	1.1818	0	12
W/2	20	24	31	1.2	0	54
SD1	14	17	22	1.2143	0	7
SD2	17	21	27	1.2353	0	5
3/W	20	25	32	1.25	0	3
D	3	4	5	1.3333	0	0
LDR	19	27	33	1.4205	0	3
LD	16	23	28	1.4375	0	5
G/2	13	19	23	1.4615	0	7
L	10	15	18	1.5	0	11
LA	17	26	31	1.5294	0	16
π/2	7	11	13	1.5714	0	22
ΦA	10	16	19	1.6	0	54
ΦP	28	25	53	1.6071	0	0
ΦB	11	18	21	1.6364	0	35
LVC	14	23	27	1.6429	0	21
GA/2	7	12	14	1.7143	1	1
4/W	18	30	35	1.6666	0	3
2/SB	11	19	22	1.7273	0	16
√3	15	26	30	1.7333	0	4
MA	4	7	8	1.75	1	1
MB	12	22	25	1.8333	0	20
M	8	15	17	1.875	0	0
MC	12	23	26	1.9167	0	19
2Q	4	8	9	2.0	0	54
2Q	9	18	20	2.0	0	53
2Q	13	26	29	2.0	0	20
2QA	13	27	30	2.0769	0	10
2QB	9	19	21	2.1111	0	10
2VC1	5	11	12	2.2	1	2
√5	9	20	22	2.2222	0	29
WA	10	23	25	2.3	0	30
W	5	12	13	2.4	0	0
WB	10	25	27	2.5	0	28
WC	11	28	30	2.5455	0	28
Φ^2	5	13	14	2.6	0	53
√7	11	29	31	2.6364	0	5
2D	6	16	17	2.6667	0	54
√8	6	17	18	2.8333	0	17
G	12	35	37	2.9166	0	0
3Q	6	18	19	3.0	0	16
π	7	22	23	3.1429	0	45
√10	6	19	20	3.1667	0	45
2LVC	7	23	24	3.2857	0	21
GA	7	24	25	3.4286	0	0

Examples of this type of analyses can be found, among others, in Ranieri (2000) for prehistoric Turkmenistan, in Malgora (2001) for ancient Egypt, in Ranieri (2002) for ancient Sardinia, in Ranieri (2005; 2011) for Etruscans, in Ranieri (2007) for the Myceneans, in Ranieri (2008b) for Mesosoamerica, in Ranieri & Polcaro (2009) for the Levant, in Ranieri (2014) for Greek temples, in Labianca et al. (2008), Ranieri & Monaco (2014), Monaco et al. (2010) for Roman architectures.

We would like to stress that the technique to use Pythagorean (or Quasi-Pythagorean) Triads of integers to obtain right angles in buildings applies to all ancient civilizations (and it is known to have been in use since recent times) and therefore cannot be considered as typical of Mithraea.

Figure 2 shows, for eight only of the 16 Mithraea, the obtained geometrical schemes, numerical combination and related length units. The large majority of the Mithraea are constructed inside re-adapted spaces. Nonetheless, as it can be seen in Figure 2 and in Table II, the very space of the Mithreum was carefully designed, for all the 16 Mithraea, with the use of precise squaring combinations of numbers (Pythagorean and Quasi-Pythagorean triads of integers) by utilizing the same length units, that is Roman feet (pr), cubits (cr) and palms (pa =1/2cr). Of course, the finding of Roman length units is to be considered as normal.

The results of the geometrical CAD analysis are reported in Table II.

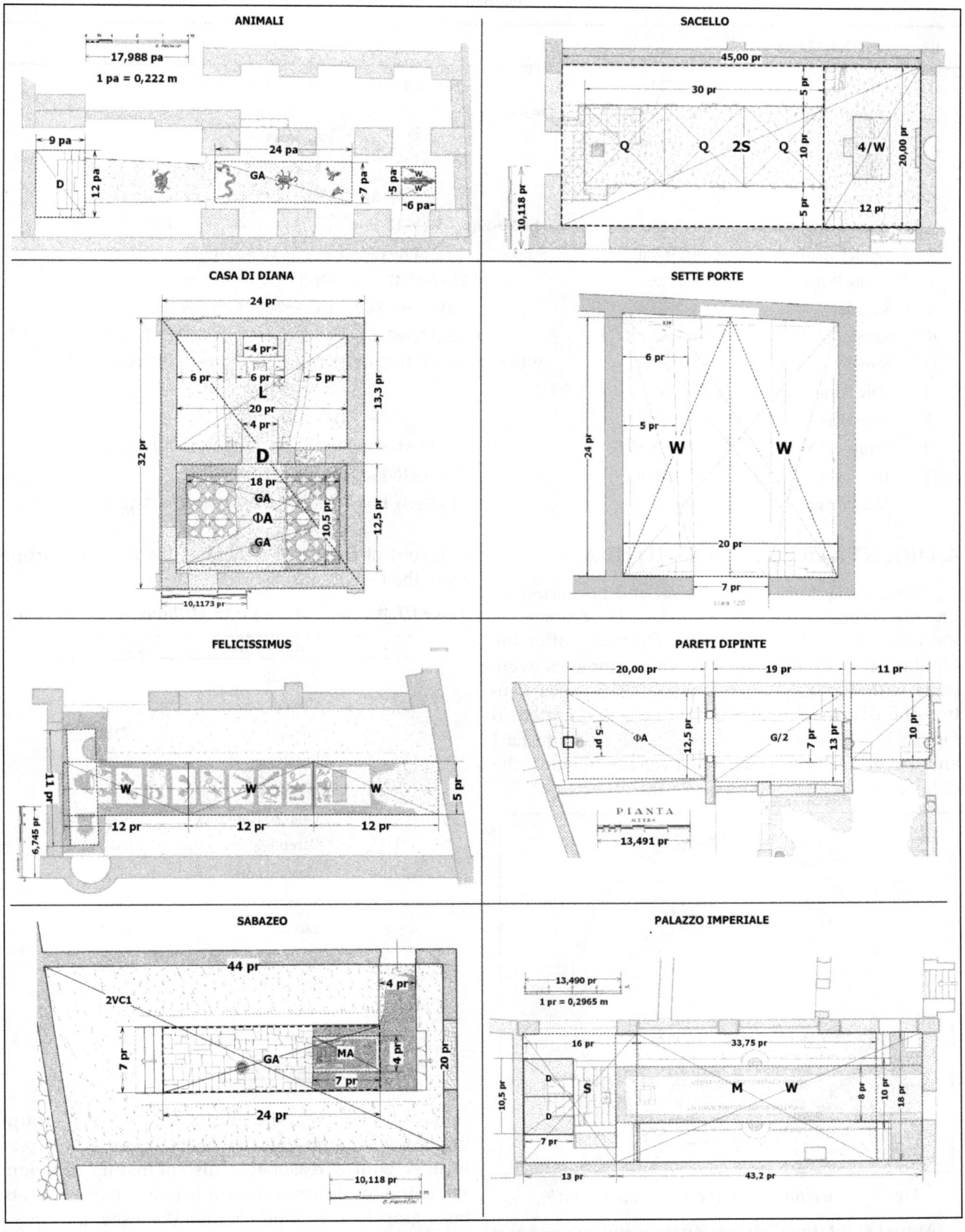

Figure 2. Geometrical schemes, numerical combinations and related length units resulting for some Mithraea (in Table II the complete results)

Table II. Pythagorean and Quasi-Pythagorean Triads of integers identified in the Mithraea and the corresponding length units

ID	MITHRAEUM	LENGTH UNIT (±0.0005)	TRIADS
1	Animali	pa=0.2220m	**D=3-4-5** (x3), **GA=7-24-25**, **W=5-12-13** (x1/2)
2	Casa di Diana	pr=0.2965m	**D=3-4-5** (x9), **L=10-15-18**(x4/3), **ΦA=10-16-19**(x5/4), **GA=7-24-25**(x3/4)
3	Felicissimo	pr=0.2965m	**W=5-12-13**
4	Fructosus	pr=0.2965m	**D=3-4-5** (x5)
5	Menandro	pr=0.2965m	**D=3-4-5** (x3), **MA1=15-27-31**
6	Palazzo Imperiale	pr=0.2965m	**S=8-9-12**(x2), **W=5-12-13** (x18/5), **M=8-15-17**(x9/4), **D=3-4-5** (x7/4)
7	Pareti Dipinte	pr=0.2965m	**ΦA=10-16-19**(x5/4), **G/2=13-19-23**
8	Planta Pedis	pr=0.2965m	**2D=6-16-17**, **18-39-43**(x1/3), **2M=8-30-31**(x1/2)
9	Sabazeo	pr=0.2965m	**2VC1=5-11-12**(x4), **GA=7-24-25, MA=4-7-8,**
10	Sacello	pr=0.2965m	**2S=24-54-59**(5/6), **Q=10-10-14, 4/W=36-60-70** (x1/3)
11	Serpenti	pr=0.2965m	**WB=10-25-27** (x8/5), **Q=12-12-17**(x4/3), **Q=7-7-10, L=10-15-18**(x2/5)
12	Sette Porte	pr=0.2965m	**W=5-12-13** (x2)
13	Sette Sfere	pr=0.2965m	**D=3-4-5** (x2)
14	Terme del Mitra	pr=0.2965m	**W=5-12-13, M=8-15-17**(x3/2), **G=12-35-37, D=3-4-5** (x3)
15	Tre Navate	pa=0.2220m	**L=10-15-18**(x2), **MC=12-23-26**(x5/2)
16	Aldobrandini	cr=0.4440(2/5)	**D=3-4-5** (x4), **D=3-4-5** (x3), **GA/2=7-12-14, Q=7-7-10**

4. ORIENTATION OF THE MITHRAEA

Being Ostia Antica at sea level and bordering on the sea, azimuth data are sufficient to characterize the orientation of the Mithraea. On place, after the identification of the axes, the measurements were done with a professional compass (±0.5° uncertainty). The distribution of the azimuths is reported in Figure 3. Azimuths are intended from North to East and are meant from the interior (altar) outward: the obtained values are reported in Table III.

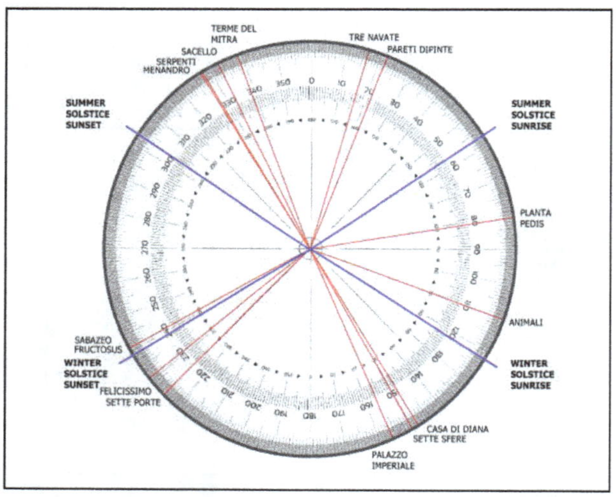

Figure 3. Distribution of the Mithraea azimuths.

At first sight, the distribution does not present any preferential orientation. In fact, the orientations of the Mithraea are bound to the orientations of the pre-existing structures, which follow the topographic layout of the street system of the town departing from the *Decumanus*.

Table III. Azimuth values for 15 Mithraea of Ostia Antica.

ID	Mithraeum	Azimuth (°)
1	Animali	110.0
2	Casa di Diana	148.0
3	Felicissimo	231.0
4	Fructosus	241.0
5	Menandro	327.5
6	Palazzo Imperiale	156.0
7	Pareti Dipinte	22.0
8	Planta Pedis	81.0
9	Sabazeo	241.0
10	Sacello	333.0
11	Serpenti	328.0
12	Sette Porte	225.0
13	Sette Sfere	150.0
14	Terme Del Mitra	339.0
15	Tre Navate	16.5

In order to reveal the presence of possible grouping in the azimuth data, a Cluster Analysis was performed with a reasonable discriminating criterion: 10° is the maximum distance between two elements belonging to the same cluster. We used the open-source MATLAB script made by Yarpiz (2015)[1] as an implementation of the DBSCAN data clustering algorithm proposed by Ester et al. (1996). From our

[1] www.yarpiz.com/255/ypml110-dbscan-clustering

analysis four different groups emerge: the first between 225° and 241° corresponding to Mithraea 3, 4, 9, 12; the second between 327° and 339° corresponding to Mithraea 5, 10, 11, 14; the third between 16° and 22° corresponding to Mithraea 7, 15; the fourth between 148° and 156° corresponding to Mithraea 12, 13, 6.

Further studies will investigate if these groupings can be related to possible hierophanies inside the mithraea, such as, for example, the one discovered in the so-called "Neo-Pythagorean Basilica" of Porta Maggiore in Rome (Labianca et al., 2008), produced by alterations to the building design (such as opening of light shafts or similar). Level of preservation does not allow this study in most cases. However, this possibility is worth to be further analyzed.

5. ORIENTATION OF THE *DECUMANUS*

The main part of the *Decumanus* extends from the entrance to the older *Porta Marina* of the *Castrum* where it bifurcates in two branches. The upper branch is supposed that brought to the ancient fluvial harbour (Goiran et al., 2014) while the other to the coastal *Via Severiana* (Fogagnolo and Valenti, 2005; Bruschi, 2015). We performed a series of measurements with a palmar Trimble GPS (which averages 100 measures per point) along the main part of the *Decumanus*. As the *Decumanus* is not a unique straight line, the coordinates of four points were measured to take account of this (Table IV).

Table IV. Coordinates of the four points along the Decumanus as taken with the averaging GPS

	Latitude N			Longitude E		
Point	Degr.	Min.	Sec.	Degr.	Min.	Sec.
P1	41	45	24.96	12	17	40.98
P2	41	45	20.22	12	17	29.76
P3	41	45	16.32	12	17	21.30
P4	41	45	13.26	12	17	14.70

The first point (P1) corresponds to the *Porta Romana* entrance; the second (P2) is in correspondence of the theatre, the third and the fourth (P3 and P4) are at the known entrances of the ancient *Castrum* (Figure 4), the first settlement of the colony of Ostia. The bearings of the three segments are reported in Table .

By means of Stellarium[2] we determined the Solsticial azimuth for year 400 BC:

Winter Solstice Sunset (WSSS) 237.941°
Summer Solstice Sunrise (SSSR) 56.746°

As it can be seen in Table , the bearings of the three segments between the four points diverge from the bearing of WSSS from a maximum value of 2° 32′ 5″ to a minimum of a mere 0° 11′ 53 for P3-P4 just corresponding to the *Castrum*.

The Cluster Analysis performed on the Azimuth data of the Mithraea confirms that their orientations follow the topographic layout of the street system. In facts, the first group (the one between 225° and 241°) point out that *Fructosus* and *Sabazeum* orientation differs very little from the *Decumanus* bearing, while the third and fourth groups are orthogonal to it. The alignments of the second group appear to conform to the direction of the deviated portion of the *Decumanus* toward *Porta Marina*.

Table V. Bearings of the three segment between the four points: Castrum corresponds to P3-P4

SEGMENT	BEARING	DIFF. (WSSS)	DIFF. (SSSR)
P1-P2	240.4756	2° 32′ 5″	3° 43′ 47″
P2-P3	238.2835	0° 20′ 33″	1° 32′ 15″
P3-P4	238.1389	0° 11′ 53″	1° 23′ 35 ″

The obtained results are not surprising because solar orientations are not uncommon in Roman *Castra* (see, e,g., Magli, 2008; Gonzales Garcia & Magli, 2014; Bertarione, 2014), although in many cases it appears that the practical mentality of the Roman overruled the symbolic principles. The winter solstice night was of particular importance in the Republican era, being the central point of the festivities dedicated to *Sol Indiges*, the archaic solar god (Torelli, 1990).

It is manifest that this orientation, in itself, has nothing to do with Mithraism, since Ostia was founded at the beginning of the 4th Century BC, while the Mithraism spread in Rome half a millennium later. However, we will suggest in the conclusion why this circumstance should have some interest in the present context.

6. DISCUSSION AND CONCLUSIONS

Our study on the Mithraea of Ostia Antica shows that:

- Pythagorean and Quasi-Pythagorean triads of integers were used in the arrangement of the available space for the Mithraea;
- all the Mithraea are based on the same length units;
- the distribution of the axes does not present any preferential orientation but follow the topographic layout of the street system built departing from the *Decumanus*.
- Finally, the *Decumanus* of Ostia Antica is oriented toward the Winter Solstice Sunset.

As conclusion, we would suggest a few considerations justifying an apparent higher density of Mithraea in Ostia Antica with respect to Rome.

[2] www.**stellarium**.org/it/

Between the Second and the Third Century AD, the epoch of its maximum expansion, Ostia reached a population of 50000 people, while it is evaluated that in the same epoch Rome had from 450000 to one million inhabitants (Storey, 1997; Dyson, 2010).

We know that at least 16 Mithraea were present in Ostia, but, following Vermaseren (1956), two further mithraea were excavated in Ostia during the 19th Century and are presently lost. Bakker (1994) suggests a total of 24 mithraea in Ostia. Considering that a community of about 20 believers used each Mithraeum (Ries, 2004), these figures suggest a diffusion of the Mithraism on about 0.65% - 1% of the population, a percentage that is not far from the estimate given by Ries (2004) from historical and archaeological evidences. By assuming that the same percentage of Mithraicism believers was present in Rome at the same epoch, they should have been between about 4500 and 10000 and use more than 200 Mithraea. At present, the estimate of the number of Rome's Mithraea by various scholar varies between 700 and 2000 (Dyson, 2010), depending from the assumed diffusion of this religion.

However, we presently know only nine Mithraea in Rome, while hints from various sparse archaeological findings suggest the presence of a few tens of others (40, following Dyson, 2010). We have to increase fivefold this number in order to reach the minimum estimate of Mithraea in Rome. Considering that the area of the city of Rome is the subject of intensive excavations, for various reasons, all supervised by law by an archaeologist, it is very unlikely that still at least 150 undiscovered mithraea are present in Rome.

Figure 4. The solsticial orientation of Ostia Antica. P1: porta romana; P2: the theatre; P3 and P4: the entrances of the Castrum.

The relatively smaller number of mithraea in Rome with respect to Ostia should be explainable by a number of hypothesis:

- The diffusion of Mithraicism in Ostia was wider because of sociological reasons. Actually, Ostia population included some seventeen thousand slaves (Dyson, 2010), mostly coming from Near East, the possible area of origin of Mithraicism. However, the percentage of slaves on the Rome population was more or less the same, as well as their origins.
- Mithra mysteries were particularly diffused in Roman Army and Ostia have a strong presence of the Roman military fleet. On the other hand, Rome too had, at the same epoch, an extremely high military presence.

- The destruction of Mithraea when the Christianism became the state religion of Roman Empire was more intense and effective in Rome than in Ostia. Actually, Rome was the major center of the Christianism and it is quite possible that the fighting of the pagan cults was stronger than elsewhere. However, Ostia is just a 20 km from Rome and there is no reason to assume that there was there a higher tolerance for the Mithra cult.

All these factors could have contributed to left such a higher presence of mithraea in Ostia with respect to Rome, but cannot totally explain it.

We thus suggest a further factor.

It has been reported that Mithraea in the northernmost regions of the Empire are generally East-West oriented (Campbell, 1968; Merkelbach, 1984). Madlenović (2009) have interpreted the identical orientation of two Roman mausoleums at Gamzigrad and Šarkamen in Dacia as related to Orion, to the Winter Solstice and in turn to Mithras. Beck (1976) stress the role of the East-West orientation of the mithraeum in Ponza to understand the symbolism of the zodiac represented on the ceiling of the cult niche.

In Rome and Ostia, portions of free land for constructing Mithraea *ex novo* were not easily (and/or cheaply) available. Mithraea occupy thus readapted spaces in pre-existing buildings, thus following the orientation of the streets bound to the topographic pattern. Rome had at that epoch (and still has) a street system that is randomly oriented and centred on the *Capitolium*. On the contrary, the Mithra believers found in Ostia a solar orientation, being the town developed, since its foundation, along the *Decumanus* oriented along the Summer Solstice Sunrise - Winter Solstice Sunset direction. Actually, three of the 15 measured mithraea have their axis approximately oriented along this direction, while 7 other have their axis approximately at 90 degrees from the solstitial line: these orientations could have made easier to build an opening in the right position to generate hierophanies inside the mithraea at the solstices.

However, it should also be possible that placing a Mithreum in Ostia, no matter which his specific orientation, could have had a symbolic value just for the symbolic orientation of the whole town.

ACKNOWLEDGEMENTS

The authors would like to thank the Dr. Dragana Mladenivic (Southampton University) for useful comments and suggestions and all the personnel of the *Soprintendenza Speciale per il Colosseo, il MNR e l'Area Archeologica di Roma – Ostia Antica*, especially the following Doctors, Dr. Angelo Pellegrino, Dr. Maria Rosaria Barbera, Dr. Cinzia Morelli, Dr. Paola Germoni, Dr. Flora Panariti, Dr. Orietta Mantovani, Dr. Marco Sangiorgio, Dr. F. Pandolfi, Dr. A. Marano, Dr. M. Mola for their support and technical cooperation.

REFERENCES

Bertarione S. (2014), *Ecco come nasce la Roma delle Alpi*, Archeologia Viva, XXXIII, pp. 70-72.

Bakker, J.T. (1994), *Living and working with the gods: studies of evidence for private religion and its material environment in the city of Ostia (100-500 AD)*, Warszawa, JC Gieben.

Beck, R. (1976), Interpreting the Ponza Zodiac: I. *Journal of Mithraic Studies* 1, pp. 1-19.

Beck, R. (1998), The Mysteries of Mithra: a New Account of their Genesis, *Journal of Roman Studies*, 88, pp. 115-128.

Beck, R. (2006), *The Religion of the Mithras Cult in the Roman Empire: Mysteries of the Unconquered Sun*, Oxford, Oxford University Press.

Bruschi A. (Ed.) (2015), *Portus, Ostia Antica, via Severiana. Il Sistema archeologico paesaggistico della linea di costa di Roma imperiale*, Roma, DiAP Print.

Campbell L.A. (1968), *Mithraic Iconography and Ideology*. Leiden, Brill.

Dyson S. L. (2010), *Rome a living portrait of an Ancient City*, Baltimore, The John Hopkins University Press.

Ester M., Kriegel H.P., Sander J., Xu X. (1996), *A density-based algorithm for discovering clusters in large spatial databases with noise.* In Simoudis E., Han J., Fayyad U. (eds.). Proceedings of the Second International Conference on Knowledge Discovery and Data Mining (KDD-96). Portland, AAAI Press, pp. 226–231.

Fogagnolo S., Valenti M. (2005), *Via Severiana*, Roma, Istituto Poligrafico e Zecca dello Stato.

Gaudenzi, S., Preite Martinez, M., Bernardini, M.L., *et al.*, (2012), *Studio interdisciplinare sui mitrei di Ostia Antica*, In V.F. Polcaro (ed.) Atti del XII Convegno SIA, Albano Laziale, Italy, 5-6 October 2012, Napoli, La Città del Sole, pp. 201-217.

Goiran J.P., Salomon F., Mazzini I., Bravard J.P., Pleuger E., Vittori C., Boetto G., Jonatan Christiansen J., Arnaud P.,Pellegrino A., Pepe C., Sadori L. (2014), Geoarchaeology confirms location of the ancient harbour basin of Ostia (Italy), 41, pp. 389-398.

Gonzales Garcia C. & Magli G. (2014), *Roman City Planning and Spatial Organization,* in Clive Ruggles (ed.) Handbook of Archaeoastronomy and Ethnoastronomy, pp. 1643-1650.

Gordon R.L. (1976), *The sacred Geography of a Mithreum: the example of Sette Sfere*, Journal of Mithraic Studies, 1, pp. 119-175.

Labianca L., Sciortino I., Gaudenzi S, Patané A., Polcaro V.F., Ranieri M. (2008), *An archaeoastronomical study of the neo-Pythagorean basilica at Porta Maggiore in Rome,* Proceedings of the SEAC 2007 and Oxford VIII Conference Astronomy and cosmology in folk tradition and cultural heritage, Archaelogia Baltica, Klaipeda University Press, 10, 195-199.

Madlenović D. (2009), Astral Path to Soul Salvation in Late Antiquity? The Orientation of Two Late Roman Imperial Mausolea from Eastern Serbia, *American Journal of Archaeology*, 113, pp. 81-97.

Magli G. (2008), On the orientation of Roman towns in Italy, *Oxford J Archaeol*, 27, pp. 63–71.

Malgora S. (2001), *L'uso dei numeri e la ritualizzazione delle strutture cerimoniali nella topografia monumentale di Saqqara,* Thesis, Bologna University.

Merkelbach R.(1984), *Mithras*, Kőnigstern Verla Anton Hain Miesenheim GmbH.

Monaco M., Gaudenzi S., Ranieri M., (2010), *Among the circles: a geometrical analysis of the Teatro Marittimo in Villa Adriana,* Proceedings of the 18th SEAC 2010 Conference Astronomy and Power, Gilching, Germany, in press.

Ries J. (2004), *Il culto di Mitra dall'India vedica ai confine dell'Impero Romano*, Milano, Jaka Book.

Ranieri M. (1997), Triads of Integers: how space was squared in ancient times, *Rivista di Topografia Antica Journal of Ancient Topography*, VII, pp. 209-244.

Ranieri M. (2000), *The Development of Architectural Forms of Prehistoric Turkmenistan,* South Asian Archaeology, Proceedings of the 14th International Conference of the European Association of South Asian Archaeologists, Roma, Is.I.A.O., II, pp. 593-615.

Ranieri M. (2002), *The stone circles of Li Muri: geometry alignements and numbers,* Proceedings of the 14th SEAC Conference Ligths and Shadows in Cultural Astronomy, Isili, Sardinia, pp. 58-67.

Ranieri M. (2005), *La geometria della pianta del tempio urbano di Marzabotto (Regio I – Ins. 5), in* Culti, Forma Urbana e Artigianato a Marzabotto. Nuove prospettive di ricerca, Ante Quem, Dipartimento di Archeologia Università di Bologna, Bologna University Press, 11, pp. 73-87.

Ranieri M. (2008), *A geometrical analysis of Mesoamerican pre-hispanic architecture: squaring triads, numbers, length units and the calendar,* Proceedings of the SEAC 2007 and Oxford VIII Conference Astronomy and cosmology in folk tradition and cultural heritage, *Archaeologia Baltica,* Klaipeda University Press, 10, pp. 183-194.

Ranieri M. (2008b), *The Pythagorean geometry of the Atreus tholos tomb at Mycenae,* Proceedings of the SEAC 2007 and Oxford VIII Conference Astronomy and cosmology in folk tradition and cultural heritage, *Archaeologia Baltica*, Klaipeda University Press, 10, 2pp. 11-213.

Ranieri M., Polcaro A. (2009), *Early Bronze Age geometry in the Dead Sea region at Tel Arad and Bab edh-Dhra' numerical squaring, harmonic compositions and length–units,* Proceedings of the 17th SEAC 2010 Conference From Alexandria to al-Iskandariya: Astronomy and culture in the ancient Mediterranean and Beyond, Alexandria, Egypt, in press.

Ranieri M. (2011), *Himera and Pyrgi: the diagonals and the alignement of the temples,* Proceedings of the 19th SEAC 2011 Conference Stars and Stones: Voyages in Archaeoastronomy and Cultural astronomy, *BAR International Series* 2015, Oxford, 2720, pp. 94-98.

Ranieri M. (2014), *Digging the archives: the orientation of Greek temples and their diagonals,* Proceedings of the 21st SEAC 2013 Conference Astronomy: Mother of Civilization and Guide to the Future, Athens, *Mediterranean Archaeology & Archaeometry*, 14, 3, pp. 165-174.

Ranieri M., (2014), *Geometrie Pitagoriche e circolari nella pianta del Pantheon adrianeo,* Proceedings of the 11th S.I.A. 2011 Conference Il Dentro e il Fuori del Cosmo, Bologna, pp. 15-27.

Sauer, E. (2012), *Mithras and Mithraism* in The Encyclopedia of Ancient History, Oxford, Jhon Wiley and sons pp. 4551-4553.

Storey G. R. (1997), The population of ancient Rome, *Antiquity*, 71, pp. 966-978.

Torelli, M. (1990), Riti di passaggio maschili di Roma arcaica. *Mélanges de l'Ecole française de Rome. Antiquité*, 102 (1), pp. 93-106.

Turcan R. (1975), *Mithras Platonicus*, Leiden, Brill, coll. Études Préliminaires aux Religions Orientale dans l'Empire Romaine. 47.

Turcan, R. (2000), *Mithra et le mithriacisme*, 2nd edn., rev., Paris. Les Belles Lettres.

Vermaseren, M.J. (1956), *Corpus Inscriptionum et Monumentorum Religionis Mithriacae*, The Hague, Martinus Nijhoff.

24

THE GREAT COMET OF 1858: A ROAD SIGN TO THE STARS

Christiaan Sterken

University of Brussels, Pleinlaan 2, 1050 Brussels, Belgium

**Corresponding author: Christiaan Sterken (csterken@vub.ac.be)*

ABSTRACT

During an expedition to the Texas–Mexican border area in September–October 1858, the Belgian astronomer Jean-Charles Houzeau (1820–1888) admired "a beautiful comet". At that moment, he was not aware that the comet had already been discovered in June 1858 by Giovanni Battista Donati. Houzeau sent detailed descriptions of his whereabouts to his family and to a colleague at the University of Brussels.

Donati's comet, with its curved tail extending 40 degrees across the sky, became a big news event, and inspired visual artists worldwide. Many paintings and sketches were produced. Some of these artworks show quite literal transcriptions of the surroundings, and even have scientific overtones, whereas others are more artistic than exact. The comet also inspired jewelry artisans, and poets: some lyric poems include — just like some artworks do — elements that refer to mid-nineteenth century scientific developments.

A basic ephemeris analysis shows that the cometary passages allegedly attributed to the same comet in the time span 104–1858 AD cannot be assigned to a single comet. An analysis of a drawing by William Hayes Hilton leads to the conclusion that the artist could not possibly have painted the landscape and the starscape at the same moment and at the date suggested by the position of the comet with respect to the star Arcturus. A similar conclusion is reached for a painting by British artist William Dyce. The artistic representations of this comet indirectly contributed to familiarization of the mid-nineteenth century public with some specific stars and constellations. As such, Donati's comet showed the road to the stars.

KEYWORDS: Comet Donati, Comet Tuttle, Comet Halley, nineteenth century literature and art, Jean-Charles Houzeau, comet mania, Abraham Lincoln

1. INTRODUCTION

Jean-Charles Houzeau de Lehaie (1820–1888) was a Belgian astronomer at the Royal Observatory of Belgium. During the social upheavals of 1848[1], Houzeau took a republican stand and was dismissed. He then left Belgium and travelled around Europe, working in various libraries and archives. In 1854 he was recalled to Belgium for survey work on the triangulation of the Kingdom, but when this project was interrupted in 1857 he left for New Orleans, in search for his revolutionary dream of freedom.

His principal reasons for leaving his native country were his dissatisfaction with Belgian politics, his views that Europe was too parochial, and his desire to discover a country with more individual freedom and with a more egalitarian society. Houzeau adhered to Fourierism, i.e., the set of ideas first put forward by the French utopian socialist Charles Fourier (1772–1837).

On May 24, 1858, he arrived in San Antonio, Texas, where he stayed until 1862. His original plan was to return home after several months, but he was caught in the American Civil War, and his final time overseas turned out to be almost 20 years. During that period he wrote hundreds of letters to his parents, to his spouse and to his colleagues at the university in Brussels.

In a letter to his parents, after returning from a field trip to the Rio Grande near the Mexican border, he wrote on October 17, 1858 "*I have discovered and admired a beautiful comet, that almost was lost in the fire of the sunset …*".[2] (Houzeau 1994). Two weeks later, in a subsequent letter to his parents, he mentions that the head of "*la belle comète de Donati*"[3] is near the star Arcturus, and that the tail passed over σ and ζ Bootes. In this, and in later letters, he also mentions that already in the prairie, he knew that this comet was not the comet of 1556 (C/1556 D1) dubbed "Charles V". The Great Comet of 1556 was nicknamed the comet of Charles V because, when the Emperor saw it for the first time, he cried: "*By these signs, then, my fate summons me*" (Aspaas, 2018; Heward, 1899): he regarded the comet as a command from Heaven to resign the Crown, and thus he retired to a retreat in Spain.[4]

2. COMET DONATI IN THE PRESS

The comet became a true media event in newspapers and popular magazines of the times, and entailed a fascination of the public, and even led to some comet mania. The newspaper *Daily Alta California* of October 8, 1858, though, gives a lot of exact information – including coordinates and orbital elements – that had been provided by Detroit Observatory (Ann Arbor). The newspaper explains that the comet will not be visible after October 16, that the comet visits us for the first time, that its orbit is parabolic and that the comet will thus never return, and that the comet is not a harbinger of pestilence, famine and evil.

Houzeau's preoccupancy with the issue of the identity of the comet of Charles V and Donati's comet can be easily understood. Ten years earlier, John Russell Hind (1823–1895), assuming that the comet of 1556 had already been seen in 1264, calculated that 292 years after its 1556 apparition, it would visit us again in 1848 (Hind, 1848). Hence, some people speculated that the comet of 1858 was the somewhat late Charles V comet. In 1857, a paper *The Great Comet, Now Rapidly Approaching, Will it Strike the Earth?* (Anonymous, 1857) appeared that warned for a coming "*monstrous body*". That comet was first noted in 1264, the paper said, and again in 1556, and now it would make its appearance between May 1857 and January 1858. The writer alluded to a probability of collision, and the consequent "*End of all Things*". The appearance of the comet of 1264 is connected to the death of Pope Urban IV (1195–1264). *The Illustrated London Almanack* (1847) reports, on page 65, that "*The chroniclers of the time … connect the appearance of the comet with the death of Pope Urban, who fell sick (they assert) on the very day when the comet was first seen, and died at the exact time it disappeared – viz., on the last day of September, 1264!* ".

The times of "passage" listed in Hind's paper lead to a straightforward ephemeris of that alleged comet, and yields a period of 290.3 ± 0.2 years for the six passages from 104 to 1556 AD. Figure 1 shows the seven cometary passages allegedly attributed to that same comet. The dashed line through the filled symbols is the linear ephemeris. This result looks amazingly accurate, yet it stands for one of the greatest pitfalls of ephemeris calculation: the period precision is entirely due to the long time-baseline of almost 1500 years and is totally illusory. The lower broken line through the open symbols stands for the differences between observed and calculated times of passage (the so-called *O–C* diagram). The trend of the

[1] The "revolutions" of 1848 were a series of political upheavals throughout Europe.

[2] "*… je n'ai pas été sans découvrir et sans observer une belle comète, qui se perdait presque dans les feux du couchant, …*" .

[3] *Comet C/1858 L1 was first observed by Giovanni Battista Donati (1826–1873) on June 2, 1858.*

[4] *Heward cites an anonymous English treatise of 1618 : "In the time of Charles the Emperor, surnamed the Great, a blazing star appeared, in the contemplation whereof the Emperor, having his eyes earnestly bent upon the star and considering profoundly thereupon, at length was wrapped into a great astonishment touching the significance of the same; and sending for a philosopher named Eginard, reasoned with him to and fro about the star, saying in conclusion that the appearing thereof did threaten unto him some dire calamity."*

observed *minus* calculated times of passage clearly shows that comet Donati cannot possibly be identified with any of the other comets. A totally different diagram would emerge for a truly periodic comet: Figure 2, for example, shows the *O–C* diagram for the last four perihelion passages of comet Halley. The labels refer to the observational technique employed for rediscovery and reveal that almost every point in this diagram has been obtained with a detector that differs from the one related to the previous data points. The dashed line is a third-degree polynomial fit. However, each observational entry has an accuracy that is an order of magnitude better than the preceding point, whereas the same also holds for the computational techniques. The diagram also shows the date of rediscovery (expressed in days before perihelion) and is a vivid illustration of the fact that the observational technique directly influences the computational result: the earlier the rediscovery occurs (as, for example, in 1982), the longer is the orbital arc that enters the orbit calculation and the more accurate the *C* term in *O–C*. The polynomial fit is also reproduced in Figure 1; the seemingly straight line is plotted from the moment when Halley saw "his" comet for the first time to the comet's latest apparition in 1985. The different character of the *O–C* curves for comets Halley and Donati is evident.

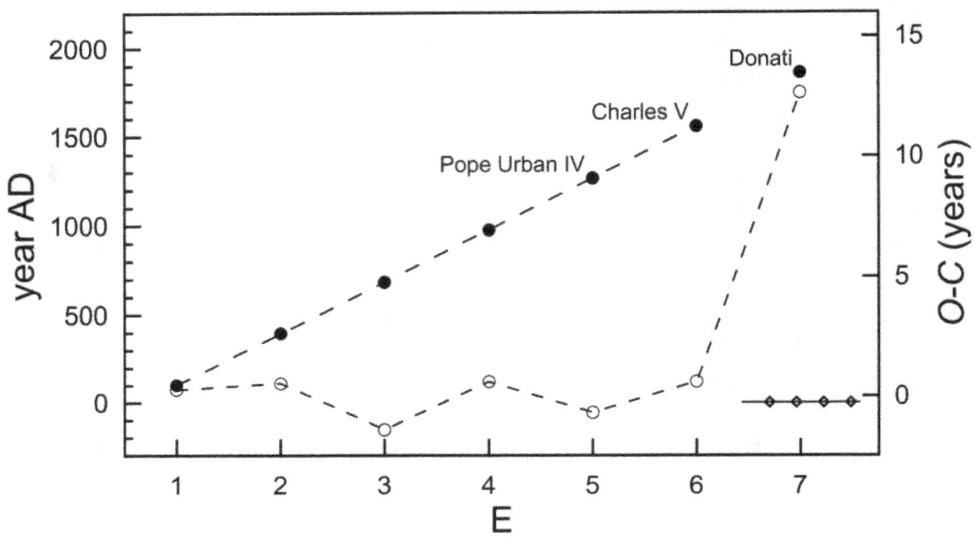

Figure 1. Seven cometary passages allegedly attributed to the same comet since AD 104, versus passage number E. The lower broken line through the open symbols refers to the vertical axis on the right, and represents the differences between observed and calculated times of passage. The four ◊ symbols are the O–C values for Halley's comet based on a linear ephemeris, and the line is the fitted polynomial of third degree in Figure 2. The line extends from 1682, when Halley saw "his" comet for the first time, to the comet's latest apparition in 1985.

New calculations of the orbit of the comet of 1556 by Martinus Hoek (1834–1873) led him to the conclusion that "*the non-identity, that now definitely follows from the non-appearance of the comet that was expected in 1858, also follows from the now obtained orbital elements*"[5] (Hoek, 1861). For a detailed discussion of the orbits of these medieval comets, see Carter (2017), who draws the conclusion that astronomers have selected only the medieval data that affirmed their presupposed theory, in fact, Halley's belief that all comets were periodic.

In a letter of July 28, 1860 to his university colleague Nicolas-Constant Schmit (1832–1879), Houzeau mentions (Schmit, 1860) that in 1857 already—thus four years before Hoek—he published his own opinion that comet Charles V was not the one seen in 1264, and that the comet that he saw the same month (discovered by H. P. Tuttle in June 1860), was also not the one of 975. Houzeau (1857) clearly stated the importance of the study of the cometary paths: "*The return of the comet of Charles V is one of those mystifications that shine with a varnish of science, but that do not have a more realistic foundation for it. It is not true that the comet observed in the times of Charles V, in 1556, is the same one that appeared in 1264. This one is also not the same as the one of 975. The hypothesis of the identity of these three stars is not sustainable from the moment when one takes the trouble to study their respective paths in the sky.*" [6]

[5] *"Die Nicht-Identität, die jetzt am sichersten aus dem Ausbleiben des im Jahre 1858 erwarteten Cometen hervorgeht, würde aus den jetzt erhaltenen Elementen ebenso gefolgert werden."*

[6] *"Le retour de la comète de Charles-Quint est une de ces mystifications qui brillent d'un vernis de science, mais qui n'en ont pas pour cela de fondement plus réel. Il n'est pas vrai que la comète observée du temps de Charles-Quint, en 1556, soit la même qui etait apparue en 1264. Celle-ci n'est pas non plus la même que celle de 975. L'hypothèse de l'identité de ces trois astres n'est pas soutenable, du moment où l'on se donne la peine d'étudier leurs marches respectives dans le ciel."*

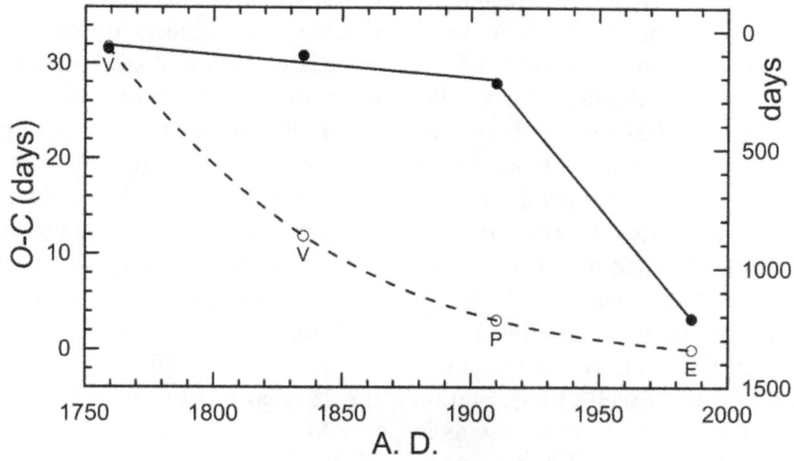

Figure 2. O–C diagram for the last four perihelion passages of comet Halley (○, left y axis). Labels V stand for visual observations, P for photographic, and E for electronographic data. The dashed line represents a polynomial of third degree. The right y axis (•) shows discovery date in days before perihelion. Full lines are linear fits. Source: Sterken (2000).

And in his letter to Schmit, Houzeau adds that *"soon the time will come that the facts will vindicate me, and I recall this note with some pleasure, because it went against the mainstream by which almost all astronomers let themselves be swept along – by laziness or by extreme precaution".*[7]

3. COMET DONATI IN VISUAL ART

Comet Donati was the first comet ever to be systematically photographed for scientific purposes. George Phillips Bond (1825–1865), at Harvard College Observatory, published a monumental treatise with dozens of drawings and engravings of the comet (Bond, 1862). Olson and Pasachoff (1998) present more than half a dozen reproductions of paintings and drawings by British artists. Many of these paintings are interesting because they show the position in the sky with respect to the constellation Ursa Major, and to Arcturus, the brightest star in the northern celestial hemisphere. Gasperini et al. (2011) outline and discuss the relations and interconnections between a scientific discovery, the artistic movements of the period, and the different social environments in a worldwide context.

A most interesting drawing was made by William Hayes Hilton (1829–1909), who made a stunning sketch of Donati's comet over a stagecoach. This sketch (Figure 3) was dated 1858, but the exact date of the drawing was not given. The picture, besides some faults in perspective, is intriguing in many aspects. The timing can be estimated from the relative position of the comet's head with respect to Arcturus: Plate XII of Bond (1862), with the comet's head West of Arcturus, was taken on October 4, and Plate XIII, with the head directly East, was taken exactly 24 hours later. Figure 4 shows the path of the comet from September 12 to October 17, 1858, as well as the path of the Moon, and the moonphases from Last Quarter to New Moon. The dashed lines are links to guide the eye. The ⊙ symbol gives the position of the Sun on October 5. The brightest stars of Ursa Major and of Bootes are also shown. The scene depicted in the drawing appears to be on a flat section of desert with a hill in the background. Ahnert (2014) situates the scene near Ewell's Stage Station (elevation 4550 ft, GPS location 32.1231, -109.6505) that was constructed about 4 miles south of present-day town of Dos Cabezas (Arizona), between Apache Pass and Dragoon Springs.

The *Daily Alta California* newspaper of November 20, 1858, cites the Arkansas *Fort Smith Times* of October 27 that mentions the arrival of the Overland Mail California at Fort Smith (AR): *"on Tuesday evening the stage that left San Francisco, on the 4th inst., arrived, with five through passengers... there were 3 through passengers from San Francisco, to wit W. H. Hilton, W. H. Ware and E. Janen."* The stage had been detained two days at Kern river (CA) by high water, and 4 hours at Boggy, in the Choctaw Nation (Atoka County, Oklahoma).

Waterman L. Ormsby, the only through passenger on the first westbound stage of the Butterfield Overland Mail gives the travel schedule for September 16 (Ormsby, 1968): the cumulative distance from San Francisco to Fort Smith is 2233.5 miles and the total travel time allowed is 485 hours or about 20 days. Taking into account the two days delay, Hilton's reported arrival time in Fort Smith was pretty exact. *The New York Times* of October 14, 1858 gives a most detailed itinerary of the 2765 mile route from San Francisco to St. Louis that was covered in 596.5

[7] *"Voilà bientôt le temps où les faits vont me donner raison, et je rappelle cette note avec quelque plaisir, parce qu'elle allait alors contre le courant où presque tous les astronomes se laissaient entraîner, – par paresse ou par extrème prudence."*

hours, thus at an average speed of 4.6 miles/hour (7.5 km/h). The stretch from Fort Yuma (CA) to El Paso (TX, formerly Franklin) via Tucson (AZ) is quite linear and yields a daily average travel distance of 100 miles. That means that the coach arrived in Dragoon Springs (32.0378, -110.0780) about 82 hours after leaving Tucson, or 12.74 days after leaving San Francisco, thus on October 17 around midnight. Since the remaining distance to Ewell's Stage Station is about 30 miles, that place was probably reached in the early morning hours of October 17.

It is evident that, if the scenery really corresponds to the Dragoon Springs-Ewell's Stage Station area, the drawing could never have been made in mid-October because at that moment the comet's head was already very low, and far away from Arcturus, with the Moon just past First Quarter. Other anomalies in Hilton's artwork are the position of the Full Moon North of Ursa Major, the location of the Milky Way, the shape of the comet's tail, and the large number of faint stars that cannot possibly be seen during the twilight hours. It is obvious that Hilton did not make his drawing *in situ*, but that he either made a sketch of the comet shortly after leaving San Francisco, and later combined it with a landscape obtained in southern Arizona, or that he copied the comet and the starry night from other sources.

The moment of the depiction, with the comet at its greatest radiance on October 5, had been much publicized because of the passage of the cometary head past Arcturus, and also because that moment brought quite a change in the orientation of the tail (Anonymous, 1858). Clerke (1893) describes this very clearly: *"But the most striking view was presented on October 5, when the brilliant star Arcturus became involved in the brightest part of the tail, and during many hours contributed, its lustre undiminished by the interposed nebulous screen, to heighten the grandeur of the most majestic celestial object of which living memories retain the impress."*

Hilton's sketch shows the consequence of constructing a painting from multiple studies over a period of time of several weeks to months. It is striking that numerous landscape painters have achieved visually accurate starscape and landscape combinations. Like, for example, James Poole (1803-1886) with his magnificently accurate large-size oil on canvas painting that shows all realistic elements related to the comet and the stars, including a reflection in a pond (Poole, 2018). There also is William Dyce's (1806-1864) *Pegwell Bay – a Recollection of October 5th 1858,* that portrays a magnificent landscape with a rather modest comet (Dyce, 1858).

Also Cornelis van der Grient (1827-1918) produced a most realistic etching in 1858-1860, (van der Grient, 2018) that really mirrors Bond's picture of October 10, 1858. Concerning the engravings in Bond's work, i.e. positive effects upon a dark background, he calls them *"almost a distinct branch of art"* in which it is not permitted to produce effects at the sacrifice of precision, *"otherwise the scientific value of the representation is entirely lost"*.

An offshoot of the nineteenth-century cometary appearances was visible in fashion, in particular in the exquisite jewels of the crowned heads in Europe that featured "star-diamond" gems. Franz Xavier Winterhalter's (1805-1873) famous 1865 portrait of the 28-year old Empress Elisabeth of Austria (in the Sisi Museum in the Hofburg, Vienna) wearing a set of diamond stars in her hair is a marvelous example of comet-inspired decorative art .

4. COMET DONATI IN POETRY

The comet inspired many poets: see Gasperini et al. (2011) for some examples from the English and French literature. An interesting poem in Dutch is by Frans Friederich Christiaan Steinmetz (1827-1897), who blends scientific thoughts with religious beliefs. He mentions Ursa Major and Arcturus (Steinmetz, 1862), though wonders about the godly law to which the comet conforms its path. He argues that scholars may very well be able to calculate the comet's distance and orbit, but are just unable to explain anything. The poet is convinced that the comet will return after centuries, hence he will see it from the Heavens—as he writes, from the *"Empire of Love and Light"*. In the same vein, Harme Bevoort (1801-1875), in a poem written on October 9, 1858, calls the comet *"Donati's star"* and explains that it follows the same orbit since *"sixty centuries"* (Bevoort 1858). Bevoort's religious overtones, however, seem in stark contrast with his October 1830 call to arms of the Dutch archers against the separatist southern provinces whose armed revolution led to the creation of Belgium in 1830. (Bevoort 1830).

Bevoort's verse line is a reference to the Ussher chronology of the history of the world, constructed from a literal interpretation of the Old Testament (Ussher, 1658). This timeline, as well as similar chronologies by Kepler, Newton and others, places the first day of creation in the fourth millenium BC. It is also referred to by Edgar Allan Poe (1809-1849) in his "prose poem" *Eureka* (Poe, 1848): *"I can afford to wait a century for readers when God himself has waited six thousand years for an observer."* Both above-mentioned poems incorporate views of the developing nineteenth-century mechanistic universe, in conjunction with social thought in the first half of the nineteenth century that was almost always associated with a religious vision of human destiny (amidst a Victorian crisis of faith).

5. COMET DONATI AND LINCOLN

During the Illinois campaign for the U. S. Senate in 1858, Abraham Lincoln (1809–1865) and Stephen Douglas (1813–1861) held a series of seven oratorical contests, and one such debate took place in "Egypt" (Jonesboro, Illinois) on September 15, 1858. Lincoln observed the comet from the porch of the Union House hotel as he relaxed during the evening before the debate (Olson 2013). That was a quite early observation, at the moment that the comet had a tail of about five degrees long, with a "small star" visible through the axis of the tail (Bond, 1862). Agnes Clerke (1842–1907) observed that "*Not before the early days of September was it generally recognised with the naked eye, though it had been detected without a glass at Pulkowa, August 19. But its growth was thenceforward surprisingly rapid, as it swept with accelerated motion under the hindmost foot of the Great Bear, and past the starry locks of Berenice*" (Clerke, 1893).

6. DISCUSSION

The appearance of Comet Donati was of a very distinct geometry that allows dating of almost any visual artwork in the blink of an eye. Houzeau's very first description of the comet can be dated as October 4 or 5, about the same time that William Hayes Hilton and other artists depicted or described the sky. But in a letter to his parents of December 7, 1858 Houzeau (1994) specified that he saw the comet for the first time on September 19, though did not spot it on September 17 (the 18th was clouded).[8] From Bond (1862) we also know that on September 15, the tail was about 5 degrees long, and it thus follows that Lincoln could never have observed the comet in all its magnificence. The same applies to William Dyce's painting of which the production date is unknown. The non-dominant comet makes it look as if the artist recorded a viewing of around mid-September, as it was seen by Abraham Lincoln and observed by Houzeau. Since there exists a sketch of the scene made in 1857, evidently without the comet (Dyce, 1857), it is clear that Dyce had the same approach as Hilton, that is, constructing a painting from previous studies over a period of time. Pointon (1978) suggests that Dyce chose to paint the day on which the comet appeared at its most brilliant, but in reality it shows detailed strata of chalk cliffs below a just visible trail of the comet.

The appearance of Donati's comet led to a unique description of sightings by a self-exiled established astronomer who lived as a "frontierman" in complete isolation, without any access to libraries, scholarly books or journals. In contrast to Houzeau, self-taught artist Hilton had no astronomy knowledge. And he was a soldier, miner, cattle drover, stock broker and rancher in Texas, Mexico, Chile, Arizona and California (Bliss, 1963). Dyce, on the other hand, was more knowledgeable in recent developments in Victorian science, and combined facets of geology and astronomy in his artwork.

The discoveries of the mid-19th century, especially in geology and astronomy, revealed the great age of the Earth and the vastness of space. The revolutionary theories of Charles Darwin (1809–1882) and Alfred Russel Wallace (1823–1913) clearly influenced the visual art of the times. Exactly one year after the comet passed out of sight, Darwin published his *On the Origin of Species*, the Introduction of which he began by saying "*the origin of species – that mystery of mysteries*" (Darwin, 1859). But the mastermind of that phrase on speciation was no one less than John Herschel (1792–1871), who wrote to Wallace "*Of course I allude to that mystery of mysteries, the replacement of extinct species by others*' and '*the origination of fresh species, could it ever come under our cognizance, would be found to be a natural in contra-distinction to a miraculous process*" (Herschel, 1837).

Houzeau, during his five-year residence in the Texas prairie, had — just like Darwin and Wallace — intensely observed animal life, as follows from his letter of June 26, 1864, to his brother Auguste Houzeau de Lehaie (1832–1922): that the "'*mens*' [mind] *of mammal and man differs only in quantity and not in quality*" (Houzeau, 1994).[9] The study was published almost thirty years later (Houzeau, 1872), and was reviewed by (Wallace, 1872) who stated "*If this work had appeared a few years ago it would have created for its author a considerable reputation. Even now, had it been written in Europe after a careful study of all the best authorities on the subject, it might have been made a very valuable and important treatise.*"

Houzeau's saga clearly demonstrates that isolated individual scientists, whether in exile or in a technologically less-developed society, who cannot participate in the activities of the global scientific community – whatever their creative abilities, genius and hard work – may not succeed in making major contributions to scientific knowledge.

---

8 "*... j'ai aperçu la comète pour la première fois le 19 sept. J'étais alors dans la prairie, sans communications avec le monde intellectuel. La soirée du 18 avait été couverte; le 19, avant la fin du crépuscule, j'ai remarqué l'astre, qui eut attiré mon attention le 17 s'il avait eu autant d'éclat. A cette époque la comète ne se dégageait pas du crépuscule, dans nos régions. Je vous ai dit à peu près tout le reste. Je l'ai suivie pendant 48 jours, jusqu'au 5 novembre inclusivement. On la soupçonnait encore les jours suivants. ... Dans la 1ère décade d'octobre, plus d'un degré de la queue était visible à l'heure où paraissaient les tertiaires du Bouvier. Dès le 30 sept. je notais que le noyau paraissait en même temps qu'Arcturus, ou peu s'en fallait.*"

9 " *... que le 'mens' d'un mammifère et de l'homme ne diffère qu'en quantité non en qualité.*"

Figure 3. Butterfield Overland Mail stage wagon in Arizona, by William Hayes Hilton, 1858. Courtesy The Huntington Library, San Marino, California.

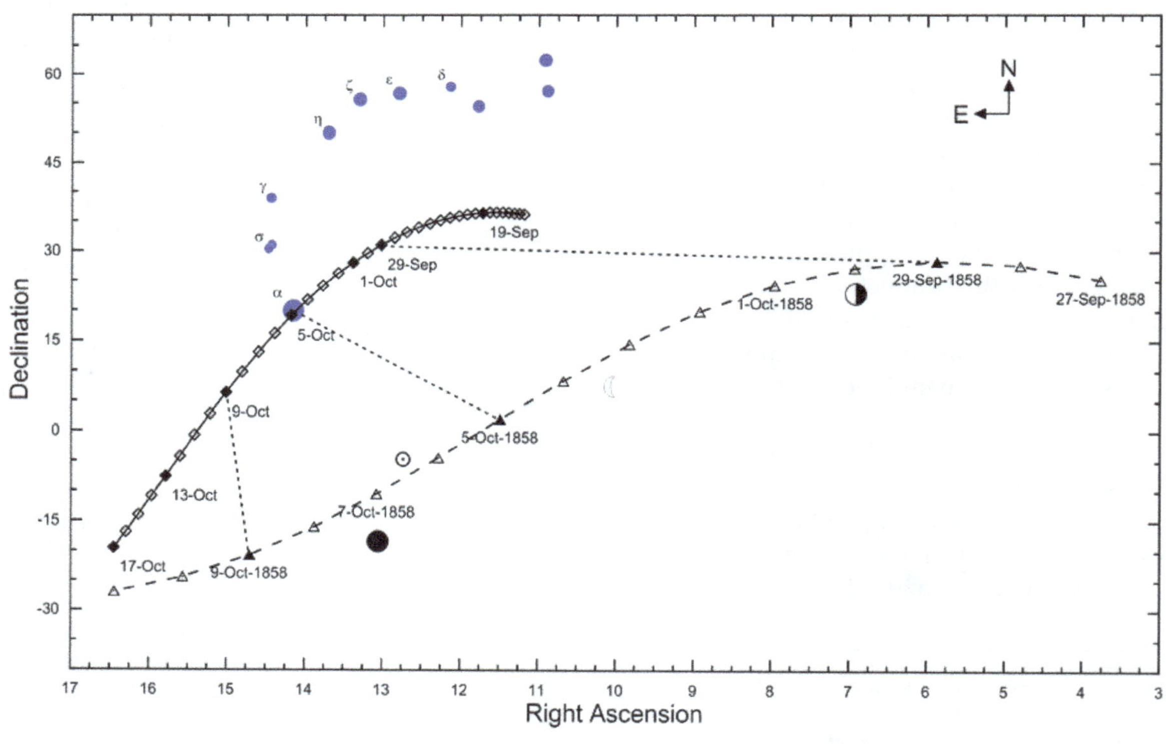

Figure 4. Path of the comet (◊) from September 12 to October 17, 1858. The path of the Moon (Δ) and the moon phases from Last Quarter to New Moon are also shown. The dashed lines are links to guide the eye. The ⊙ symbol gives the position of the Sun on October 5. The filled bullet symbols represent the brightest stars of constellations Ursa Major and Bootes, the symbol sizes are proportional to the brightness of the stars.

ACKNOWLEDGEMENTS

I thank Arlo Landolt (Louisiana State University, Baton Rouge, LA) for proofreading this manuscript, and Per Pippin Aspaas (University of Tromsø, Norway) for the translation of the Latin verse in Section 1.

REFERENCES

Ahnert, G. (2014, Sep 8). Personal Communication.

Anonymous. (1857). *The Great Comet, Now Rapidly Approaching, Will it Strike the Earth?* London.

Anonymous. (1858). De komeet van Donati. *De Tijd. Noord-Hollandsche Courant.* Amsterdam, NL.

Aspaas, P. P. (2018, Jan. 29). "His ergo indiciis me mea fata vocant", Personal Communication.

Bevoort, H. (1830). Retrieved Feb. 01, 2018, from http://home.hccnet.nl/k.fijma/hbged208.html

Bevoort, H. (1858). Retrieved Feb. 01, 2018, from http://home.hccnet.nl/k.fijma/hbged3140.html

Bliss, C. B. (1963). *Sketches in the Southwest and Mexico, 1858-1877.* Los Angeles: Dawson.

Bond, G. P. (1862). Account of the Great Comet of 1858. *An. Harvard College Obs., Vol. 3*, 1–372.

Carter, C. (2017). I Still Continue Convinced. *JHA, Vol. 48(2)*, 180–206.

Clerke, A. M. (1893). *A Popular History of Astronomy During the 19th Century.* London: A. & C. Black.

Darwin, C. (1859). *On the Origin of Species by Means of Natural Selection.* Oxford: J. Murray. Retrieved from Darwin .

Dyce, W. (1857). *Pegwell Bay by William Dyce.* Retrieved Feb. 01, 2018, from Art Fund.

Dyce, W. (1858). *Pegwell Bay, Kent – a Recollection of October 5th 1858.* Retrieved Feb. 01, 2018, from Tate UK n01407: http://www.tate.org.uk/art/artworks/

Gasperini, A., Galli, D. and Nenzi, L. (2011). The worldwide impact of Donati's comet on art and society in the mid–19th century. *Proc. IAU Sym. No. 260* (pp. 340-345). Cambridge: CUP.

Herschel, J. F. (1837). Letter of Feb. 20, 1836 to Charles Lyell. In C. Babbage, *The ninth bridgewater treatise a fragment* (pp. 202-204). London: Murray.

Heward, E. V. (1899). Comet–lore. *The Anglo-Saxon review, Vol. 6*, 28–42.

Hind, J. R. (1848). *On the Expected Return of the Great Comet of 1264 and 1556.* London: Hoby.

Hoek, M. (1861). Schreiben des Herrn Prof. Hoek, Directors der Sternwarte in Utrecht an den Herausgeber. *Astronomische Nachrichten, 55*, 215–217.

Houzeau, J.-C. (1857). La Comète de 1857. *Revue Brittanique, édition Franco-Belge*, 9–15.

Houzeau, J.-C. (1872). *Études sur les facultés mentales des animaux comparées à celles de l'homme.* Brussels: Manceaux.

Houzeau, J.-C. (1994). *Jean-Charles Houzeau: lettres adressées des Etats–Unis à sa famille.* Brussels: Elkhadem, H. et al. (Eds.), Centre National d'Histoire des Sciences.

Olson, D. W. (2013). *Celestial Sleuth.* New York: Springer.

Olson, R. M. and Pasachoff, J.M. (1998). *Fire in the Sky.* Cambridge, UK: CUP.

Ormsby, W. L. (1968). *The Butterfield Overland Mail.* San Marino: The Huntington Library.

Poe, E. A. (1848). *Eureka: A Prose Poem.* New York: G. P. Putnam.

Pointon, M. (1978). A study of William Dyce's Pegwell Bay. *Art History, Vol. 1*, 99–103.

Poole, J. (2018). *Donati's Comet.* Retrieved Feb. 01, 2018, from Sheffield Museums: http://www.museums-sheffield.org.uk/ Accession number K1908.66

Schmit, N.-C. (1860). *BE.ULB-A&B-ARCH. 014PP.* Brussels: Archives Université Libre de Bruxelles.

Steinmetz, F. (1862). De komeet van Donati. Eene herinnering aan het jaar 1858. *Vaderlandsche letteroefeningen*, 399–400.

Sterken, C. (2000). The *O–C* Diagram. *Variable stars as essential astrophysical tools, ed. C. Ibanoglu* (p. 529). Boston: Kluwer Academic Publishers.

Ussher, J. (1658). *The Annals of the World.* London: E. Tyler.

van der Grient, C. (2018). *De komeet van Donati.* Retrieved Feb. 01, 2018, from Rijksmuseum Amsterdam: https://www.rijksmuseum.nl/en/search?q=RP-P-OB-55.999&ii=0&p=1

Wallace, A. R. (1872). Book Review. *Nature*, 469–471.

25

ARCHAEOASTRONOMY AND BEDOUIN STAR-LORE IN THE ROCK ART OF THE NEGEV DESERT

George F. Steiner

Negev Rock Art Center, Israel
(georgesteiner@gmx.net)

ABSTRACT

The archaeological record of the Late Neolithic – Chalcolithic – Early Bronze Age of the Negev Desert exhibits prevalent east to west orientations. This is understood in literature as an expression of preoccupations that characterized emerging pastoralist elites, namely: after-life beliefs, mortuary cult and ancestor worship. Such archaeological remains are generally explained as astronomical alignments and are tentatively related to the position of the setting sun on the day of the summer solstice. Orientation seems to be also the central theme in the oral traditions of contemporary pastoralists. While the material remains exhibit orientation in space, the oral traditions, which are illustrated at their best in star-lore, exhibit an orientation in time: the cyclic renewal of seasons is observed in the east to west passage of stars and asterisms. As material and spiritual expressions of the beliefs that characterize pastoral nomads, the archaeological record and star-lore seem to be closely related. However, due to polar shift and the precession of equinoxes, contemporary star-lore orientates itself differently from its Chalcolithic – Early Bronze Age forerunner, therefore it cannot reflect the spatial orientation exhibited by *tumuli* fields, walls, *masseboth* and other remains from the said periods, except in a very approximate manner. A significant precession of equinoxes occurred in the early phases of the Middle Bronze Age. The event apparently left a deep mark on cultures worldwide and it was likely paralleled by shifts in symbolism. Moreover, in the Negev, the precession was also accompanied by a climatic deterioration. Nomadic mythology and star-lore had to re-adjust to the new coordinates that superseded an apparently perfect previous order. The majority of the rock art corpus in the Negev is dated - based on stylistic considerations - to the period that preceded the shift. However, a few of the engravings attributed to the Early Bronze Age become meaningful only when related to the changes that occurred during and after the precession of the equinoxes. Other petroglyphs reflect precisely Chalcolithic – Early Bronze Age realities, but their symbolic implications outlived the astronomical context in which they were conceived and are still meaningful to the Bedouin pastoralists of today.

KEYWORDS: Aldebaran, Antares, astronomical alignments, equinoxes, petroglyphs, polar shift, precession, star-lore

1. INTRODUCTION

The correlation between the positions of the stars Antares and Aldebaran and the dominant east to west orientation of numerous Chalcolithic - Early Bronze Age archaeological sites was tentatively introduced in a previous paper (Steiner 2016), to which this article is a sequel. The literature describing these complexes is very cautious in its approach and it generally associates the prevalent orientations with the position of the setting sun on the day of the summer solstice (Rosen, 2007). For example, the sacred precinct on the Saharonim Plateau in the Ramon Crater is related to a black hill to the west and to the sun setting behind it on the day of the June solstice. The orientation exhibited by the complex and the large number of shrines and burial mounds (tumuli) found there are interpreted as reflecting a preoccupation with after-life, dead cult and ancestor worship of an emerging pastoral elite. However, an azimuth deviation of 2° to 8° from the point where the sun sets is observed and explained with the polar shift that occurred since the Chalcolithic-Early Bronze Age, the archaeological periods to which the complex is dated (ibid.). Another site, the Zin-Mehia alignment, seems to follow the pattern of the Ramat Saharonim precinct (Steiner, 2010; 2016). Judging by the suggested age of the oldest rock engravings found there (Anati,1986; 2015) the alignment was conceived at the same time. The azimuth deviations do not correspond to those measured at the Ramat Saharonim precinct, and a correlation of this site with the setting sun of the summer solstice is very unlikely.

Even when the combined effects of polar shift and the precession of the equinoxes is taken into consideration, the orientation of the alignment only approximates that presumed to have had determined the layout of the Ramat Saharonim structures. However, while researching the Zin-Mehia alignment, the author has noticed a correlation between the orientation of the site and the nocturnal/heliacal rising/setting of the stars Antares and Aldebaran, at the beginning of June and the end of November (Steiner, 2010; 2016). When the first petroglyphs at the site were carved into the rock, at approximately > 4,000 BP (Anati, pers. comm.), the two stars marked precisely the location of the September and June equinoxes (Berger, 1976). Because these two red stars also seem to play a role in renewal, transformation and other concepts related to dead cult and after-life beliefs (cf. Steiner, 2016), their association with archaeological sites - that apparently reflect the same preoccupations - should be reassessed.

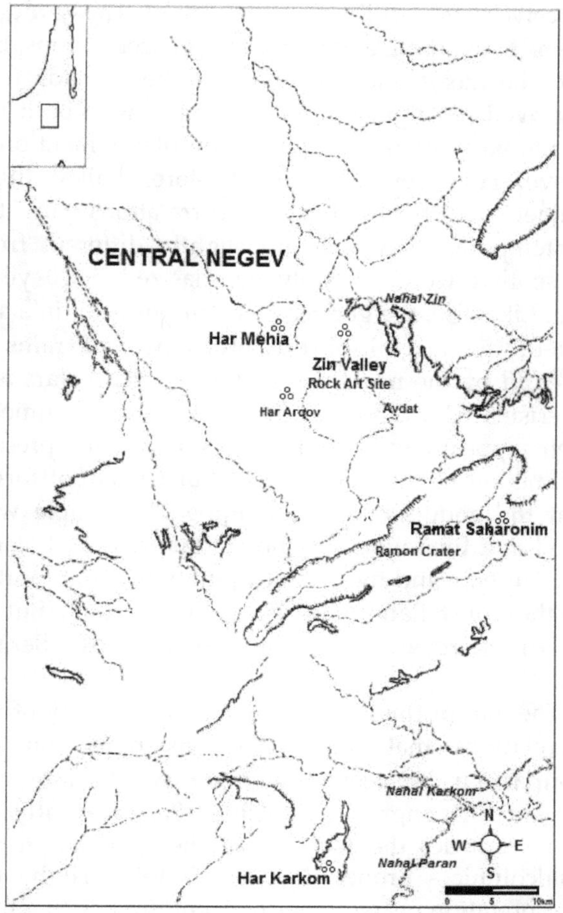

Figure 1. Map of the Central Negev (Israel) showing the sites mentioned in the text

In the same paper (ibid.), a link between an iconic figure that is featured in the rock art of the Negev, southern Jordan and northern Hejaz, and the two stars was also proposed. One of the locations where the engraving is found is the Zin-Mehia alignment. Moreover, when this 'icon' is related to modern Bedouin star-lore, it is possible to place it into the sky, with the help of clues provided not only by oral traditions, but also by the contexts in which it appears, the orientation of the rock faces on which it is engraved, the position of the stars that mark the seasons, and the shapes of the constellations. The suggested position of the icon in the night sky, and the graphic concept behind the composition - with its accent on symmetry - seems to further emphasize the relationship between Antares and Aldebaran on one hand, and between the markers of the rainy and dry seasons on the other. Its location in the circumpolar area, to the left (from the perspective of the stars) of the constellation which is traditionally identified as the 'seat of God' (Tomkins, 1971; Magli, 2015), led the author to float the conclusion that the icon was the symbol of a 'goddess' associated with rebirth, cyclic renewal and transformation, in the desert. These concepts are related to what was perceived as fertility, in the sown. The goddess seems to symbolize not only the preoccupations, but also the ideals and the hopes of the nomads who engraved her figure in the rock, and who built the east to west oriented funerary complexes mentioned above. However, Bedouin star-lore (Bailey, 1974; Steiner, 2010) establishes correlations in the contemporary sky that are slightly different from those that were probably emphasized 5,000 years ago. Likewise, the goddess may fit perfectly in a sky where the arrival of the much-awaited rains is foretold by the nocturnal rising of certain stars and asterisms whose position in heaven and the time of their appearance are determined by the present inclination of the Earth's axis, but let us not forget that the goddess was worshipped by people who lived at a time when the north-star was not Polaris, but Thuban, and the vernal equinox was not located in the constellation Pisces, as it is today, but in Taurus (Dreyer, 1953; Pannekoek, 1961; Berger, 1976).

The aim of this paper is to test the validity of the correlations that the author has based on the comparison between contemporary star-lore and ancient rock engravings (Steiner, 2016). An attempt to reconstruct the location of the goddess in the Chalcolithic - Bronze Age sky is followed by the interpretation of two other rock engravings to which she seems directly related. The dating of these petroglyphs (Anati, 1985; 1986; 2015) is re-evaluated in the article and they are attributed to the period when the changes induced by the polar shift and the precession of the equinoxes occurred. Moreover, the shortcomings of still prevalent dating criteria (set by Anati, 1968a; 1968b; 1972; 1974) are highlighted, and their origins explained (Bednarik and Khan, 2005). Techniques such as archaeometallurgy (as in Steiner, 2016) and archaeoastronomy (in this paper) are recommended as fitting sub-disciplines of archaeological science that bear the potential to overcome the said shortcomings in dating and to place certain petroglyphs in a more realistic chronological context. The suggested archaeoastronomical approach and the dates derived from it are supported by the motifs depicted, which seem to reflect the new realities in the distribution of the seasons and the orientation of the stars. The author argues that the marked increase in the complexity of rock engravings during this period should be attributed to the need to explain and illustrate the changes, and to embed them in the human consciousness, via ritual. The motifs depicted seem to express the fear that the sun may vanish and the magic employed to restore it is also illustrated in the engravings. Zoomorphic figures that represent the constellations Taurus and Aries are drawn as successive guardians of the sun, chased by heroic hunters (Rotblum, 2010), whose purpose is to re-establish the order that governed the world before the precession of the vernal equinox (Magli, 2015). As the polar and equinox precession was followed by an arid period with heavily felt repercussions on pastoral economy (Anati,1985; Steiner 2010), the nostalgia for the previous order is understandable. The readiness of the nomad to accept metallurgist values, precisely in this period (Amzallag, 2008; 2009 Steiner, 2010; 2016) may also be rooted in his anxiety: the disappointment in the immutability of the laws governing the heavens, added to the spiritual affinity with copper-related symbolism, induced the nomad to look for more abstract, but at the same time more reliable alternatives. The heroic hunters who restore the cosmic order are associated in many mythologies with coppersmiths (Amzallag, 2008). The abandonment of the old stellar and lunar religion for the sake of an abstract deity, that is, the replacing of the God of Abraham with that of Jethro (ibid.) may be understood as the voicing of the disappointment caused by the changes that took place at the beginning of the Middle Bronze Age.

2. THE CONTEMPORARY SKY

2.1. Equinoxes and Seasons

The stories of Bedouin star-lore focus on asterisms stars and constellations which, in their turn, are associated with the seasons (Bailey, 1974). The nocturnal (at dusk) or heliacal (at dawn) rising of these predict the arrival or the end of the rainy and dry periods of the year (ibid.). The knowledge is based on observations that describe a sky that has been relatively stable for the last two millennia. The area around Polaris marks the position of the north celestial pole since immemorial times. The seasons are defined by equinoxes and solstices that occur against the same background of stars, year after year. The 1° shift of the constellations located on the 360° ecliptic (the apparent path of the sun in the sky) in 72 years is virtually imperceptible in one lifetime, and the accumulated 27° shift in the last 2,000 years, although noticeable, does not interfere too much with the distribution of the seasons. The spring equinox still occurs when the sun is in the constellation Pisces in March. The Bedouin relies on what he sees and the fact that scientists call the spring equinox "point zero Aries" does not impress him. Since 67 BC the vernal equinox, that is, when the sun crosses the celestial equator into the northern hemisphere at the point where the ecliptic and equatorial planes intersect each other, occurs in the constellation Pisces. At the diametrically opposite intersection of the two planes, the autumn equinox, that is, the point on the ecliptic where the sun descends into the southern hemisphere, is in the constellation Virgo, since 780 BC. Before these dates, however, the sun crossed into the northern hemisphere in the constellation Aries and left it in the constellation Libra, due to an astronomical phenomenon called precession of the equinoxes (Pannekoek, 1961; Dreyer, 1953; Berger, 1976). Pisces, Aries, Virgo and Libra are only four of the twelve constellations located along the ecliptic, which form the zodiac. Because the zodiac was conceived as a celestial coordinate system meant to measure ecliptic longitude at the time when the vernal equinox still occurred in Aries, and because at the time of its conception the phenomenon of precession was unknown, longitude 0° is still located in Aries (instead of Pisces); longitude 90°, or the summer solstice in Cancer (instead of Gemini, or rather Taurus, to where it shifted in 1989); longitude 180° in Libra (instead of Virgo, where the autumn equinox actually occurs); and longitude 270°, the winter solstice, in Capricornus (instead of Sagittarius) (ibid.). In today's astrology, the imaginary zodiac became distinct from the stars (Evans, 2008; Parker, 2010). In astronomy, ecliptic longitude and nomenclature are still in use, although not related to the 30° arcs of the zodiacal signs, but to the real longitudinal extension of the constellations, as defined in 1930 by the IAU (Hilton et al., 2006). The desert nomad does not know much about the ecliptic, the celestial equator, precession and the zodiac, but solstices and equinoxes are important to him, as they mark the transition between seasons (Bailey, 1974; Steiner, 2010).

2.2. *Winter and Summer Constellations; Bedouin Star-lore*

To the Bedouin, the sky appears as being parted in two: the winter half and the summer half, that is, the part visible between October and May and its November to April counterpart. The first interval is associated with luck and plenty, while the second period with misfortune and hardship (Bailey, 1974; 'Id al Wadj, pers. comm.). The blessed part of the year is foretold by the rising of Canopus (as Suhayil in Arabic, in the constellation Vela) at daybreak, in the south, approximately two weeks after the flowering of the Canopus reed (Urginaea maritima). The first chilly winds start blowing at dawn, a release from the heat of the summer. In the second half of October the nocturnal rising of the Pleiades in the east, which coincides roughly with the disappearance of Antares from the sky, marks the beginning of the rainy season, called in the honor of the asterism Wasm-ath-Thuraya, which means "the sign of the Pleiades". The rains that fall in the 25-27 days marked by the asterism would bring forth the wadi-(riverbed) vegetation in the following spring. The next 25-day season begins with the rising of Aldebaran at sunset, also in the east, towards the end of November. The rains of Aldebaran would make the slopes above the wadis bloom in the following spring. The second half of December is the beginning of the third 25-day rainy interval, in which the rains of Betelgeuse fall. These are important to the vegetation that would sprout on the plateaux. The real cold weather sets in together with the nocturnal rising of Sirius in the east, in mid-January. It lasts for forty days and its Arabic name, Al-Arbainiyah, means just that. However, despite the discomfort, it is a very much-anticipated season because, during it, the seeds gather strength to burst into life at the end of the period marked by Sirius, when Ar-Rabi, that is, spring arrives (ibid.). The sun and its warmth make the vegetation sprout, the kids and lambs are born at this time of the year and the desert turns green. In anticipation of the blessings, the Bedouin become impatient with Orion, which dominates the sky during the winter months, and wish him gone. Sirius is seen as a Saluki dog that would chase Orion for forty days and forty nights, to allow the spring constellations to take its place in the sky ('Id al-Kashkhar and 'Id al-Wadj, pers. comm.). Orion would sink slowly under the western horizon to appear again, just as gradually, as Bootes, in the March sky. During the first days of May, when the Pleiades, Aldebaran and Orion do not rise anymore, and when Sirius also stops hanging above Canopus, the unlucky part of the year sets in (ibid; Bailey, 1974; Steiner, 2010). The extremely hot weather is followed by the nocturnal rising of Antares, towards the end of May. The rains of Antares bring forth poisonous vegetation, which replaces what is still left of the fodder that sprouted because of the rains of Aldebaran. The season known as As-Smayih, that is, summer sets in. Once a month, the moon passes through Sagittarius, Scorpius and Libra, for seven consecutive nights. These nights are dreaded by the Bedouin, who associates them with extreme bad luck (Bailey, 1974). When Bootes slowly starts disintegrating, the desert nomad sighs with relief: soon, the Canopus reed would bloom again…

Interestingly, at the Nabataean temple at Khirbet Tannur, one of the most intriguing items found was a 2nd Century AD Nabataean zodiac that seems to emphasize the partition of the year according to the considerations mentioned above. The symbols of the Nabataean zodiac conform to their Roman and Hellenistic counterparts – let us not forget that at this stage the Nabataeans were already a settled population that had assimilated Hellenistic beliefs and values. However, the most significant difference in the Nabataean zodiac is the arrangement of the order of the houses within the zodiacal circle. The Nabataean zodiac found at Khirbet Tannur is extraordinary in its two opposite and completely separate halves - Aries to Virgo, and Pisces to Libra. Both halves start and culminate at the spring and autumn equinoxes (Liritzis et al., 2015). This may very well reflect the persistence of a nomadic desert star-lore that was inherited from the Nabataeans' pre-Hellenistic ancestors. Moreover, the depiction of the goddess Allat in the center of the zodiacal circle (ibid.) supports very well one of the main suggestions of this paper, namely, that a goddess-figure that appears at a number of rock art sites in the Negev Desert may arguably be associated with engravings that seem to depict the opposite - "blessed" vs. "cursed" - halves of year.

2.3. *The Goddess in the Sky*

There are two rocks in the Negev, almost 100 km away from each other (in the dry riverbed of Nahal Zin, on the already mentioned Zin-Mehia alignment and in Nahal Zihor, in the southern Negev), on which the same figure is engraved. It is an abstract image, but at the same time strikingly anthropomorphic: an oval, with a line drawn as its extension are transformed into an ankh-like figure because of a second line that intersects that under the oval. The "arms" formed this way are similarly engraved on both rocks: the right arm (from the observer's perspective) is a two-horned serpent and the left arm ends in an exaggerated three-fingered hand (which is connected to a zoomorphic figure in Nahal Zin). The arms alone are a well-known symbol of mediation between the worlds, which is often used to represent the shaman, whose role was precisely that (Lewis-Williams and Dowson, 1988). Variations of the petroglyph are also found on Har Karkom (where the "ankh" is drawn beside a hand, or only the arms are engraved in the rock, without the "body"), in southern Jordan (where the central figure is a male, and the right arm holds a snake instead of being one itself), in eastern Jordan (where it is very similar to that found in Nahal Zihor) and in Spain (where the figure is markedly feminine, with her arms blended with the figures of a male and female wild sheep, but where the serpent is only an element of the petroglyph and not the arm of the central figure).

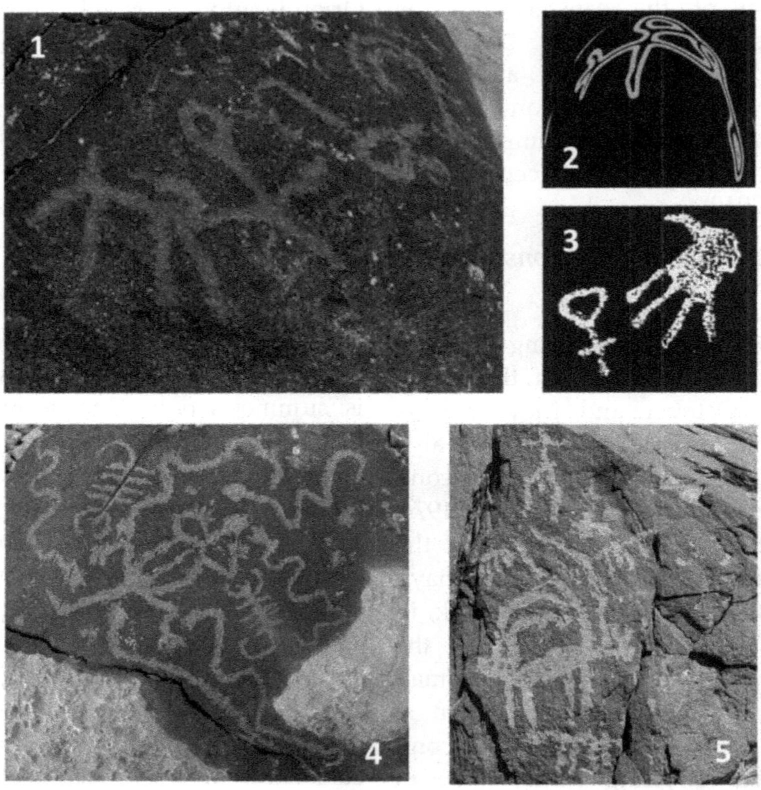

Figure 2. The petroglyphs discussed in the text: 1 Nahal Zin; 2 Nahal Zihor; 3,4,5 Har Karkom. (photographs and sketches by the author)

The last-mentioned engraving resembles very much the drawing on the 8th century BC Lachish ewer, where the central icon is not anthropomorphic, but it is represented as a six-branched menorah. The two latter are representations of the fertility goddess Astarte (as inherited from Phoenicia, before she became the abstract Tanit of the Carthaginians) and the Canaanite Ashera. The unmistakable "maleness" of the Jordanian petroglyph is explained in a previous article (Steiner, 2016) that describes the goddess. The Negev engravings are presumably also icons of the goddess. She was associated with fertility in the sown (Phoenicia and Canaan), which would mean rebirth, renewal and transformation in the desert (the southern Levant), where fertility was a rather utopian concept, associated with Paradise (Steiner, 2010; 2016). In this case, the goddess may also be associated with after-life which, in its turn, can be related to concepts like renewal and rebirth. Her presence on the Zin-Mehia alignment also seems to illustrate these concepts (ibid). Her zoomorphic companions seem to be male (or adult) and female (or kid) ibex, or gazelle (Nahal Zin), and Barbary sheep (Spain), and they are symmetrically placed to the left and to the right of the goddess, who in this case is the ''tree of life,'' on which the animals feed.

The concepts of transformation, rebirth and renewal are also well-illustrated in the stories about the stars (Bailey, 1974). The passage of seasons, which is the main theme of nomadic star-lore, is likewise east to west oriented: the position of the equinoxes, the setting or rising of marker stars and the correlations between stars like Antares and Aldebaran, which are complementary to each other (they cannot be observed at the same time in the sky, they are both red, and they mark diametrically opposed directions and seasons). Antares is always associated with the west and the after-world, whereas Aldebaran with the east and with life (Van der Waerden, 1953; Tomkins, 1971; Magli, 2015). The goddess seems to be responsible at the same time for the concepts that determined the spatial east to west orientation of archaeological alignments (rebirth, after-life), and the temporal east to west orientation of star-lore (renewal, passage). As archaeological remains like alignments, tumuli fields and walls display characteristics related to astronomy (Steiner, 2016), a parallel between star-lore and after-life beliefs does not seem out of place. In this case, the common denominators would be the figure of the goddess (which symbolizes renewal and rebirth) and the stars Antares and Aldebaran (which define the orientation of renewal and rebirth in physical and ritual space).

In many rock engravings, a marked affinity to the starry night sky may be recognized. This does not mean that they are ''star maps,'' but that the stories told by petroglyphs are the same with those related by the stars. Because archaeological alignments are understood even by mainstream scholars as related to astronomical orientations (Avner, 1984; Haiman, 2000; Granot, pers. comm.), rock art associated with these - or dated to the period in which the alignments were conceived and built - should be related to the same astronomical factors that influenced the orientation of the alignments. In this context, the goddess, who is clearly related to the prevalent east to west symbolic orientation of renewal, should also be found between the stars, with respect to the directions that seem to have had influenced the archaeological alignments. Indeed, her position in heaven describes precisely her attributes. Her shape would locate her to the left of Draco (from the perspective of the stars): the head almost on the north-star and the body along the axis of the constellation Cepheus (Steiner, 2016). The main star of this constellation is Alderamin, or Adh Dhira al Yamin in Arabic, which means "the right forearm." The palm at the end of the forearm and its exaggerated fingers would fit perfectly the constellation Cassiopeia, which is also called Khaf, "the palm," by the Bedouin. It is also known as Khaf al Habib, or "the henna-dyed hand," and it is thought to represent the hand of Fatima, the daughter of Muhammad (Kunitzsch, 1959). The opposite (left, snake-) arm would be in this case the tail of Draco (or the neck of the camel - see below) and the two-horned head of the snake would correspond to the asterism at the end of the tail (neck), with its two prominent stars, Rastaban and Eltanin.

Draco is at the center of the circumpolar area and it can be seen all year long. Because of its reliability, it was imagined in the antiquity as the ''seat of God'' (Tomkins, 1971; Magli, 2015). In Bedouin star-lore this is still mentioned, although in a figurative way. The constellation is not God, but a reminder (sign-lesson, or ayah) of His existence: it is the figure of a ten-month pregnant she-camel that was miraculously brought forth from the rock by the Thamudic prophet Salih, in order to prove the existence of God to the unbelievers (Poret, 2010 [Qur'an, surah 8]). Indeed, the shape of the constellation resembles that of a camel. The goddess is positioned at the left of this constellation, like Ashera is always at the left of Ba'al or Yhwh and like the broad female masseboth are in most cases at the left of the narrow male ones (Avner, 1990; 2001). These stelae were thought to contain the essence of deities and were erected starting with the 10th millennium BC, to biblical times (ibid.). The constellations and asterisms on which the body of the goddess is traced are also in the circumpolar area today and are visible all year long.

The location of the goddess becomes even more significant when the directions to which her arms are pointing are considered: the snake-arm points straight to Antares, while the middle finger of the hand that is Cassiopeia points to Aldebaran. The arc described by her spread fingers encompasses all the "lucky" stars and asterisms of the winter sky: Hamal, the Pleiades, Aldebaran, Capella, Betelgeuse and Sirius. As the two arms resemble the symbol of the shaman in his role as mediator, it may be stated that the goddess is mediating between the world of Aldebaran and that of Antares, that is, between east and west. The tailed tumuli described in The Goddess and the Copper Snake (Steiner, 2016) seem to have the same role, not only because of their orientation, but also because of their shape which, if drawn on paper (or rather on rock), would also resemble the symbol of the ritual specialist. Although the head of the goddess is roughly positioned under the north-star, it points straight to the brilliant Regulus in the constellation Leo. In the opposite direction, the axis of her body may be extended to the prominent star Fomalhaut, in Pisces Austrinis. These four stars were once called the "archangel stars": Michael (Aldebaran) watcher of the east, Oriel (Antares) watcher of the west, Raphael (Regulus) watcher of the north and Gabriel (Fomalhaut), watcher of the south (Van der Waerden, 1953; Pannekoek, 1961). To the ancient Persians they were known as the Four Royal Stars. Moreover, 5,000 years ago they marked the equinoxes and the solstices: Aldebaran the vernal equinox, Antares the autumn equinox, Regulus the summer solstice and Fomalhaut the winter solstice (ibid.).

The markedly symmetric figure of the goddess may be called an iconic image, not only because it is found in more than one location, but also because of its apparent relatedness to the concepts of renewal, transformation, rebirth and after-life (Steiner, 2010; 2016). The symmetry also emphasizes the conceptual and physical east to west orientation, which is clearly associated with the categories listed above. Furthermore, its location between the four cardinal stars link the icon directly to the prevalent orientations exhibited by archaeological alignments, and also to the distribution of the seasons, which is the central motif of nomadic star-lore.

2.4. The Goddess Today

Another detail related to the proposed location of the goddess is that in the contemporary sky, which is defined by the present axial tilt and the stars which are the backdrop of the equinoxes, she stretches her arms precisely over those 90° of ecliptic longitude that are situated between the winter solstice and the spring equinox (see Figure 3.). This means that she must have been related directly to the increasingly longer days that characterize the period between the December solstice and the March equinox, and to the renewal of life that is heralded by the latter. Hence, her tentative identification with Ashera becomes justified. In the illustration below, the grey dotted lines that intersect each other in Polaris delineate the 90° arcs on the ecliptic between today's solstices and equinoxes. Note the position of the goddess between the winter solstice and the spring equinox, as described above. Instead of zodiacal signs the constellations on the ecliptic are illustrated with petroglyphs from Har Karkom (when a correlation between the two can be determined), or with the first three letters of the names of the constellations. The black lines, circles and arcs should be ignored meanwhile, as they are meant to illustrate how the sky looked 5,000 years ago. The illustration is not drawn to scale or with mathematical precision, just like petroglyphs are not, because they are meant to tell stories and not to be astrological maps.

The zoomorphic composition at the left seems to be a combination of the constellations Orion, Taurus and Aries (note the three dots, which were incised in the rock with the clear purpose not to leave any doubt about what the petroglyph was supposed to convey – it is Orion's Belt). All the "lucky" stars mentioned in the text are located in and around this engraving. It would be very tempting and quite logical to interpret it as an illustration of the winter sky. In a measure it is, only that the petroglyph is dated to a period (Early Bronze Age [Anati, 1985]) when due to the axial tilt and the different backdrop of stars against which the equinoxes occurred, the seasons were different than today's, hence they could not have been foretold by the marker stars used as references by contemporary nomads. An Early Middle Bronze Age dating would already qualify it as an illustration of an almost contemporary winter sky. The same applies to the partial reproduction of a petroglyph at the right side of the figure, which depicts the constellations Scorpius, Libra, Virgo and Bootes (the latter is easily identified in the original rock engraving, due to a few details, which leave no doubt about the figure being the constellation [see Steiner, 2016]). These constellations dominate the summer sky and because of their negative image in Bedouin star-lore (Bailey, 1974), and because they represent animals associated with the scorching heat of summer, it would be tempting, again, to interpret the petroglyph as an illustration of the summer sky. Without taking into consideration the age of the engraving (dated to EBA [Anati, 1985]), such an interpretation would be very logical: Bootes, the bird-like lizard, sinks slowly behind the horizon immediately after the autumn equinox, devoured by snakes and scorpions toward the end of summer. The stripe on the lizards back only seems to emphasize its relatedness to the equinox that divides the two seasons. This is in a way the correct interpretation of the engraving, only that instead of a seasonal story, the composition rather relates a one-time EBA/MBA event, and the snakes and scorpions are more like the biblical "sarafim" than common reptiles and arachnids. Just like in the case of the composition diametrically opposed to it, in which the ibex and the ram are not common animals, but mythical heroes or villains. However, both compositions are within those quarters of the ecliptic plane that border the one occupied by the goddess. Symbolically seen, their interpretation is meaningful, as both petroglyphs are clearly related to the goddess: her snake-arm points to the heart of the scorpion (Antares) and the middle finger of the opposite hand to the bull's eye (Aldebaran). The length of her body is on the line that unites Regulus and Fomalhaut. Although the Nahal Zin goddess is a relatively recent engraving (Iron Age, or even later, [ibid.]), its Nahal Zihor replica, judging by the patina covering it, was incised into the rock at an earlier period. The composition was in all probability conceived to illustrate the realities of the Early to Middle Bronze Age I sky. The depiction of the same icon after almost a millennium, when the equinoxes and the seasons were already different from those of the Early Bronze Age, only emphasizes the importance of the goddess. The continuity of motifs may also be observed in the case of other engravings at the Zin-Mehia site. The orientation of the alignment, and the age of the earliest engravings on it, point to a Chalcolithic-Early Bronze Age origin (Anati, pers. comm.). Apparently, the site did not lose its importance, and the numerous petroglyphs engraved during the Iron Age, and even later, show a continuity of traditions in which the goddess or, rather, the concepts that she represented, played a key-role.

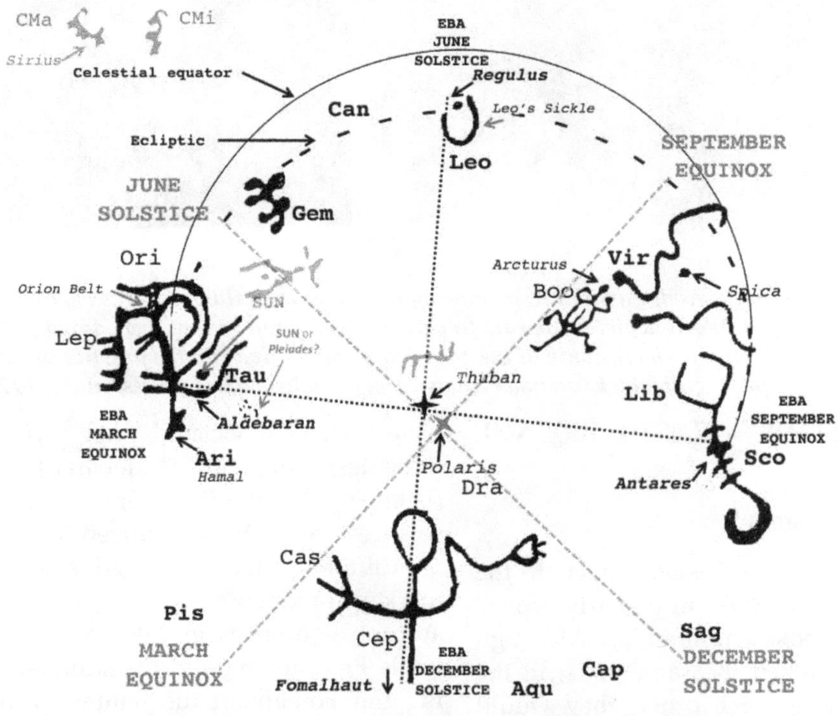

Figure 3. Contemporary (grey) and Early Bronze Age (black) equinoxes and solstices correlated with petroglyphs suggested to represent the winter (left, Har Karkom) and summer (right, Har Karkom) constellations. A "goddess" icon was inserted (center, Nahal Zihor) in order to illustrate her role in the transition from season to season. (drawing based on photographs by the author)

Indeed, the ethnicity and economy of all the documented inhabitants of the area - from Shasu nomads, Nabataeans and Thamudic tribes, to the present-day Janabib and 'Azazme Bedouin, who were all Semitic nomadic herders of North Arabian origin who made the Negev Highlands their home and marked its rocks with their signs attesting their territorial claims (Khan, 2000) - points toward a parallel, conceptual continuity. On the rocks bearing their tribal signs (wusum), they also engraved their stories and 'magic,' in order to attract blessing and abundance to the claimed grazing and watering grounds. "Bedouin" dominated the ethnographic landscape even through the Idumaean, Nabataean,

Roman and Byzantine stable periods (Fabian, 2005). Throughout much of the past millennium and a half since the end of the Byzantine era, the Negev Desert accommodated such pastoral nomads of various tribal identities whose migratory paths took them from Arabia across the Negev, into the Sinai and beyond, and back again to the Sinai and Negev, where wusum engraved during the first period of occupation were recognized as valid territorial markings, and to which new ones were added as a re-confirmation of ancestral rights (Abu Rabi'a, 2001; Galilee et al., 2013; Isaac, 1992; Sharon, 1975; Khan, 2000. NOTE: a recent publication on Arabian rock art brings up all these issues, MAA (2017)). This pattern seems to be also the case for the present inhabitants of the area, the Janabib, whose oral traditions narrate a >1,400 years history (Bailey, 1989). The Nahal Zin "goddess" petroglyph is in Janabib territory, and the site where it was engraved was - judging by the presence of Bronze Age, Iron Age, Nabataean, Thamudic and Janabib engravings - continuously used for fertility-related ritual practices (Mailland, 2015; Steiner, 2016). The same phenomenon was observed in the case of other Arabian rock art sites where, according to R.G. Bednarik (2017, pers. comm.), certain Bedouin groups have apparently defied the restrictions imposed by Islam for 1,400 years and are continuing to the present time to create rock art and use pre-existing rock art for ceremonial purposes.

Figure 4. Left: a petroglyph from the Avdat area that, judging by its degree of patination, was engraved recently, during the Bedouin period. However, the motif depicted appears frequently in engravings that were dated to the Early Bronze Age. (photograph by the author) Right: The Bedouin of the Negev have never given up the practice of leaving their marks on rock faces. (photograph by Emmanuel Anati 1954; republished in Expression 15, 2017)

3. THE CHALCOLITHIC - EARLY BRONZE AGE SKY

3.1. Equinoxes and Seasons

Despite the preliminary conclusions above, to the contemporary Bedouin the Chalcolithic-Early Bronze Age sky would have looked unfamiliar. Although they would have recognized the same stars, in the same relative positions to each other, they would have heard different stories about them. The north-star would not have been Polaris, as it is today, but Thuban in the constellation Draco, which is only one fifth as bright as Polaris. Thuban was the star closest to the celestial pole from 3,942 BC to 1,850 BC, when the much brighter star Kochab (Ursa minor) superseded it (Burnham, 1978). Bootes, which was closer to the pole than it is today, was a permanent feature in the sky, while Cassiopeia disappeared seasonally (ibid.). Even the familiar stars would have risen and set at different times and the rains would have been associated not with Ath-Thuraya (Pleiades), Imjayid (Aldebaran) and Al Jawza (Betelgeuse), but with other stars. The equinoxes and solstices would have occurred against the backdrop of different stars, too. Aldebaran and Antares marked precisely the location of the spring and autumn equinoxes in 3,044 and 3,052 respectively, while Regulus marked the summer solstice in 2,345 BC and Fomalhaut the winter solstice in 2,582 BC (Berger, 1976). These dates fall within the archaeological period known as the Early Bronze Age. The vernal equinox, which is considered the most important not only in astrology, but also to the nomad, was associated with the constellation Taurus until 1,865 BC (Middle Bronze Age), when it passed into Aries. The autumn equinox, on the other hand, stayed within the constellation Scorpius until 729 BC (Iron Age), when it passed into Virgo (to the ancients, the constellation Libra was part of Scorpius, or to be more precise, the stars of Libra were imagined to be the pinchers of the scorpion – the equinox passed from Scorpius into what we know today as Libra at around 1,850 BC). The June solstice passed from Leo into Cancer in 1,458 BC (Late Bronze Age) and the December solstice from Aquarius/Pisces Austrinus into Capricornus at around the same time (ibid.).

The backdrop of stars against which equinoxes occur and the location of the celestial pole are complementary phenomena. The equinoxes are located in the points where the celestial equator intersects the ecliptic. When the axis of the Earth changes its inclination, the equatorial plane of the Earth moves with it. As the celestial equator is just the Earth's equator projected onto the celestial sphere, it moves as the Earth's equatorial plane moves, and the intersection with the ecliptic also moves with it. The positions of the poles and equator on Earth do not change, only the orientation of the Earth against the fixed stars (ibid.; Dreyer, 1953; Evans, 1998).

3.2. *Antares and Aldebaran*

To the nomad, the stars are part of everyday-life. The dry air of the desert makes them appear bigger and closer, and their passage through night, season and year is observed and memorized with care. The symmetry displayed by the Early Bronze Age sky must have impressed the nomad, whose skyward orientation in matters related to faith was described and analyzed elsewhere (Steiner, 2016). The association of the four cardinal stars with the equinoxes and solstices only strengthened the belief that the seasonal cycle of renewal and life was not incidental, but carefully planned and a part of the cosmic order. Closely related to renewal and life are rebirth and after-life. Placing the four categories within the coordinates of such a meaningful sky seems logical. When around 3,000 BC the equinoxes coincided precisely with the locations of the two red stars Aldebaran and Antares (see above), which are on the two ends of the Milky Way (which in many beliefs was thought to be the path taken by the soul after death [Magli, 2015]) and were associated not only with the east and the west, but also with life (vernal equinox, rebirth, Aldebaran) and death (autumn equinox, decay, Antares), the significance of the sky only gained in importance. In the Early Bronze Age, Antares was not the brilliant red star of today: at some times it was brighter, while at other times it seemed to fade (Burnham, 1978; Parker, 2010). This may also be one of the reasons of its association with a world very different from Aldebaran's, where life was supreme. The color of the stars was also important, as red was associated, since times immemorial, with transformation and renewal (Wilmsen, 2009). The ochre-painted bones found in prehistoric secondary burials hint at the relatedness of the color either with rebirth, or with after-life beliefs (Gilead, 2002). The magic that surrounded copper reflected the fascination with the physical properties of the red metal, which also illustrated, in a palpable way, the concepts of transformation and renewal (Amzallag, 2008, 2009; Steiner, 2010, 2016). As Antares and Aldebaran are precisely those that from all the fixed stars are the closest to the orbit of the moon, occultations were a common phenomenon – the moon was another symbol of renewal (Bastoni-Brioschi, 1998; Mailland, 2015; Wachtel, 2014), and the association of the said stars with it must have had a marked symbolic meaning.

Antares and Aldebaran are equally distanced from the celestial pole, which at that time was the star Thuban. They are seen in the night sky during cardinally opposed seasons, and when one of them sets at dawn, its counterpart rises. While Aldebaran follows the path of the sun, Antares follows that of the moon, and vice versa. They can never be seen at the same time, as life and death do not belong together. The arms of the goddess mediate between their worlds, which never meet, except in the twilight zone to which only shamans or priests have access. Until around 1,850 BC the constellations to which the two stars belong (Taurus and Scorpius) marked the spring and autumn equinoxes, while the summer and winter solstices were associated with Regulus (Leo) and Fomalhaut (Pisces Austrinis) (Berger, 1976).

3.3. *Archaeological Alignments*

Although Regulus and Fomalhaut were equally important stars, the symbolic implications of Aldebaran and Antares, especially with respect to renewal, life, rebirth and after-life, seem to have been more significant. Placing these four categories within the coordinates of Aldebaran and Antares and taking into consideration that these were precisely the concepts reflected in the orientation of the archaeological alignments dated to the period (Steiner, 2016 and references therein), the placing of the alignments within the same ritual coordinates seems justified. The prevalent east to west orientation of tumuli lines and rock walls may be interpreted as the reflection of the symbolism ascribed to the two stars. Late Chalcolithic-Early Bronze Age archaeological complexes, in which masseboth, shrines, rock art sites and tumuli are distributed along the same orientation, may therefore be related to the positions and the roles of Antares and Aldebaran at the time when the complexes were conceived. The continued ritual significance of the alignments in periods when the equinoxes ceased to be marked by the two stars, and the presence of petroglyphs that despite this fact still emphasize the complementary character of Aldebaran and Antares may be explained with the survival of the traditions that link the two stars with after-life beliefs. Even that since 1,850 BC they are not associated with the spring and autumn equinoxes, their relative position to each other made them into the prominent stars of winter and summer, and their symbolic meaning remained actual. The complementary character of the two red stars continued to define the cycles of seasonal renewal and decay, thus the parallel concepts of life, death and rebirth are still associated with them (ibid.). The fact that the image of the goddess was engraved on a boulder on the Zin-Mehia alignment during the Iron Age, and that the alignment still had a ritual meaning in relatively recent periods (as illustrated by the numerous Thamudic drawings and inscriptions from the 3rd century AD [Tsafrir, 1996]) reflect the continuity of the traditions that were initially inspired by the two stars.

3.4. *The Chalcolithic - Early Bronze Age Goddess*

The symmetry displayed by the icon identified as the goddess seems to mirror the symmetry of the Late Chalcolithic-Early Bronze Age sky, in which her image fits perfectly. This is also recognizable in the illustration in Fig.3. With Thuban as the celestial pole, the intersection of the ecliptic with the celestial equator was located within the constellations of Taurus and Scorpius, on the two opposite ends of the Milky Way. Aldebaran and Antares marked the location of the vernal and autumn equinoxes, while Regulus and Fomalhaut marked the highest and lowest latitudes of the sun, with respect to the equatorial plane. The celestial equator during the Early Bronze Age was the constellation Hydra (Dreyer,1953; Berger, 1976; Parker, 2010). In the illustration, the black circle is the celestial equator and the black interrupted arc represents the ecliptic. The outstretched arms of the goddess encompass precisely the 180° of the ecliptic between its two diametrically opposed intersections with the celestial equator, and point to Aldebaran (the March equinox) and Antares (the September equinox). Her head and body are accurately aligned with Regulus (the June solstice) and Fomalhaut (the December solstice). She is parting the dome of the sky projected on the ecliptic plane into four 90° slices which, in their turn, correspond to the four seasons. However, the four seasons that characterize temperate zones are not so sharply differentiated at subtropical latitudes. They rather represent transitions than seasons. The arms of the goddess differentiate between the dry and the rainy parts of the year, which are separated by the vernal and autumn equinoxes. We do not know much about the seasonal distribution of rainfall during the Early Bronze Age, and the change in the axial tilt of the Earth that occurred since that period was paralleled by changes in climate (Anati, 1985), which probably shifted the months associated with the arrival of the rains. However, the Negev was still in the northern subtropical climatic zone, which is defined by rainy winters and dry summers. That Antares and Aldebaran had a key-role in the transition between the two important seasons is also emphasized by the equidistant position of the goddess between them. Because the image of the goddess fits perfectly in the Early Bronze Age sky, it can be assumed that the figure was conceived around 3,000 BC. However, with Thuban as the pole star, the palm of the goddess, Cassiopeia, sank under the horizon during the dry months. Its designation as the "cutoff hand" and its association with Fatima's hand in contemporary star-lore (Kunitzsch, 1959) may be an ancient inheritance, which shows the esteem in which the constellation was held. Symbolically seen, the appearance of the hand in the sky must have been interpreted by the ancients as the sign that foretold the imminent arrival of the lucky constellations, at which the fingers were pointing. The rising of Cassiopeia was probably associated with plenty and blessings, very much like Fatima's hand (also known as "khamsa") is today. The other two petroglyphs, tentatively placed at the right and the left side of the central figure, are also dated to this period (Anati, 1985; 1986). However, it is difficult to guess their symbolic meaning in an Early Bronze Age context. They evidently represent the stars and constellations associated with the ecliptic longitudes at which the equinoxes occured, but they do not tell a story. Although the dot between the horns of the animal that clearly represents Taurus depicts the position of the sun at the vernal equinox, which would date the engraving to precisely this period, this would still not explain the presence of the third zoomorphic figure, to the right of the bull. The drawings cannot be interpreted as references to the marker stars of seasons either, as the arrival of the rains in different climatic conditions was most likely foretold by other asterisms. These petroglyphs are carefully conceived complex compositions and not mere graffiti. If they were conceived to illustrate banal, everyday aspects, there should be more such engravings, but this couple is unique in its complexity and artistic presentation. The dynamism that characterizes the two compositions seems to express more than simple statements about the whereabouts of certain stars.

4. POLAR AND EQUINOX PRECESSION

4.1. Period

- Thuban was the star closest to the celestial pole from 3,942 BC until 1,850 BC when it was superseded by the much brighter star Kochab (Ursa minor).
- Aldebaran marked precisely the location of the spring equinox in 3,044 BC.
- Antares was associated with the autumn equinox in 3,052 BC.
- Regulus marked the summer solstice in 2,345 BC.
- Fomalhaut corresponded with the location of the winter solstice in 2,582 BC.
- The vernal equinox was associated with the constellation Taurus until 1,865 BC, when it passed into Aries.
- The autumn equinox passed from Scorpius into what we know today as Libra at around 1,850 BC.
- The June solstice passed from Leo into Cancer in 1,458 BC.
- The December solstice passed from Aquarius/Pisces Austrinus into Capricornus at around 1,450 BC.

Although these dates were already mentioned (Berger, 1996; Evans, 1998; Hilton et al., 2006), their listing together evidences that the astronomical stability that characterized the Late Chalcolithic and the Early Bronze Age (3,200 to 2,200 BC), came to an end during the Middle Bronze Age (2,200 BC to 1,550 BC) and the early phases of the Late Bronze Age (1,550 BC to 1,200 BC). The stable period was characterized by the millennia-long association of certain constellations with the equinoxes and solstices, which resulted in the reliability of the seasons. The beliefs, symbols, myths and star-lore, which emerged during this period mirrored the astronomical realities, to which material vestiges like oriented walls, alignments, tumuli lines, masseboth, etc., also confirmed (for a discussion see Avner, 1984; 2001; Haiman, 2000; Steiner, 2010; 2016). Beside the astronomical stability, of which the desert nomad did not know much, there was a parallel "orderliness" displayed by the relative position to each other of the stars and asterisms that were associated with the equinoxes and solstices. This was likely perceived as an evident proof of the beliefs, and the perfect symmetry depicted in the night sky was understood as divinely designed. However, the apparent position of the sun relative to the backdrop of the stars at some seasonally fixed time, say the vernal equinox, slowly regresses a full 360° through all twelve traditional constellations of the zodiac, at the rate of about 1 degree every 71.6 years. This is known as the precession of the equinoxes, which in its turn is related to axial precession or, in other words, when the axis of the Earth changes its inclination, the equatorial plane of the Earth remains perpendicular to it, and the points of intersection with the ecliptic, where equinoxes occur, slowly precess (Berger, 1976). The positions of the poles and equator on Earth do not change, only the orientation of the Earth against the fixed stars. When Thuban was succeeded by Kochab as the pole-star, around 1,850 BC, the equatorial plane moved progressively, together with Earth's axis. The effect was the shift in the position of the equinoxes, as perceived by an observer on Earth, and around 1,850 BC the backdrop of stars against which the equinoxes occurred, was different (ibid.).

4.2. Effects and Fears

Any change that alters a perfect state is to the worse. The symbolic implications of the Earth's new orientation against the fixed stars were catastrophic. Order was exchanged for chaos, not only on the spiritual level, but also in aspects related to everyday life (Velikovsky, 1945). The precession triggered out an arid period (Anati, 1985) that would last until the Iron Age (1,200 BC). The Chalcolithic - Early Bronze Age - Middle Bronze Age I were characterized by high population densities in the Negev, which would be followed by a demographic hiatus (ibid.). The changes in climate started to be felt already at the close of the Early Bronze Age, which was a period of upheaval, from Egypt to Mesopotamia (Velikovsky, 1945). There are many catastrophist scenarios that describe this period, and there are also hypotheses that explain the sudden deterioration in climate and the precipitous end of the Egyptian Old Kingdom with a sudden, non-gradual polar shift (ibid. et passim). Interestingly, out of billions of stars, only one is called by that name: kochab in Hebrew and kawkab in Arabic both mean "star." This may serve as an indicator that the polar shift was a gradual process and the change in the inclination of the Earth's axis did not happen overnight. Thuban was not suddenly displaced by Polaris as the celestial pole, but gradually, with the star Kochab as an intermediate pole. The name was probably given to the star during the Middle Bronze Age I or after, when it was simply referred to as "the star." It must have been also made responsible for the changes and maybe it was forbidden to speak out its real name. In Bedouin star-lore Polaris is the killer of the hero called An-Na'ash, whose coffin, Ursa major, is carried by his three mourning daughters (Bailey, 1974). It seems plausible enough that the origin of the legend goes back to the days when Thuban lost its role in heaven. Thuban did not "jump" to Polaris, just like the equinox did not jump from Taurus to Aries. Actually, Thuban, Kochab and Polaris are not the celestial poles, but the closest stars to it. 2,000 years ago, Polaris was still not the pole-star, although the equinox had already passed into Pisces. The latter is a constellation that occupies many longitudinal degrees. Although in astrology the zodiacal signs occupy only 30° on the ecliptic, in astronomy the length of the constellations varies (Hilton et al., 2006). Therefore, the passage of the vernal equinox from one zodiacal sign into the other does not imply that the celestial pole jumps from one star to another. The change in obliquity is known to attract climatic effects. However, these are gradual and cumulative, and are expressed by the lengthening or the shortening of the seasons in the two hemispheres. Today for example, summers are longer in the northern hemisphere (Berger, 1976). It seems that a critical point was reached at the beginning of the Middle Bronze Age, which led to a sudden deterioration in climate, the consequences of which were catastrophic (Velikovsky, 1945).

The symbolic consequences were just as far-reaching as the climatic ones. The changes in climate were probably correlated with the escape of the sun from between the horns of Taurus to Aries. The fear of losing it definitively must have been very real, together with the need to restore it and to re-establish the order that ruled the world prior to its escape, that is, before chaos took over on earth and in the sky (Rotblum, 2010). The ram was most likely thought to be the thief who stole the sun and who must therefore be hunted down. The epical hunt is apparently the theme of many Middle Bronze Age myths, worldwide (ibid.). It is depicted in the rock art of regions between which there were no cultural exchanges at that time. In the Negev, where the effects of the equinox shift were devastating, the story of the event must have also been carved in the rock.

4.3. Rock Art and Mythical Re-orientation

A closer look at the two petroglyphs that appear at the location of the Early Bronze Age equinoxes in Fig. 3. - in the light of the implications described above - may explain their complexity and dynamism. The zoomorphic figure seems to be a combined image, in which three animals are depicted as one. The ibex may be identified as Orion, because of the three dots that represent Orion's Belt, and because of its relative position to the horned animal facing the opposite direction, which is evidently Taurus. The dot between the horns of the latter - in this context - does not depict Aldebaran, but the position of the sun at the time of the Early Bronze Age vernal equinox. The third animal does not display the impressive horns of the ibex or the bull, but it has smaller horns and markedly ovine features. Judging by its relative position to Orion and Taurus and by its size, it may be identified as Aries, the ram. Above it, in front of the dot between the horns of the bull, there is another dot surrounded by concentrically arranged lines. Although it could be interpreted as a depiction of the Pleiades in the context of contemporary star-lore, the detail is differently arranged and it is rather a representation of the sun. The rock engraving is apparently the story of the sun passing from Taurus into Aries around 1,850 BC. The patina covering it shows that the composition was drawn as it is, at the same time, and that the ram and the new position of the sun are not later additions. Thus, it must be dated to the Middle Bronze Age - this would mean that it is more recent than what its age was suggested to be (Anati, 1985; 1986; 2015).

Whereas the bull, the ram and the two suns tell the story of the equinox shift, Orion's role in the composition is not so clearly defined. It must be more than a simple reference to the position of the bull and the ram. The fact that it faces the opposite direction and that at its feet a fourth animal (constellation Lepus, Al-Arnab, the hare) does the same, implies that its presence is not incidental, but a part of the story. In the original engraving, there is an anthropomorphic figure with a dot between its legs, placed above the combined image of ibex, bull, ram, and hare. The latter is perceived as a symbol of fertility in contemporary star-lore (Bailey, 1974). The dot between the legs of the anthropomorphic figure and between the horns of the bull are complementary, and they illustrate the plenty associated with the spring equinox and the role of the ritual specialist's magic in the unfailing annual return of the plenty. Before the precession, Orion was on the path of the sun during the period of plenty that followed immediately the equinox. This may explain its presence in the composition, where the symbolism of the hare and the magic of the shaman try to spell Orion back to its place, with the hope that the bull, the ram and the sun would follow it. There are many engravings in the Negev that depict a hunt that seems related to the interpretations above. An animal - predominantly the ibex - with a dot between its horns is shown to be chased by two dogs, or hunted by a figure with a bow and with ideograms around its head, which would identify it, again, as the ritual specialist. This only shows that the engravings do not depict common hunt scenes (Eisenberg-Degen and Rosen, 2013; Mailland, 2015). The animal is sometimes chased in the direction of the precession, in which case the engravings would illustrate the phenomenon. Gemini (the twins) and Canis major/minor (the dogs, of which the larger is Sirius) are apparently involved in the hunt. When the ibex is chased in the direction opposite to that of the precession, the petroglyphs seem to voice the wish to see Orion back on its proper longitude on the ecliptic, only a few degrees away from its intersection with the equatorial plane, as it was the case before the sun escaped from between the horns of the bull. Perseus may in this case be identified as the hunter/shaman who initiates the chase, which is rather a magical, figurative hunt. In today's context, the hunt expresses the wish to precipitate the arrival of spring and the plenty associated with it, which means that the dogs chase Orion towards Aries and Pisces, in other words, they chase the winter away – Bedouin star-lore describes this hunt ('Id al-Kashkhar and 'Id al-Wadj, pers. comm.). The second petroglyph in the illustration is placed diametrically opposed to that described above, at the location of the Early Bronze Age autumn equinox on the ecliptic. Like its counterpart, it also relates the story of the precession. At the time when Thuban was close to the celestial pole, the constellation Bootes - identified as the lizard in the engraving - was a constant feature in the sky, given its proximity to Thuban. Today it disappears shortly after the September equinox and it only returns after the March equinox. During the summer months, the constellation is almost overhead. The disappearance of such a prominent constellation from the circumpolar area was associated with the advance of the equinox from Scorpius to Libra, which occurred at the same time with that from Taurus to Aries. The snakes in the petroglyph can be very easily reconstructed as curves drawn along the main stars of Virgo (Steiner, 2016). The advance of the equinox on the ecliptic led to the banishment of the lizard from the circumpolar area and to its seasonal exile below the horizon. The figure of the lizard may be interpreted either as that of a villain responsible for the changes, or as that of a victim. In the first case Bootes would be a "fallen angel" banished from the circumpolar area (the seat of God) because of his guilt in triggering out the chaotic conditions that followed the precession of the equinoxes. The role of the scorpion and the snakes would be positive in this interpretation. If the lizard were a victim, the snakes and the scorpion would be the evil characters of the story. However, considering the relatedness of Antares to Aldebaran and the association of Scorpius with the autumn equinox during the Early Bronze Age, the seasonal exile of Bootes may be understood as a well-deserved punishment for some kind of involvement in the precession, which was probably the subject of a long-forgotten Early Bronze Age story of which this rock engraving is the sole reminder. Again, it must be made clear that Bootes did not disappear from one autumn to the next, but was slowly "dismembered," through the decades. Poles and equinoxes do not "jump", but shift gradually. Another consequence of precession, parallel to the gradual disappearance of Bootes was that the "cut-off hand" of the goddess - the constellation Cassiopeia - became a constant feature in the sky. This did not happen overnight either, but it was a slow process. The increase in the popularity of the goddess during this period may have its roots in this interpretation (Achrati, 2003; Cornelius, 2000), which may also explain why an icon conceived during the Early Bronze Age would still be depicted in relatively recent petroglyphs. The constellation Draco, that is, the seat of God, the goddess, and the complementary character of Antares and Aldebaran are the only reminders of the Early Bronze Age order. The symbolic implications of the goddess are even more pronounced today: the arc described by her fingers encompasses progressively - between November and May - the constellations which were associated with the vernal equinox during the last 6,200 years, from Gemini to Pisces, while at the same time, the fingers also point at the lucky stars associated with plenty between late October and December. Between May and October, the snake-arm describes an arc from Sagittarius to Virgo, the ecliptic degrees associated with the autumn equinoxes, from prehistoric to modern times. Her outstretched arms - in the context presented in this paper - still mediate between the two celestial regions.

In conclusion, the two other petroglyphs were most likely engraved immediately following the changes triggered out by the precession of the equinoxes, during the Middle Bronze Age. Their dating by Anati (1985) to an earlier period is challenged by the stories that they apparently tell. They were conceived and incised in the rock during the last phases of human habitation in the Negev, shortly before the demographical hiatus that would last until the Iron Age. Their location on Mount Karkom, one of the most sacred mountains in the Negev (ibid.), seems to be more than incidental. On the mountain, there are many other petroglyphs that were engraved around 4.200 BC, which corresponds to the beginning of the Chalcolithic. The date marks not only the start of an archaeological period, but also that of the long-lasting astronomical interval in which the four cardinal stars and the constellations to which they belong were associated with the equinoxes and the solstices. The early engravings on the mountain most likely illustrate the process of mythic re-orientation to the "orderly" and symbolically meaningful Chalcolithic sky. Pastoralism became an established economy during a period of astronomical stability (Close, 2002; Haiman, 2002), to which its beliefs and symbols were adjusted and in which they were also mirrored (Rosen, 2007; Steiner, 2016). Given the longitudinal extension of both Taurus and Scorpius, the order lasted for almost two millennia and a half. The position of Antares and Aldebaran in these constellations and their symbolically complementary character only added to the significance of the cosmic order. At the beginning of the Middle Bronze Age, the visual effects and the climatic implications of the precession brought the long-lasting equilibrium to a relatively abrupt end (Berger, 1976). With chaos setting in, a mythological re-orientation became imperative. Apparently, the cult of a goddess associated with plenty and fertility gained in popularity during this period (Cornelius, 2014), when due to the arid conditions precipitated by the changes, these concepts became idealized and imagined to exist only in a better after-world.

5. CONCLUSION AND RECOMMENDATIONS

Even in the most cautious approaches to the subject, the prevalent east to west orientation displayed by purportedly Chalcolithic - Early Bronze Age tumuli lines, walls and other archaeological remains is recognized as being related to astronomical alignments, that is, solstices and equinoxes. Mortuary cult, ancestor worship and after-life beliefs are referred to when the orientation of the alignments is explained. As the function of rock art is also recognized to be ritual, when there are petroglyphs on such archaeological alignments, they must be regarded as graphic illustrations of the concepts and beliefs that inspired the orientation of the latter. Hence, rock art may also be approached from an archaeoastronomical perspective. However, the engravings cannot be placed haphazardly into the sky, and other factors than mere shape must also be considered. Contemporary nomadic star-lore may prove to be a useful instrument, as the cyclic renewal of seasons - which is apparently its main theme - may be related to the concepts of life, death, rebirth and transformation which, in their turn, are the concepts behind the topographical layout of the archaeological alignments. In a sense, star-lore also displays a prevalent east to west orientation: the cyclic renewal of seasons and life is associated with marker stars, which rise in the east and set in the west, or constellations, which are located either in the western or in the eastern celestial hemispheres, depending on the seasons in which they can be observed and on their location with respect to the points on the ecliptic where the equinoxes occur.

An iconic figure that is engraved on a rock situated on one of the aforementioned alignments and two complex, and carefully executed petroglyphs - which are recognizable as depictions of constellations - were selected and placed into the sky. Not only that they fit perfectly, but they also seem to illustrate the stories related by Bedouin star-lore. The iconic figure - which was tentatively identified as a goddess - seems to define the seasons, and her arms point to the celestial areas where the two other petroglyphs were placed. Moreover, she seems to emphasize the relationship between the red stars Antares and Aldebaran which, besides being symbolically related in tradition, also have a key-role in the summer, respectively in the winter sky. The fingers of the goddess point at the marker stars of the rainy season. However, the two petroglyphs were dated to the Early Bronze Age, while Bedouin star-lore describes the contemporary sky and seasons. Due to axial and equinox precession, the Early Bronze Age climate was different than that described in Bedouin star-lore and the rains were likely predicted by other marker stars. Placing the icon and the two other petroglyphs into the Chalcolithic – Early Bronze Age sky, the former only gains in importance, whereas the two petroglyphs, although fitting the location of the vernal and autumn equinoxes, do not seem to state more than this mere fact. The arms of the goddess point to Aldebaran and Antares, the stars that marked the March and September equinoxes around 3,050 BC and her body is perfectly aligned between Regulus and Fomalhaut, which were associated with the summer and winter solstices. Her head is in the vicinity of Thuban, the polar star of the period. Her position leaves no doubt about the icon being conceived during the period discussed, and the fact that she also fits in the contemporary sky seems to be incidental, although her suggested location between Aldebaran and Antares is still meaningful. Thus, the initial presumption of the paper, namely that the icon was related to the concepts of renewal and passage in the case of star-lore, and to rebirth and after-life in the case of astronomically oriented alignments seems solid enough. The identification of the icon as a goddess associated with after-life beliefs during the Chalcolithic and Early Bronze Age is supported by its evident relatedness to the stars Aldebaran and Antares, which in their turn played important symbolic roles in the same beliefs. The other two petroglyphs become meaningful only when placed into the Middle Bronze Age I sky, in which they suddenly come alive. They illustrate the precession of the equinoxes, which occurred around 1,850 BC.

The archaeoastronomical approach adopted in this paper contradicts the Early Bronze Age date that was attributed to the two petroglyphs. This may be a sign that the still widely accepted and employed dating techniques - based on a stylistic analysis of a small number of photos randomly taken by inexperienced travellers in the 1950s and extrapolated to all of Arabian and southern Levantine rock art should be reconsidered. The application of archaeometry is strongly recommended, especially because it is encouraged by the International Federation of Rock Art Organizations (IFRAO), of which the local Negev Rock Art Center (NRAC) is the newest (58th) member.

The precession was accompanied by heavy consequences in the Negev and the demographical hiatus between Middle Bronze Age II and the Iron Age may be attributed to it. The climatic catastrophe was accompanied by a symbolic catastrophe, perceived as the dissolution of the cosmic order into chaos. The Chalcolitic – Early Bronze Age sky was characterized by a perfect order, which is mirrored in the symmetry displayed by the goddess. The equinoxes and solstices were marked by the four cardinal stars, with the Milky Way stretching between Taurus and Scorpius, and the constellation Hydra as the celestial equator. This order lasted for a very long period, given the longitudinal extension of the constellations Taurus and Scorpius. The prevalent orientations exhibited by the archaeological alignments of the period reflect this order, in which the symbolic and astrological correlations between Aldebaran and Antares played a key-role. Thus, it is proposed that rather this relationship, than the direction of the sunset on the day of the summer solstice should be considered in the study of such alignments, and that the role played by a goddess in the beliefs that determined the construction of these alignments should also be researched. Her worship at these complexes may be indicated by the prevalence of broad female-masseboth over tall and narrow male-stones. As masseboth continued to be erected until relatively recent times, and also because symbols and icons conceived in the Early Bronze Age would be engraved in the rock thousands of years later, the continuity in the archaeological data reflects a parallel continuity in beliefs. Apparently, the goddess still played a key-role during the Iron Age, as evidenced by the petroglyph in Nahal Zin. As Aldebaran and Antares are closely related to the goddess, their symbolic meaning outlived the astronomical context in which they rose to prominence, which is still recognizable in their central role in Bedouin star-lore and in the orientations observed by archaeological alignments, in relatively recent times. The persistence of the Ashera cult is the most palpable example of the continuity of beliefs crystallized in prehistoric periods.

In this paper, the connection of the present Bedouin population of the Negev Highlands to the rock art corpus found in their tribal lands becomes evident. Therefore, any decision regarding the research and protection of the said rock art should involve the elders of the Janabib and 'Azazme tribes, who should be recognized as the traditional owners of the petroglyphs. The IFRAO Code of Ethics (www.ifrao.com/wp-content/uploads/2014/06/ethic.doc) is very specific on this, and the literature that addresses rock art traditions in the southern Levant also highlights the connection of contemporary pastoralists to the rock art of their cultural and genetic ancestors. Having in mind that the inauguration of a National Park meant to familiarize the public with the rock art of past and present nomads is considered by the Negev Rock Art Center, following the ethical lines of IFRAO is strongly recommended in its planning. Moreover, given that rock art only offers a restricted spectrum of the culture that created it, a National Park should offer the public an encompassing picture of the traditions that are mirrored in the symbolism of the said rock art. I would recommend that the planned National Park should focus on the preservation and presentation of Bedouin nomadic culture as a whole, including herding, traditional dwelling and various cottage industries. Such an approach would be beneficial not only to the Janabib and 'Azazme Bedouin – to whom being part of such a protected area could offer recognition and legal status on their traditional lands - but also to the Negev Highlands Regional Council which, because of its interest in promoting tourism, would only profit from the revenues of a Bedouin Heritage National Park.

ACKNOWLEDGMENTS

I would like to express my gratitude to Professor Robert G. Bednarik who has attracted my attention to the origin of the shortcomings in the prevalent dating techniques of the rock art of Arabia in general, and of the southern Levant, in particular. I am also indebted to the elders of the al-Kashkhar and al-Wadj clans of the Janabib tribe, who have accepted me as kin and shared their traditional stories and knowledge, around desert campfires under the stars. Special thanks to Dr. Liora Kolska Horwitz, head of research at the Negev Rock Art Center.

REFERENCES

Abu-Rabi'a, A. (2001) Bedouin century: education and development among the Negev tribes in the twentyeth century. Berghahn, New York.

Achrati, A. (2003) Hand and foot symbolisms: from rock art to the Qur'an. *Arabica*, vol. L, No.4, pp. 464-300.

Amzallag, N. (2008) *The copper revolution: smelters from Canaan and the beginning of civilization.* Hameara, Shani-Livna.

Amzallag, N. (2009) From metallurgy to Bronze Age civilizations: the synthesis theory. *American Journal of Archaeology*, Vol. 113, pp. 497-519.

Anati, E. (1968a) *Rock art in Central Arabia,* Vol. 1. *The 'ovalheaded people of Arabia'.* Bibliothèque du Muséon, Institut Orientaliste, Louvain.

Anati, E. (1968b) *Rock art in Central Arabia,* Vol. 2. *Part I: Fat tailed sheep in Arabia; Part II: The realistic dynamic style of rock art in Jebel Qara.* Institut Orientaliste, Louvain.

Anati, E. (1972) *Rock art in Central Arabia,* Vol. 3. *Corpus of the rock engravings, Parts I and II.* Institut Orientaliste, Louvain.

Anati, E. (1974) *Rock art in Central Arabia,* Vol. 4. *Corpus of the rock engravings, Parts III and IV.* Institut Orientaliste, Louvain.

Anati, E. (1985) Has Mt. Sinai been found? *Biblical Archaeology Review*, Vol. 11, p. 4.

Anati, E.(1986) *The mountain of God.* Rizzolli International Publications, New York.

Anati, E. (2015) *The rock art of the Negev and Sinai.* (3rd ed.), Atelier Editions, Capo di Ponte.

Anati, E. (2017) Editorial notes. *Expression,* Vol. 15, pp. 2-3.

Avner, U. (1984) Ancient cult sites in the Negev and Sinai Deserts *Tel Aviv,* Vol. 11, pp. 115-131.

Avner, U. (1990) Masseboth sites in the Negev and Sinai and their significance. In J. Aviram (ed.), *Proceedings of the Second International Congress on Biblical Archaeology in Jerusalem*, pp.166-181.

Avner, U. (2001) Sacred stones in the desert. *Biblical Archaeology Review,* Vol. 27, pp. 30-41.

Bailey, C. (1974) Bedouin star-lore in Sinai and the Negev. *Bulletin of the School of Oriental and African Studies,* Vol. 14, pp. 35-80.

Bailey, C. (1989) The Janabib tribe in the Negev: facts and folkloric traditions. *Notes on the Bedouin*, Vol. 20, pp. 9–21.

Bastoni-Brioschi, R. (1998) Arte rupestre: Har Karkom e il dio Sin. In F. Mailland (ed.), *Archeologia e Mito,* pp. 21-38. Atelier Editions, Capo di Ponte.

Bednarik, R. G. and Khan, M. (2005) Scientific studies of Saudi Arabian rock art. *Rock Art Research,* Vol. 22, No.1, pp. 49-81.

Bednarik, R. G. (2017) Scientific investigations into Saudi Arabian rock art: a review. In *Mediterranean Archaeology and Archaeometry*, Volume 17, No. 4 (Rock art special issue); Al Ghanim, K. (ed.), The rock art of pre-Islamic Arabia - part I., pp. 43-59.

Berger, A. L. (1976) Obliquity and precession for the last five million years. *Astronomy and Astrophysics,* Vol. 51, pp. 21-42.

Burnham, R. (1978) *Burnham's Celestial Handbook.* Dover, New York.

Close, A. E. (2002) Sinai, Sahara, Sahel: the introduction of domestic caprines to Africa. In R. Kuper (ed.), *Africa Prehistorica*, Vol. 14. *Tides of the Desert*, pp. 459-470. Köln.

Cornelius, I. (2004) *The many faces of the goddess: the iconography of the Syro-Palestinian goddesses Anat, Astarte, Qedeshet, and Asherah, c.1500-1000 BCE.* Academic Press, Fribourg.

Dreyer, J. L. (1953) *A history of astronomy from Thales to Kepler.* (2nd ed.) Dover, New York.

Eisenberg - Degen, D. and Rosen, S. A. (2013) Chronological trends in Negev rock art: the Har Michia petroglyphs as a test case. *Arts*, Vol. 2, pp. 225- 252.

Evans, J. (1998) *The history and practice of ancient astronomy.* Oxford University Press, Oxford.

Fabian, P. (2005) (Unpublished PhD thesis) *Avdat during the establishment of the Provincia Arabia: the Roman military camp and its position in the eastern defense system of the Empire.* Ben-Gurion University of the Negev, Beersheba.

Galilee, E., Kark, R. and Kressel, G. M. (2013) (in Hebrew) Like footprints in the desert: reflection of tribal boundaries and immigration processes in Negev Bedouin cemeteries, 1800–1966. *Hamizrah Hahadash*, Vol.52, pp. 116–137

Gilead, I. (2002) Religio-magic behaviour in the Chalcolithic period of Palestine: smiths and shamans come from the same nest. In I. Oren and S. Ahituv (eds.), *Magic and craftsmanship in 5th millennium Palestine – Aaron Kempinski Memorial Volume,* pp.103-128. Beer Sheva University Press, Beer Sheva.

Haiman, M. (2000) (in Hebrew) "The K-line" at Har Romem in light of the survey of the map of Har Romem. *Atiqot,* Vol. 39, pp. 21-29.

Haiman, M. (2002) (Conference paper) Enclosed structures, pens, and pastoralism in the Negev and Sinai Deserts - Past and Present. *ASOR Annual Meeting*, Toronto.

Hilton, J. L. et al. (2006) (IAU home page, .pdf download) Report of the International Astronomical Union Division I Working Group on precession and the ecliptic.

Isaac, B. (1992) *The limits of Empire: The Roman army in the East.* Oxford University Press, Oxford.

Khan, M. (2000) *Wusum, the tribal symbols of Saudi Arabia, Parts I and II.* Ministry of Education, Riyadh, Kingdom of Saudi Arabia.

Kunitzsch, P. (1959) *Arabische Sternnamen in Europa.* Otto Hrasowitz Verlag, Wiesbaden.

Liritzis, I., Al-Otaibi, F. M., Castro, B. et al. (2015) Nabatean tombs orientation by remote sensing: provisional results. *Mediterranean Archaeology and Archaeometry,* Vol. 15, No 3, pp. 289-299. doi: 10.5281/zenodo.33835

Lewis-Williams, J. D. and Dowson, T. A. (1988) The signs of all times: entoptic phenomena in Upper Palaeolithic art. *Current Anthropology,* Vol. 29, pp. 201– 245.

MAA (2017) Arabian Peninsula Rock Art in Pre Islamic Era (Part I), Kaltham al-Ghanim (editor) *Mediterranean Archaeology and Archaeometry,* SPECIAL ISSUE Vol. 17, No.4, (www.maajournal.com).

Magli, G. (2015) *Archaeoastronomy: introduction to the science of stars and stones.* Springer International Publishing, doi 10.1007/978-3-319-22882-2_8

Mailland, F. (2015) Ibex, crescent and swastika as symbols of a lunar god in the rock art of the ancient Near East and Central Asia. *Expression,* Vol. 10, pp. 48-53.

Pannekoek, A. (1961) *A history of astronomy.* Dover, New York.

Parker, J. (2010) *The astrologer's handbook.* Alva Press, Brunswick, NJ.

Poret, R. (transl.) (2010) *Der Koran.* Kohlhammer Verlag, Stuttgart.

Rosen, S. A. (2007) The rise of desert cult: the sacred precinct at Ramat Saharonim, Central Negev. *Bulletin of American Schools of Oriental Research,* Vol. 346, pp. 1-27.

Rotblum, Y. (2010) *Heavenly art.* Lulu Independent Publishing Platform.

Sharon, M. (1975) The political role of the Bedouin in the sixteenth and seventeenth centuries. In M. Maoz (ed.), *Studies on Palestine during the Ottoman Period.* Magnes Press, The Hebrew University of Jerusalem.

Steiner, G. (2010) *Smiths and prophets: the curse of Cain on history.* CSIPP | Renetius, Zurich

Steiner, G. F. (2016) The goddess and the copper snake: metallurgy, star-lore and ritual in the rock art of the southern Levant. *Expression,* Vol. 12, pp. 73-91.

Tomkins, P. (1971) *Secrets of the Great Pyramid.* Harper Colophon Books, New York.

Tsafrir, N. (1996) New Thamudic inscriptions from the Negev. *Muséon,* Vol. 109, pp. 1-22.

Van der Waerden, B. L. (1953.) A history of the zodiac. *Archiv für Orientforschung,* Vol. 16, pp. 27-64.

Velikovsky, I. (1945) *Ages in chaos: a reconstruction of ancient history from the Exodus to King Akhnaton.* Doubleday, New York.

Wachtel, I. (2014) 'Jethro cairn' – an Early Bronze Age site in the Upper Galilee, Israel: function and significance. *9th International Congress on the Archaeology of the Ancient Near East.* June 9 - 13, Basel, Switzerland.

Wilmsen, E. (2009) *Tsodilo Hills: Copper bracelet of the Kalahari.* Michigan University Press, Ann Arbor, MI.

26

SURVEY, ARCHAEOASTRONOMY AND COMMUNICATION: THE MAUSOLEUM OF THEODORIC IN RAVENNA (ITALY)

M. Incerti*, G. Lavoratti and S. Iurilli

Architecture Department of Ferrara (Italy)

Corresponding author: manuela.incerti@unife.it

ABSTRACT

The Mausoleum of Theodoric (520 ca.), one of Ravenna's most globally renouned monuments and UNESCO heritage site, has been object of archeoastronomical research in the past years. This paper examines the orientation of the building as well as other elements of the architecture, especially the small openings of the building, to investigate their possible archaeoastronomical significance. The architectural survey was carried out with a laser scanner device (Faro-Focus 3D, software Scene) while the textures were extracted using digital photomodelling software. Starting from the archaeoastronomical analysis, a 3D model was developed to check astronomical phenomena and allow the communication of scientific content through several forms of media. This second part of the contribution is part of the experiments conducted by the research group (Incerti and Iurilli 2014), (Incerti and Iurilli 2015), focusing on new means of multimedia communication, both interactive and not. The research looks at virtual models as a means of *edutainment,* aiming to maximise the fruition of artifacts and cultural sites.

KEYWORDS: Theodoric, architectural survey, archaeoastronomy, edutainment, digital museum

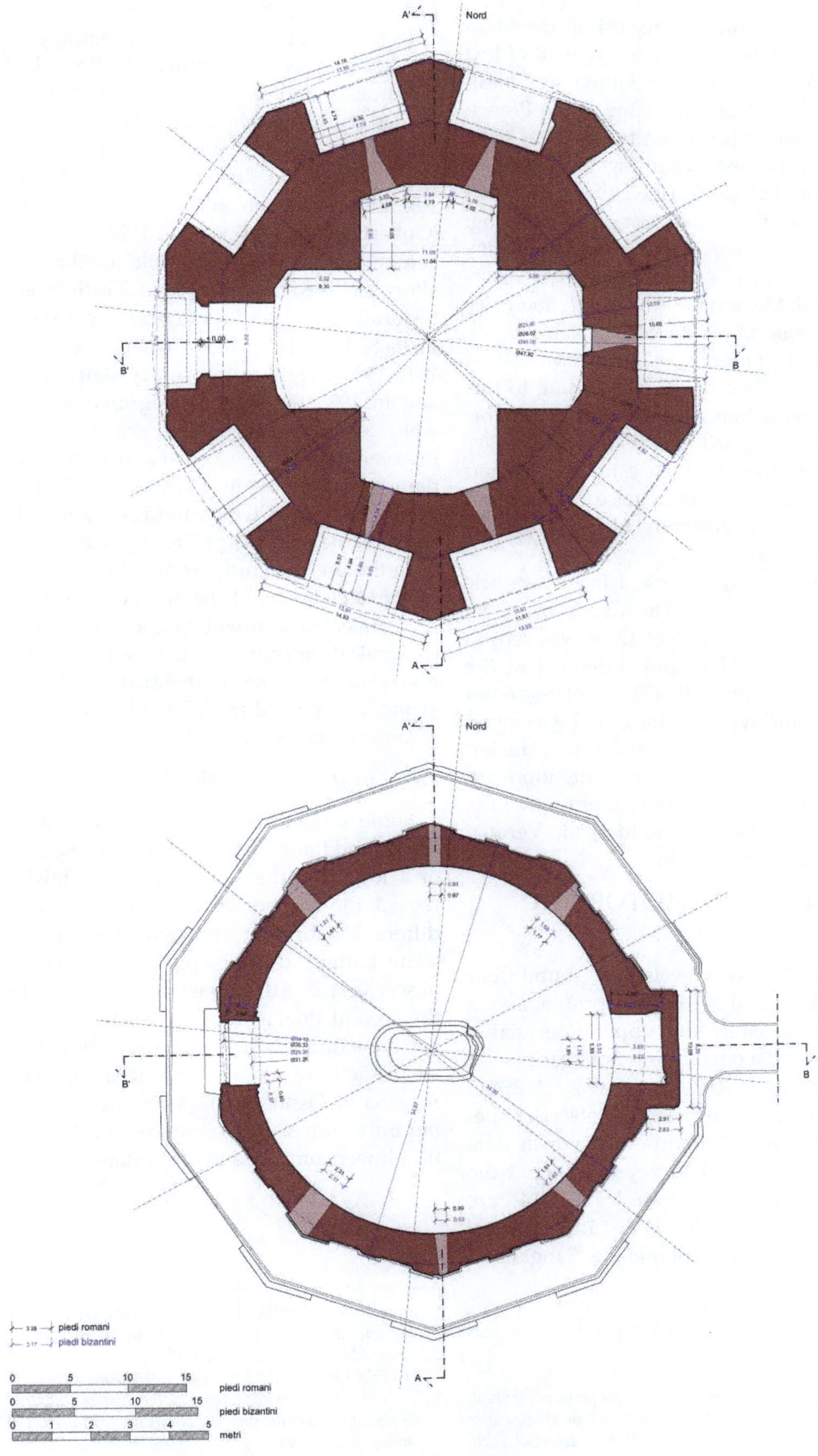

Figure 1. Mausoleum of Theodoric, Ravenna. Plans of the ground floor and first floor. Graphic elaborations of 3d survey.

1. INTRODUCTION

The most ancient information regarding the Mausoleum of Theodoric is included in a fragment of text written by an anonymous reporter known as Valesiano (*Anonymus Valesii),* from the name of the French academic Henri Valois, who edited the piece in 1636. The second part of the text, regarding the period in which Odoacre and Theodoric ruled (years between 474 and 526), narrates the foundation of the building: *"Se autem vivo fecit sibi monimentum ex lapide quadrato, mirae magnitudinis opus, et saxum ingens quod superponeret inquisivit".*[1] Meaning *"When still living, he erected a funeral monument for himself in squared blocks, a work of marvellous grandeur, and requested that an enormous rock be found to cover it".*[2] According to this document, the construction of the building was ordered by Theodoric himself before his death (30th August 526). It was then built in carefully squared blocks, a rarity for this time and practice, and covered with a "*saxum ingentem*", evidently purposefully demanded (*inquisivit*) by the King.

The commissioner of the work, King Theodoric was born around 454, son of Theodemir from the Amali family. At the young age of 12 he was sent to Constantinople as a hostage, and remained at the court of Leo I the Thracian until 472. Scholars do not agree on the terms and type of education he received in the East, in any case it is undeniable that during his kingdom he showed great attention to architecture. This is testified by the restoration of the ancient buildings of Rome, new buildings in Verona, in Pavia and, above all, in Ravenna.

2. THE ARCHITECTURE: HISTORICAL NOTES

The mausoleum is on two levels: the ground floor has decagonal plan in the external profile and a Greek cross interior, while the upper floor has a decagonal exterior and a circular interior space.

Like all monuments of Ravenna, now a UNESCO heritage site, the building has been the object of specialized studies (the most important of which date back to the last century,[3] and survey campaigns: the oldest, also published by Heidenreich (108-114), are those of Giuliano da Sangallo (1445-1516),[4] Baldassarre Peruzzi (1481-1536),[5] Antonio da Sangallo il Giovane (1483-1546)[6] and Giovanni Battista da Sangallo il Gobbo (1496-1552).[7]

In its long history, the small, central plan building was supported by other bodies, including a lighthouse, and later annexed to the church of a monastery (Santa Maria della Rotonda) where it was used as a "choir" (Guberti, 1952, 8-16). In 1748, a large part of the first level of the Mausoleum was still interred in an area often subject to flooding (Guberti, 1952, 17). In 1774 two flights of symmetrical stairs were built on the west side to allow access to the upper floor. Further campaigns of restoration and archaeological excavation followed in 1808, 1810, 1825, 1839, 1844 (Guberti, 1952, 19). In 1918-19 the eighteenth century stairs were removed and in 1927 the surrounding area was released as well as the first access ramp built. Further interventions were necessary in 1945 because of the damage caused by the war[8], as in 1977[9] and finally in 1998, the year in which the restoration of the stone facing stone of Aurisina[10] took place.

In the present study, some of the issues discussed relate to elements of the architecture that are possibly connected to orientation and to their archaeoastronomical meanings. It is therefore absolutely necessary to verify the authenticity and dating of the elements involved in the analysis in order to avoid misinterpretation of the data.

2.1. *The question of the "unfinished"*

Some small arches appear on the external face of the second floor, which may hint at the past presence of a loggia, perhaps lost or never finished. In this regard, the question of "unfinished" and the possible different dating of the two levels introduced by some authors do not appear to interfere with our observations. All reconstructions hypothesised for the second floor, amongst which one must remember the extremely accurate and sophisticated one of De Angelis of Ossat (very accurate graphics) (De Angelis of Ossat 1962), never involve the openings but only address the presence and shape of the portico, lower compared to the system of windows.

---

[1] *(Muratori, 1738), text available at link in bibliography ("Anonymus Valesianus: Chronica Theodericiana," n.d.).*

[2] *Translation in (Bovini 1977, 10-11).*

[3] *Amongst the most important and extensive works we point out (Haupt, 1913; Guberti, 1952; Heidenreich and Heinz, 1971) whose studies were actually carried out in 1938, and (Bovini, 1977). See also (Gotsmich, 1958; Johnson, 1988).*

[4] *(Sangallo, 1910), f. 38-37, see also the Geymuller codex, Uffizi, of which Heidenreich publishes an image (Heidenreich and Heinz, 1971, fig. 65, p. 63).*

[5] *Uffizi, Dis. 441 A (Heidenreich and Heinz, 1971).*

[6] *Antonio da Sangallo, Uffizi. Dis. 1406 A, 888 Av, 887 Ar, (Heidenreich and Heinz, 1971). On the same subject see also Uffizi. Dis. 875, 878, 896 (De Angelis d'Ossat, 1962, 97, tav. V)*

[7] *Uffizi Dis. 1563 Av, 1563 A, 1394 A, (Heidenreich and Heinz, 1971).*

[8] *The substitutions are highlighted by the different colours of the ashlars in the photographs of the time, see also (Conti and Berti, 1997).*

[9] *(Bovini, 1977, I-XV). Regarding restorations see (Piazza, 2013, 84-86) from the same volume see pages 24-25, 35-27, 56-57, 72 and especially the profile by Paola Novara pages 111-116.*

[10] *The recently identified construction material is the stone of Aurisina (Piazza et al., 1998; Bevilacqua, Fabbri, Grillini, and Iannucci, 2003).*

2.2. *The flooring*

The current flooring of both rooms, as previously mentioned, is certainly not original: in 1557 Leandro Alberti mentions traces of a mosaic floor, evidently placed in the top cell, since the bottom was buried underground.[11]

Regarding the progress of this element historians have reported a major failure of the ground on the Eastern side, which led to a drop of 14 cm in the ground floor and a 6 cm drop of the upper floor. The difference in height between the two levels led scholars to believe that an initial failure occurred during the construction of the ground floor. For this reason the upper floor was probably put in place "leveling" the already installed plan, which however, later another slight lowering. Some drawings stored in the Superintendency of Ravenna [12] show an accurate survey of the pavement slabs dated 1903, on which many measurements are noted as well as the measurement of elevation at a point indicated by the letter A. The existing pavements were put in place during the works of the biennium 1975-77 (Novara, 2013, p. 116).

2.3. *The small apse*

On the eastern side of the top space there is a small apse, whose function has been questioned by many historians: its height cannot accommodate an altar or an officiant or even the great porphyry sarcophagus (today placed at the center of the space). On the keystone of the arch is a large Latin cross, the only sculptural element of the interior, which highlights its relevance in the project. The small space, whose floor was slightly lower than that of the rest of the room is, according to scholars, contemporary to the building (Messina, 1980; De Angelis d'Ossat, 1962, 128-129).

2.4. *The sarcophagus*

According to tradition, the remains of King Theodoric were conserved in the great sarcophagus of red porphyry measuring 305x190x101 cm. The tub is characterized by four rings on the top edge and two lion heads at the bottom center of the side faces. The sarcophagus' troubled history has been well documented, its movements traced by Ambrogi (Ambrogi, 1995, 109-111), recalling its relocation to the site in 1913. There is no certainty regarding the original orientation of the object, which is however considered by scholars to be consistent with building, originally arranged in the East-West direction (the current one).

2.5. *The small windows*

The wall of the ground floor has a thickness of about 140 cm and is pierced by 6 splayed narrow slits, with an approximately horizontal intrados, arranged on three sides (each of which is divided into three segments). Their sizes vary in width from 11 to 25 cm, a minimum height of 40 cm, and 70 cm, and are arranged as follows:

- (1a, 2a) Northern wall: 2 small windows placed in the first and third segments (intrados lowered outward approximately 6 cm)
- (3b, 4b) Eastern wall: 2 small windows vertically aligned with the axis of the building.
- (5c, 6c) Southern wall: 2 small windows placed in the first and third segments (intrados lowered outward approximately 6 cm).

The decagonal part of the upper level ends with a series of cylindrical section bands, the highest of which is decorated by the famous "tenaglia" (pincer) shaped design. The central and receding one, about 77 cm thick, is smooth and perforated by 11 windows. Arranged roughly in the directions North-South, East-West with the two diagonals at 45 ° (directions of the compass rose), the small openings have dimensions that vary from 40 cm in height for the windows on the North-South axis, to 62 cm on the diagonal axes. Starting from the entrance, we have:

- (7-8-9-10) 4 small windows with rounded arches, combined two by two (West)
- (11) 1 small rectangular window with overhead arch between two small straight sections (North-West)
- (12) 1 small window with rounded arch (North)
- (13) 1 small rectangular window with overhead arch between two small straight sections (North-East)
- (14) 1 small Greek cross window (East)
- (15) 1 small rectangular window with overhead arch between two small straight sections (South-East)
- (16) 1 small window with rounded arch (South)
- (17) 1 small rectangular window (South-West).

The opening are almost unanimously considered contemporary to the founding of the building, excluding the rectangular South-Western one which was clearly enlarged at a later date (Guberti, 1952, 59; De Angelis d'Ossat, 1962, 94)[13]. In various pictures preceding the last restoration one can notice the presence of a twelfth opening positioned on the back wall of the small apse, probably added during the period

[11] *Sources in (Fagiolo, 1972, 148-149).*

[12] *Dating from 1903 numbered 2080 and 2081.*

[13] *According to (Heidenreich and Heinz 1971, 58-60) the small windows on the diagonals are not original but date back to the Middle Ages.*

when the building became a church choir (Messina, 1980, 33; Guberti, 1952, fig. 26 pp. 40 and 59).

Keeping the axes of the openings described above into consideration, the internal lighting system of the two rooms, formed by 17 openings (11 + 6) can be traced back to 12 different directions: 5 in the lower deck and 8 in the upper one. Of these, only one - the Eastern one - follows the trend of and lower system.

Below the windows is a protruding band (of about 8 cm) on which inscriptions laid out on 3 different levels have been recently found (Novara, 2013, p. 116, see also the volume at p. 85), (Piazza, 2009), investigated and restored in 2012 (results not published yet).

2.6. *The cupola*

Even the great monolith that covers the building was subject to a great number of specialized studies that have investigated physical, technological, figurative, historical and design aspects (Dyggve 1957; Bean 1972; Tabarroni 1973; Bianco Fiorin 1993).

The inner diameter is about 925 cm and the height on the springing is of about 190 cm. A large crack, which popular tradition blames on a bolt of lightning, marks the southern side where the lighthouse was built adhering to the building.

12 protruding elements are present on the outer edge of the roofing. Their triangular holes convey the image of a "royal crown". Historians have often questioned the real function of these elements and their figurative origin, (Bean 1972). The assumption is that they were used for the passage of cables and ropes necessary for its positioning, as hypothesised by Antonio da Sangallo with a drawing already published (Heidenreich and Heinz, 1971, fig. 65, p. 63), may be considered unfounded because of the enormous weight of the monolith and the common laying techniques of the time (Tabarroni 1973). What all scholars emphasize is the lack of regularity in the arrangement of the dodecagon traced by the protruding elements, for it is not aligned with any of the geometries of the building. The monolith is in fact slightly rotated in relation to the main axis of the building, which led to the unanimous conclusion of a faulty, unfixable installation due to the creation of the dangerous lesions on the southern side.

The names of the apostles and evangelists are inscribed on the vertical face of the elements in the sequence (from the door, clockwise): Lucas, Marcus, Mathias (?), Matteus, Felippus, Johannes Jacobus, Andreas Paulus, Petrus, Simeon, Thomas.[14] The elements are closed with a gable roof, almost simulating a small sarcophagus, except for one: that of San Pietro has a flat roof. This led to believe the existence of a terminal made of a different –and perhaps more valuable– material (gone missing), highlighting the figure of Peter, founder of the Church (Tabarroni, 1973, 141)

3. THE ARCHITECTURAL SURVEY

The architectural survey was carried out with a Faro focus3d scanner[15]; 30 stations were performed covering the interior and exterior. The individual clouds have been registered with the aid of spherical targets. At a later date two different photographic campaigns, needed for the reconstruction of the three-dimensional, textured model, were made: the first relating to the exterior, the second to the interior. The exterior shots were performed by a compact Lumix DMC-TZ7. Interior shots, due to matters of critical illumination, were produced with a digital SLR camera on a tripod. In this stage 406 photographs were elaborated and 188 targets used as follows:
Exterior: 123 photos, 38 targets
Exterior walkway details: 96 photos, 146 targets
Lower interior: 76 photos, 0 targets
Top interior: 111 photos, 4 targets.

4. SURVEY DRAWINGS: METHODS AND PROCEDURE

The pointcloud available, which is particularly dense and with limited bands of occlusion, was sectioned by horizontal and vertical planes in order to extract the two-dimensional canonical drawings (plans, elevations and sections) used to effectively describe dimensions and geometries. A thin slice (thickness of 1 cm) was extracted from the cloud for each cut-plane to represent the elements dissected, re-drawn in field vector by interpolation of the points and exported for the realization of raster images definitive 1:50 scale. The choice of using a "slice" of points so reduced yet still sufficiently detailed was possible thanks to the particular density of the pointcloud, which provided a high degree of detail, also on particularly elaborate portions such as shell decorations on the interior brackets. At the same time, the reduced thickness of the slice has guaranteed a higher precision on the section itself, likening it more to a geometric plan. The elements in the projection are obtained by re-drawing, in vector terms, screenshots specially extracted from the pointcloud and orthophotos produced by software for digital photomodelling. Particular attention was paid to the architectural

[14] *Regarding possible reasons for this sequence see (De Angelis d'Ossat, 1962, 129-131, fig. 9; Heidenreich & Heinz, 1971, p. 14; Fagiolo, 1972, 108, 119, 142, 149, 152-4; Tabarroni, 1973, 133-141).*

[15] *Survey M. Incerti and P. Lusuardi, dated 30/04/2015. Data handling software Scene 5.3, elaboration of data M. Incerti. We would like to thank the Compagnia delle Misure for the use of the tool as well as the Soprintendenza per i Beni Architettonici e Paesaggistici di Ravenna for the permissions accorded.*

details, for which ad hoc screenshots with appropriate definition were taken, suitable to describe even the most complex of moldings.

On the rare occlusion bands,[16] the texture has been reproduced (where possible). This has been done by integrating the gaps with specially shot photograms in order to make the final elaborate complete in all its parts. The areas treated this way have nonetheless been graphically differentiated through the use of superimposed layers and textures to emphasise the diffrent material reliability compared to the rest of the survey.

The final elaborates are therefore the result of the overlapping of parts in section and projection obtained through the procedures described above. This work format, now well established in the science world, ensures greater metric control of the architecture (through comparison and contamination of drawings obtained through different processes, distinct survey campaigns and conducted with different instruments). It also allows for detailed graphics containing material and chromatic information that a traditional survey would not have been able to render.

Plans, elevations and sections thus obtained, describing the fundamental geometry of the architectural artefact with a metric reliability that allows for the analysis reported in the next section. It also reproduces the exact wall texture, the state of conservation, the grain, color and deterioration of different materials, allowing the expert operator to perform numerous other integrated readings.

5. ARCHEAOASTRONOMIC ANALYSIS

5.1. *The orientation*

The survey of the orientation was structurally created by Giuliano Romano (Romano, 1995) who first published the Azimut value of 84,5°.[17] The building is rotated, compared to the equinoctial direction, of a considerable amount that should not be overlooked during the alignment operations. The azimut from the apse (North)[18] is aligned with the rising of the sun on the 26-27 March and the 13th of September at the time of construction (pre 526). The first date is very close to 25th of March, celebration of the Annunciation of Mary, a celebration which was already documented in the *Martirologio Gerominiano* dating back to the IV/V century.[19]

Following the correction of the slight rotation, the survey methodology involved specially processed survey drawings. By overlaying graphics to the 4 main astronomical directions (solstices and equinoxes), the small windows placed on the 45° direction were not found to be perfectly aligned in relation to the center. Despite this, it is clear that these windows allow the entry of light during the two solstitial dates. [20]

5.2. *The windows*

Only 3 of the 17 windows, those on the North side, two downstairs and one on the top floor, do not receive significant sunlight. All others are involved in important moments of the astronomical year. The behavior of sunspots on horizontal and vertical surfaces were analyzed through plans and sections, height and azimuth angles were traced, through specific software, to the ephemeris. Among the episodes we note:

- *window (14).* Cross-shaped window, the rising sun penetrates through it on the day of the Annunciation, shining on the wall placed in front, on the axis of the cell. On the day of the summer solstice, about an hour after sunrise, the light spot passes over the stone sarcophagus.
- *window (3b, 4b).* Behaves like the previous for the rising sun of the 25th March. It should be noted that the phenomenon is no longer visible due to the construction of the new upper floor access system.
- *windows (7-8-9-10).* The sun penetrates them at sunset on the days of the equinoxes, illuminating the previously mentioned emerging scripted band.
- *window (16).* At noon of the winter solstice, the light spot approximately reaches the intersection between the wall and floor. Light only enters this window around noon in the winter solstice period because of its shape.

The windows at a 45° direction still welcome the rays of the sun at dawn and dusk during solstice

---

[16] *As previously mentioned, the pointcloud and orthophotos obtained through digital photographs are very detailed and devoid of distracting occlusion areas. It was nevertheless necessary to intervene with the described procedures to limit the parts of the architecture where undercuts and non-removable elements (balustrades, fixed decor elements etc.) would not allow a complete description of the product.*

[17] *At the time it was not considered necessary to repeat the measurements because of the tools used and their precision. The investigations have therefore been carried out at a second date through the analysis of the behaviour of light in the interior in relation to the theoretical hypotheses elaborated with the aid of graphic tools.*

[18] *The horizon was untarnished because the building was situated near the sea shore.*

[19] *The festivity is documented of the 25th of March (ante diem octavum Kalendas Aprilis) in (De Rossi and Duchesne, 1894). We thank E. Spinazè for the notice. On this date the Hilaria celebrations took place in Rome in honour of the goddess Cibele, mother of the gods (Vaccaj 1927, 52-65).*

[20] *The building was the subject of an archaeoastronomical study in 2000, today still in form of a manuscript, by the scholar M. Kerner, 2000 cited in a report of the Soprintendenza per i Beni Architettonici e Paesaggistici di Ravenna (Piazza, 2009). An image of the work is published, but not consultable, showing lunar and solar symbols, perhaps referring to lunar standstills as well as solstices and equinoxes.*

dates. Phenomena of this kind may have been used for purposes of a calendar to mark the advent of a particular date of the year, or for the computation of time: in other words, to indicate a precise moment of the day.

5.3. *The dome*

The results of the survey allowed us to verify that the arrangement of the protruding elements follows the cardinal directions with rather accurate approximation. The item marked with the name of Paul, founder of the Church (the only one with a flat roof), is aligned to South. Aligned to the East is Jacobus, West Lucas, North Matteus. This condition is a possible explanation for what scholars consider an "executional mistake".

6. DIGITAL MODELS FOR RESEARCH VERIFICATION AND DISSEMINATION

The investigation described above, and the survey supporting it, generated a large amount of data. Their analysis and reading produced significant results, providing new information on the archaeoastronomic characteristics of the Mausoleum.

To be disseminated effectively, such results require a visual support that allows the storage of different types of data (morphological, dimensional and chromatic features, archaeoastronomical analysis, external metadata, video and processed media, etc...), and, on the other hand, facilitate an immediate understanding of complex phenomena by the addressee of the information.

A detailed 3d model of the entire building has been created starting from the digital survey, in order to allow a direct visualization of the luminous phenomena affecting the inner spaces on some particular dates of the year.

The model is a low-polygon mesh made of quadrilaterals (quad mesh), textured with a UV mapping method, whereas the high resolution textures are derived from the orthophotos coming from the SFM[21] survey. The model was ideated to be optimized for future use in real time applications.

The 3d maquette, oriented and placed in a Cartesian space reference, was put in relation with a light of the directional type, which simulates the parallel rays from a source similar to the sun, and is therefore the best suited to virtually reproduce the tour of sun around the mausoleum. The light is put on an animated path that reproduces the movement of the sun on the ecliptic. Every moment of animation (keyframe)[22] was created by parametrically inserting the values (position and rotation) derived from the calculation of ephemeris in "key" times. In particular, the exact time of sunrise has been inserted as the beginning of the path, and the sunset time as the end. This time range was then further divided into half hour fractions to cover the whole course of the day.

After setting the animation, the positional coordinates of the sun during the intermediate hours come automatically from the interpolation of the data provided, and can be read simply by moving a cursor on the timeline: the construction phase of the model thus becomes a moment of assessment and comparison of the calculations previously made (ephemeris).

The procedure was repeated for four significant dates (solstices and equinoxes). The model thus obtained was used as a kind of virtual laboratory for the observation of the effects of light within the burial cells, in an ideal condition as free of obstacles and external elements that, at present, occlude the passage of the sun rays.

The observations made experimentally on the model confirmed the results previously recorded from two-dimensional drawings with extreme precision. In addition, the simulation also provides an "emotional" component, increasing the impact of the observation of archaeoastronomical events. The high level of detail of the model and the photo textures, the use of volumetric lights and the general photorealism of the scene allow the user to observe the phenomenon of light in an accelerated form as it may observed inside the mausoleum.

The system is very powerful and effective from a communication point of view, capable of involving the user in an engaging learning experience.

Some short animated sequences –related to a specific stage– have currently been made; however in the future the program will include the creation of a complete interactive application[23]. It will display the behavior of light inside the mausoleum by interactively moving a cursor on a timeline, or simply by entering parameters such as date and time.

---

[21] *Structure from motion (SFM) is a range imaging technique; it refers to the process of estimating three-dimensional structures from two-dimensional image sequences, which may be coupled with local motion signals. We used this technique to assemble a first version of the 3d model of the mausoleum, from which we extracted detailed ortophotos and basic morphological data.*

[22] *A key frame in animation and filmmaking is a still-image that defines the starting and ending points of any smooth transition. It is commonly called "frame" because its position in time is measured in frames on a strip of film.*

[23] *Similar experiences carried out by our research group are described in: (Incerti, Foschi, Iurilli, and Velo, 2014; Incerti and Iurilli, 2014; Incerti and Iurilli, 2015).*

Figure 2. From left, counterclockwise: daylight simulation within the upper cell of the Mausoleum (vernal equinox, 21th of March, rising and sunset; summer solstice, 21th of June, sunrise) and full textured 3d model of the building.

7. CONCLUSION

To conclude, the archaeoastronomical investigation has certainly given significant results which have extended our knowledge of this extraordinary building into topics which had been previously unexplored. The database of measurements produced by the advanced survey offers additional insight, for example on the geometrical and numerical aspects in relation to units of measurement, and on the figure of Severino Boezio, imprisoned and murdered by Theodoric. Lastly, we tested the important contribution of digital models, both during analysis and in the transmission of complex, layered content such as historical and astronomical knowledge.[24]

Figure 3. The Mausoleum of Theodoric (Photo by M. Incerti)

[24] *Paragraphs attributed as following: M. Incerti (1-3, 5-7), S. Iurilli (6), and G. Lavoratti (4).*

REFERENCES

Anonymus Valesianus: Chronica Theodericiana (n.d.) Retrieved November 1, 2015, from http://www.thelatinlibrary.com/valesianus2.html

Bevilacqua, F., Fabbri, R., Grillini, G.C. and Iannucci, A. (2003) Il mausoleo di Teodorico: la pietra di Aurisina. Tecniche e strumenti di lavorazione. F. Lenzi (Ed.), *L' Archeologia dell'Adriatico dalla preistoria al medioevo. Atti del Convegno internazionale: Ravenna 7-9 giugno 2001*, 572–580. Firenze, All'Insegna del Giglio.

Bianco Fiorin, M. (1993) Il monolite del mausoleo di Teodorico. In: *Teoderico il Grande e i Goti d'Italia. Atti del XIII Congresso internazionale di studi sull'alto Medioevo, Milano, 2-6 novembre 1992, vol II pp. 601-610 Spoleto, Centro italiano di studi sull'alto Medioevo.*

Bovini, G. (1977) *Il Mausoleo di Teodorico.* Ravenna, Longo.

Conti, G. and Berti, M. (1997) Lacune e integrazioni murarie nel mausoleo di Teodorico a Ravenna. In *Lacune in architettura: aspetti teorici ed operativi. Atti del convegno di studi, Bressanone, 1-4 luglio 1997*, 361–378. Marghera, Edizioni Arcadia Ricerche.

De Angelis d'Ossat, G. (1962) *Studi ravennati: problemi di architettura paleocristiana.* Ravenna, Edizioni Dante.

De Rossi, G. B. and Duchesne, L. (Eds.) (1894) Martyrologium Hieronymianum. In *Acta Sanctorum Novembris, tomo II, pars prior* , 140. Bruxellis, Société Belge de Librairie.

Dyggve, E. (1957) Mausoleo di Teodorico: le origini della cupola. *Corso di Cult. sull'Arte Ravenn. e Biz. Univ. degli Stud. di Bol. Ist. di Antich. Ravennati e Biz. Ravenna.*

Fagiolo, M. (1972) Theodericus, Christus, Sol. Nuove ipotesi sul mausoleo. *Arheoloski Vestnik*, Vol. 23, 83–117.

Gotsmich, A. (1958) Il Mausoleo di Teodorico. *Felix Ravenna Università Degli Studi Di Bologna, Dipartimento Di Archeologia, Centro Di Studi per Le Antichità Ravennati E Bizantine Giuseppe Bovini.*, Vol. 3., Ser. 25, 56–71.

Guberti, V. (1952) *Il Mausoleo di Teodorico detto anche "La Rotonda"*. Ravenna, Arti grafiche.

Haupt, A. (1913) *Das Grabmal Theoderichs des Grossen zu Ravenna.* Leipzig, Seemann.

Heidenreich, R. and Heinz, J. (1971) *Das Grabmal Theoderichs zu Ravenna.* Wiesbaden, F. Steiner.

Incerti, M., Foschi, P., Iurilli, S. and Velo, U. (2014) Rilievo, conoscenza e comunicazione: il complesso di san Michele in Bosco e il Chiostro ottagonale dei Carracci. Survey, insight communication: the complex of San Michele in Bosco and the Octagonal Cloister by Carracci. C. Vernizzi and P. Giandebiaggi (Eds.), *Italian survey and international experience. 36° Convegno Internazionale dei docenti della Rappresentazione - Undicesimo congresso UID*. Roma, Gangemi.

Incerti, M. and Iurilli, S. (2014) From survey data to virtual environment. Two case studies. *SCIRES-IT*, Vol. 4 (2), 87–108. http://doi.org/10.2423/i22394303v4n2p87.

Incerti, M. and Iurilli, S. (2015) Schifanoia Palace in Ferrara a virtual storytelling for its knowledge and dissemination. J. C. Torres, H. Graf, F. Remondino, J. Barceló, P. Brunet, H. Susan and L. Duranti (Eds.), *2nd International Congress on Digital Heritage 2015. 28 sept- 2 oct. 2015, Granada* (pp. 41–49). IEEE Conference Publications.

Johnson, M. J. (1988) Toward a History of Theoderic's Building Program. *Dumbarton Oaks Papers*, Vol. 42, 73–96. http://doi.org/10.2307/1291590.

Messina, A. (1980) La cosiddetta "scarsella" del Mausoleo di Teodorico. *Felix Ravenna Università Degli Studi Di Bologna, Dipartimento Di Archeologia, Centro Di Studi per Le Antichità Ravennati E Bizantine Giuseppe Bovini.*, Vol. 1979, 117 , 29–38.

Muratori, L. A. (1738) Anonymus Valesianus. *Rerum Italicarum scriptores, XXIV, IV* (p. 640, col. X). Mediolani, Ex Typographia Societatis Palatinae. Retrieved from http://gutenberg.beic.it/view/action/nmets.do?DOCCHOICE =2307661.xmlanddvs=1446379827278~115andlocale=itandsearch_terms=DTL10andadjacency=and VIEWER_URL=/view/action/nmets.do?andDELIVERY_RULE_ID=7anddivType=

Novara, P. (2013) Mausoleo di Teodorico. A. Ranaldi and P. Novara (Eds.), *Restauri dei monumenti paleocristiani e bizantini di Ravenna patrimonio dell'umanità, Parte I. Architetture*, 114–116. Ravenna, Comune di Ravenna, Archidiocesi di Ravenna-Cervia, MIBAC.

Piazza, V. (2009) *Mauseleo di Teodorico.* Scheda Soprintendenza per i Beni Architettonici e Paesaggistici per le province di Ravenna, Ferrara.

Piazza, V. (2013) Mausoleo di Teodorico. A. Ranaldi and P. Novara (Eds.), *Restauri dei Monumenti paleocristiani e bizantini di Ravenna patrimonio dell'umanità*, pp. 84–86. Ravenna, Comune di Ravenna, Archidiocesi di Ravenna-Cervia, MIBAC.

Piazza, V., Bevilacqua, F., Grillini, G. C. and Pinna, D. (1998) Ravenna - Mausoleo di Teodorico. Esiti delle ultime indagini e progetto. G. Driussi and G. Biscontin (Eds.), *Progettare i restauri: orientamenti e metodi-indagini e materiali. Atti del Convegno di studi, Bressanone 30 giugno-3 luglio 1998*, 251–260. Venezia, Arcadia ricerche.

Romano, G. (1995) *Orientamenti ad sidera. Astronomia, riti e calendari per la fondazione di templi e città. Un esempio a Ravenna.* Ravenna, Edizioni Essegi.

Sangallo, G. (1910) *Il libro di Giuliano da Sangallo, codice vaticano Barberiniano latino 4424 riprodotto in fototipia con introduzione e note di Cristiano Huelsen.* Lipsia, Harrassowitz.

Vaccaj, G. (1927) *Le feste di Roma Antica* (II). Torino, Fratelli Bocca Editori.

ASTRONOMICAL ALGORITHMS OF EGYPTIAN PYRAMIDS SLOPES AND THEIR MODULES DIVIDER

Hossam M. K. Aboulfotouh[1,2]

[1]*Faculty of Environmental Design, King Abdulaziz University, Jeddah, KSA*
[2]*Faculty of Fine Arts, Minia University, Minia, Egypt*

Corresponding author: (fotouh@mail.com & haboulfotouh@kau.edu.sa)

ABSTRACT

This paper is an attempt to show the astronomical design principles that are encoded in the geometrical forms of the largest five pyramids of the fourth Egyptian dynasty, in Giza and Dahshur plateaus, based on using the pyramids' design-modules that are mentioned in the so-called Rhind Mathematical Papyrus. It shows the astronomical algorithms for quantifying the slopes of pyramids, with reference to specific range of earth's axial tilt, within spherical co-ordinates system. Besides, it decodes the design of the ancient Egyptian cubit rods with reference to the systems of measurement used by Herodotus and Diodorus, and proves that Herodotus never talked about Giza pyramids and he meant only the pyramids of Memphis in Dahshur. In addition, it proves that the Egyptian arrow of 22.5cm, or half the mean cubit of 45cm, is the largest common divider of the design-modules of these pyramids, and accordingly it retrieves their intended astronomical design dimensions.

KEYWORDS: Giza pyramids, Dahshur pyramids, obliquity, axial tilt, Egyptian cubit, bent pyramid, red pyramid.

1. INTRODUCTION

Methods of design and construction of the Egyptian pyramids were, and still are, in the forefront of scientific debates. In Egyptian mathematics, the so-called Rhind Mathematical Papyrus-RMP that was first translated by Thomas Eric Peet (1923), and then by Arnold Buffum Chace *et al* (1927-1929), is the oldest mathematical source book, where four of its eighty-four problems (RMP#56-59) show the design-sketches and slope-ratios of two pyramidal design models, i.e., the pyramid's height to half its side-length. In the architectonic and archaeological realms, there are five large pyramids found in Egypt from the days of the fourth dynasty, which still retain their original form, and are grouped in two sites: the three Giza pyramids, and the red and bent pyramids in Dahshur. According to Petrie (1883, 1887), the slopes of these pyramids are between 43° and 55°. Retrieving the design method that was used to set the pyramids' slopes with reference to spherical co-ordinates became the interest of modern scholars[1]. In archaeoastronomy, Belmonte and Magli (2015) proposed an idea similar to the shed design method, where they suggest that the slopes of these pyramids match the elevation angles of the ascending sun when it crosses the first vertical east-west plane and thus illuminate their four faces for few minutes, in specific days. Contrary to this opinion, previous publications by the author showed that the prime aim of the pyramids designers was to encode and preserve their scientific knowledge for the coming generations. For example, Aboulfotouh (2007) discussed the concept of shrinking contour circles that models the relative change of earth's axial tilt (obliquity), from an assumed lower limit (~21.672°), as a frame of reference, in order to retrieve the ancient Egyptian relativistic algorithms that most likely were used to reckon the tilts of the entrance passages of these five pyramids. Related to this, Aboulfotouh (2005) showed that the cotangent of edge angle of the great pyramid in Giza equals the result of subdividing the obliquity angle of time ~24.10° by the lower limit. However, the astronomical parameter of earth's obliquity-range, proposed by the ancient Egyptian designers, was never discussed in relation to the design modules of the large pyramids of the fourth dynasty, or in relation to the design modules of the four pyramids problems in RMP. Besides, in history, Herodotus (440BCE, pp.426-437) mentioned the dimensions of the four pyramids of Memphis, where one of them he called the great pyramid. Nowadays scholars of Egyptology believe that Herodotus mentioned the dimensions of the largest four pyramids in Giza Plateau; despite he did not talk about the Sphinx. In geography, Strabo (1932-1967, p.89) said that those pyramids are about 40 stadia (~ 6.3km) south of Memphis[2], which matches the location of the four pyramids in Dahshur. On the other hand, Diodorus (60BC-30AD, p.32, 63-3) mentioned the correct location of the largest three pyramids, in the Giza plateau, as 120 stadia from Memphis and 45 stadia from the Nile, but the dimensions he mentioned are less than those of the Giza pyramids, and are similar to those of the pyramids of Memphis. Perhaps he cited the dimensions given by Herodotus[3]. Today, scholars find difficulties in verifying the dimensions given by those historians with the survey data of Petrie (1883, p111), and many anti-Herodotus criticisms were published accordingly. For example, How *et al* (1913-1927) pointed out that Herodotus's figures of Giza pyramids' dimensions are inaccurate; despite, many parts of Herodotus's text indicate that he studied civil engineering. Besides, in stating the heights of the Egyptian pyramids, ancient and mediaeval historians alike were mentioning instead either the height along the inclined edge of the pyramid, or the height along the line from the crest of the pyramid to the midpoint of its side[4], which did mislead some scholars. The accurate lengths of measurement units used by Herodotus and Diodorus and their correlation to the Egyptian cubits are still unknown. Hence, the paper discusses four key points. Firstly, it shows the most likely astronomical design algorithms for quantifying the slopes of pyramids, and in relation to the pyramids-modules in RMP. Secondly, it shows that Herodotus never talked about, and defiantly never saw the Giza pyramids; and he meant only the pyramids of Memphis, in Dahshur plateau. Thirdly, it decodes the design of the ancient Egyptian cubits, and identifies the metric value of measurement units used by Herodotus and Diodorus. Fourthly, it identifies the largest common divider of the design-modules of pyramids of the fourth Egyptian dynasty, and retrieves their intended astronomical design dimensions.

2. PYRAMIDS DESIGN MODULES

Fig.1 shows the hieratic text of two of the four pyramids problems in RMP, namely: RMP#56-59. Based on reviewing its hieratic text (see: Chace *et al*, 1927-1929, plates 78-81) and the translation by Clagett (1999, pp.166-168), the data of pyramids' height h and their side-length l in cubits are: in RMP#56, h=250 and l=360, in RMP#57-58, h= (93+1/3) and l=140. However, in RMP#59, the hieratic text states that h=12, and l= 8, and not the contrary. Accordingly, the first observation is that these problems show two design models, and not only one as some scholars believed. Besides, in the hieratic text, the pyramids' slopes are not represented as a number of palms[5], but as abstract modules, written under the modular arc notation (see: Aboulfotouh, 2012, p.135). In RMP#56-58, the number of modules of pyramid's height h_m is constant and equal 7 𓏤, and the number of modules in each side of the pyramid's square base l_m is variable and is less than 7 modules in these problems. Hence, as known, $l_m/2$ indicates the slope β of the pyramid's four faces; i.e., $tan\ \beta = 2h_m / l_m$. In the hieratic text, $l_m/2$ is denoted by the Egyptian Arabic word *Saqet*[6] that we use in descriptive geometry and means the horizontal projection (run) of the inclined line. In RMP#56, $l_m/2 = (5+1/25)$, which imply $\beta = 54° 14' 46.01''$; and in both RMP#57 and RMP#58, $l_m/2 = (5+ 1/4)$, which imply $\beta = 53° 7' 48.37''$. Contrary to opinions of scholars, e.g., Gillings (1982, p.186) and Clagett (1999, p.168), RMP#59 shows different model, where the number of modules of the pyramid's side l_m is constant and equal 7; and h_m is variable, and in this case it equals (10+1/2). Hence, $tan\ \beta = 10.5/3.5$, which imply $\beta = 71° 33' 54.18''$. It is similar to the pyramidal part at the top of an Egyptian obelisk, or like the pyramid of Khonsu in Deir el-Medina, with its steep slope.

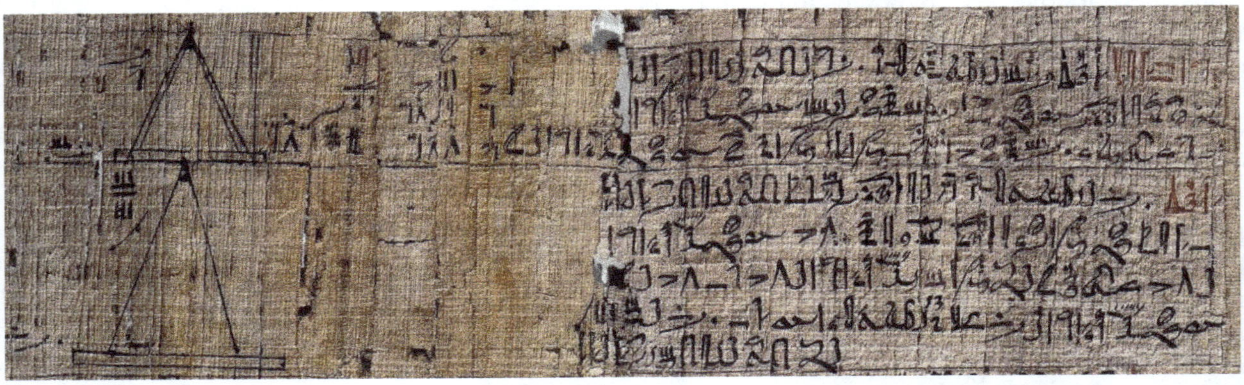

Figure 1: The hieratic text of RMP#56 upper, and RMP#57 lower; the pyramid diagram with horizontal rectangular base is explained in fig.2; and the double inclined lines in the upper diagram imply a pyramid with concave faces, where the outer line is its main slope; it is used with the permission of the British Museum.

3. PYRAMIDS DESIGN AND ASTRONOMICAL PARAMETERS

Aboulfotouh (2007) retrieved the relativistic equations for reckoning the tilt a of the entrance passages of the largest five pyramids of the fourth Egyptian dynasty. The equation of the great pyramid in Giza plateau is: $sin\ a = (sin\ \lambda * sin\ O_m)/(1-(O_i^2/O_t^2))^{1/2}$; where its $a = \sim 26.56°$. Here, λ is the latitude ~30° at which the pyramid stands. Additionally, Aboulfotouh (2002, 2005, 2014) showed that the encoded axial tilt[7] O_t (or ε) in the design of the great pyramid and the site plan of the horizon of the Giza pyramids is ~24.10°; where its designer used an obliquity range from $O_i = \sim 21.672°$ as lower limit to $O_x = \sim 24.30°$ as upper limit, and $O_m = \sim 22.986°$ as a mean. Besides, using the arcs' lengths of the obliquity-angles as radii of contour circles, Aboulfotouh (2007) showed that $O_i/O_t = cos\ \varphi$, where φ is the angle of the horizontal deviation of time that equals ~25.93° for $O_t = 24.10°$, which is used as tilt of the upper entrance passage of the 2nd pyramid in Giza plateau. The tilt of its lower entrance passage was found equal to $O_i = \sim 21.672°$.

Furthermore, using the data of Petrie (1883, p.43), Aboulfotouh (2005) showed that $O_i/O_t = h/r$, where h of the great pyramid is ~146.51m; and r is radius of the circle that encloses the pyramid's square-base, i.e., $r = (l/2^{1/2})$. For $l = 230.35$m., $r = 162.88$m. Besides, it was shown that r is also the radius of the spherical co-ordinates system of the pyramid (see fig.2) upon which its internal-design was created. That imply, $cos\ \varphi = h/r = tan\ \psi$, where ψ is the pyramid's edge angle. Using the equation: $h = r\ (O_i/O_t)$, if we intend to construct another great pyramid to encode $O_t = 23.452145°$, i.e., ε of the epoch 1900AD (Wittmann, 1979), for $r = 162.88$m, its h would be ~ 150.51m.

Fig.2 shows north-south cross section in a spherical co-ordinates system, where point C represents the earth, line Ys-Yn represents the earth's axis of rotation, line P_1-P_2 represents the plane of the ecliptic, and line Xe-Xw represents the plane of the equator. Ahoulfotouh (2005, 2014) showed that this cross section was used for creating the internal design of the great pyramid as well as the site plan of the horizon of Giza pyramids.

In fig.2, the generated main-rectangle P_1-D_1-P_2-D_2 includes the triangle cross-section of the pyramid M_1-Z-M_2 and the residual-rectangle P_1-G_1-G_2-D_2, similar to the hieroglyphic sign that denotes a pyramid △. The horizon line H-H divides the main-rectangle into two equal parts; the upper half is V_1-D_1-P_2-V_2. If we assign names for the vertical distances P_2-V_2, V_2-G_2 and G_2-D_2 as: j, k, and f, respectively; and taking into consideration that $J= f+k$, the pyramid's height $h= k+J$. Moreover, the altitude k of center of the pyramid's spherical system C (and the horizon line H-H) from its zero ground-level in the construction site; and the tilt of its entrance passage a will vary according to the latitude λ where the pyramid stands, and also according to the used O_t. The altitude k could be reckoned using the equation: $k= h- (\sin (90- \lambda- O_t)^* r)$. Therefore, this pyramid design-model is latitude and axial-tilt identifier, i.e., a space-time pyramid. In the great pyramid, $f= \sim 44.5$m., $k= \sim 51$m., and $J= \sim 95.5$m, where its k is close to the level of the highest point of its grand gallery (Aboulfotouh, 2005). If we imagine that this pyramid will not collapse and keep standing during one complete obliquity cycle, over time, it is expected that k will fluctuate between two limits. The upper limit is k_x for O_x and the lower limit is k_i for O_i; e.g., its k_i=45.45m for $O_t = O_i$ =21.672°, and its k_x =51.42m for $O_t = O_x$ =24.30°.

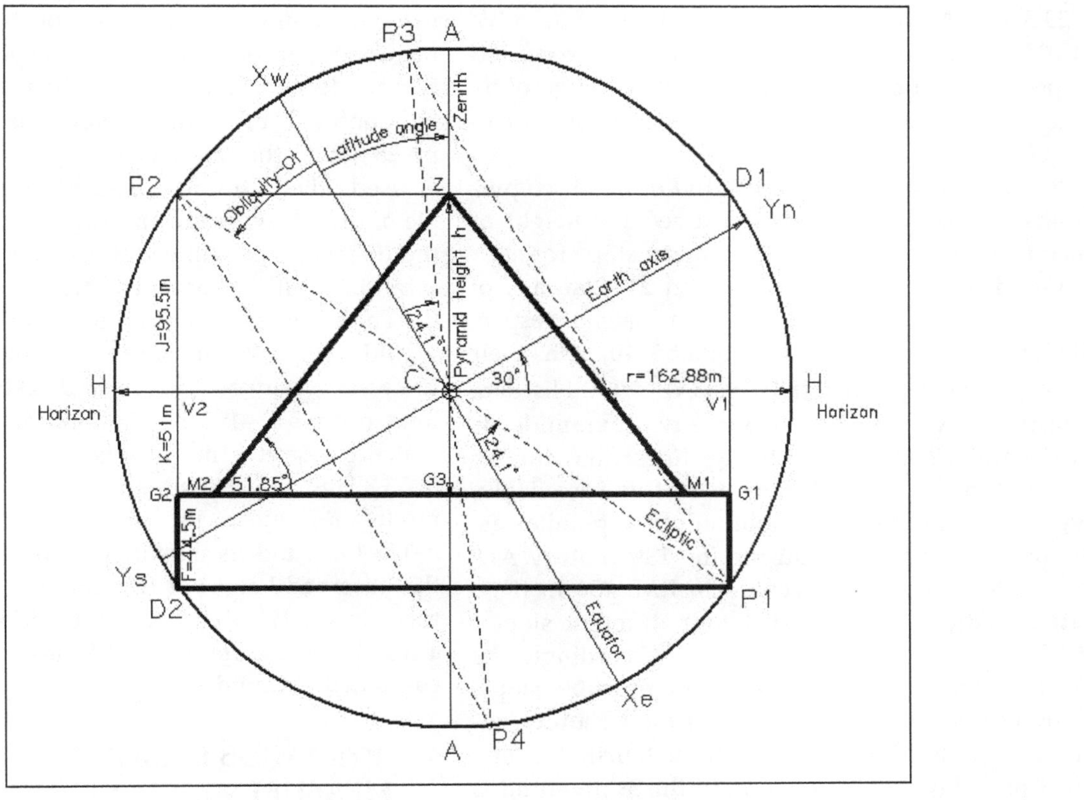

Figure 2: North-south cross section in a spherical co-ordinates system that shows the astronomical design parameters of the Great Pyramid in Giza Plateau, while looking due west.

Based on reviewing the diverse slopes of the Egyptian pyramids, the author identified four design models. Their astronomical slope-ranges are symbolized here below as: β_1, β_2, β_3, and β_4. These ranges of pyramids slopes are reckoned, using the assumption of the ancient Egyptian designers concerning the descending values of earth's axial tilt O_t (or ε), from ~24.30° as upper limit to ~21.672° as lower limit.

In the first model, since $h/r=O_i/O_t$; $l/2= r/(2^{1/2})$; and $\tan \beta_1= 2h/l$; then, $\beta_1= Atan\ (O_i^*(2^{1/2}))/O_t$. Accordingly, the lower limit of β_1 is ~51° 35' 26.81", and its upper limit is 54° 44' 8.2". Using the data of Petrie (1883 & 1887), the slopes of the three Giza pyramids and the lower part of the bent pyramid are within the range of β_1. The slopes of pyramids models in RMP#56-58 are also within this range.

In the second model, the pyramid's slope β_2 could be set equal to the edge angle ψ of the first pyramidal model. That is, by using the diagonal cross-section of the first model as the north-south cross-section of the second model, taking into consideration that, the real radius r of the circle-enclosing the pyramid's square base, and its spherical co-ordinates system, is variable. In this case, $\beta_2= Atan\ (2h/l)= Atan\ (O_i/O_t)$. Accordingly, the lower limit of β_2 is ~41° 43' 41.63" and its upper limit is 45°. Using the data of Petrie (1887, pp.27-32), the slope of the red pyramid and the upper part of the bent pyramid are within the range of β_2.

In the third model, the pyramid's slope β_3 could be set equal to the angle that complements the deviation angle of time φ, i.e., $\beta_3 = Asin\ (90°- \varphi)= Asin$

(O_i/O_t). Accordingly, the lower limit of β_3 is 63° 6' 24.77" and its upper limit is 90°. For example, β_3 = 70° 33' 41.16" for O_t= 22.986°. The slope of the pyramid-model in RMP#59 is within the range of β_3.

In the fourth model, the pyramid's slope could be set as β_4= *Asin* (O_i/φ), where φ= *Acos* (O_i/O_t). This case implies that the deviation angle of time φ is used as radius of shrinking circle, and is being referenced to the circle that its radius is the lower limit of earth's obliquity O_i (see Aboulfotouh 2007). For O_i =~21.672°, the possible range of O_t is from ~24.30° to ~ 23.3205°. Accordingly, the lower limit of β_4 is 53° 41' 34.0" and its upper limit is 90°. The recorded slope of pyramid of Unas[8] is within the range of β_4; where it is close to 56° 39' 52.77" that corresponds to O_t=24.10°.

Petrie measured the side length l of the three pyramids in Giza plateau and reckoned the height of each using what he recorded as mean slope for each pyramid. He recorded 230.35m and 215.26m as l of the great and 2nd pyramids in Giza plateau, respectively; and recorded 51° 52' ±2' and 53° 10' ±4' as their β_1, respectively (Petrie, 1883, p.43 & p.98). His data confirm the values of O_t for these two pyramids, as 24.10° and 22.986°, respectively. Regarding the 3rd pyramid, he recorded 105.5m as l of its base. However, either due to the demolition of the granite casing of the 3rd pyramid[9] during the 13th century AD, or due to its concave faces, as in RMP#56 in fig.1, Petrie (1883, p.111) recorded four different slopes: 50° 57', 50° 42', 51° 0', and 51° 58'. Accordingly, he assumed that 51° 0' ±10' is its most probable slope. None of his four records encodes the obliquity angles that are used by designers of the fourth dynasty. Most probably, the intended β_1 of the 3rd pyramid is 51° 33' 12.41" that corresponds to O_t=24.30°, as what is encoded in the tilt of its entrance passage (Aboulfotouh, 2007).

Furthermore, according to Petrie (1887, p.27&p.32) the slope of the core masonry of red pyramid is 44° 36', and he said Vyse stated this as 43° 36' 11". Perhaps, Petrie did not notice the concaveness in its four faces[10]. Despite 44° 36' is within the astronomical range of β_2, it does not imply encoding any of the obliquity values that are used by designers of the fourth dynasty. In addition, 45° is not valid because it will increase its h for l=220m. Most properly, its intended β_2 is 43° 18' 52.7" that corresponds to O_t=~22.986°, as what is encoded in the tilt of its entrance passage (Aboulfotouh, 2007). Besides, Petrie (1887, pp.28-30) measured the slope of the lower part of the bent pyramid and recorded diverse values from 54° 36' to 55° 1'. Most probably, the intended design slope of the lower part is the upper limit of β_1, i.e., 54° 44' 8.2". He recorded 189.45m as l of its base, and concerning the horizontal-projection of the slope of its lower part, he recorded two distances: 32.7m and 33.0m. Concerning the slope of the upper part of the bent pyramid, he recorded 43° 0' and in the west side, he said, "it is perhaps 42° 39' higher up". Most probably the intended slope of the upper part is 42° 54' that encodes O_t=~23.3205°, and identifies the lower limit of the deviation of angle of time φ, i.e., at this limit φ=O_i=~21.672°.

By reviewing the design of the bent pyramid, it seems that its designer combined the first and second models in one pyramidal form. The final product shows only the lower part of the model β_1, and the upper part of the model β_2. Despite the rest of the two models are not observed, their r, l, and h could be identified, using the survey data of Petrie (1887, pp.28-30). In this special case, apparently, its designer used other altitude k_φ related to the deviation angle of time φ, where $k_\varphi = r - (\sin(90 - \lambda - \varphi) * r)$. In fig.2, the distance, AZ will equal k_φ, when O_t is put equal φ; and k_φ will be measured from the ground point G_3. The altitude of the bent line[11] of the bent pyramid would be ~46.5m, for the lower horizontal-projection, between 32.7m and 33.0m, and the intended β_1= 54° 44' 8.2". The altitude of the bent line matches k_φ=46.51m of its model β_1, for r=h =133.875m, λ = 29.788°, and φ = 0.0°. Besides, in its model β_2, most probably its intended radius r=158.19m, and its intended h= 103.95m, where it is the observed height of this combined pyramid; and the observed l of its base= 133.875*$2^{1/2}$= 189.33m. Table.3 shows the intended design data for this combined pyramid.

4. PYRAMIDS DESIGN AND MEASUREMENT UNITS

In the realm of architecture, designers state the dimensions of buildings in three ways: (i) in terms of design modules, (ii) in terms of specific measurement-unit to be used for implementation, and (iii) in terms of other measurement-unit that might was used for survey, after implementation. In the realm of megalithic construction, modern surveys, using any measurement unit, belong to the third category; and it might not imply knowing the original design module or the measurement unit used for design and implementation. In all cases, particularly after thousands of years, the results will include some deviations from the original design dimensions, in drawings. The first is at least the allowable implementation tolerance, if not exceeded it. The second is due to other vertical or horizontal force of seismic waves that might have caused, e.g., displacement, cambering, deflection, twisting, and/or tilting of all or part of the megalithic structure. The third is partial demolition due to human vandalism or by natural effects that create the second cause. In pyramids design realm, it is necessary to understand the design principles of the ancient Egyptian cubit rods. Sarton (1936, pp.399-402) noted out that the cubit rod of 28 digits is 52.5cm, and referring to plate-I by Lepsius (1866, p.18), from right to left, he said the first 16 digits are 18.75mm long, the eight following are 17.19mm long, and the last four are 21.87mm long. Correlating these numbers, the last two figures should be 17.1875mm and 21.875mm, to become 7/6 and 11/12 of the mean-digit δ (18.75mm) respectively. Hence, its design length[12] is exactly 52.5cm. Besides, starting from the right side, the first fifteen mean-digits are subdivided to create mini-intervals, starting from 1/2, 1/3, 1/4,..., down to 1/16 of δ=18.75mm, respectively. By reviewing the design of four types of ancient Egyptian cubit-rods of the same size, it seems that all cubit-rods contain three types of digits; one of them is the mean-digit δ (18.75mm). The diversity remains in changing the sizes of the other two types of digits: above and below the mean δ, as well as their number and location in the cubit-rods[13]. Fig.3 shows the design components of the ancient Egyptian measurement rod of 52.5cm, the so-called the royal cubit of Torino.

Besides, in some cubit rods, we see special hieroglyphic marks that identify the limit, or the domain, of lengths, e.g.: the bird's claw implies a foot; claw with sparrow imply a short foot; an arrow looks like reed pen imply half a cubit; and hand with sparrow imply a short cubit. In general, a palm is four digits; an arrow is 3 palms, a foot is 4 palms; 1.5 foot or 2 arrows make a cubit. Thus, the number of digits in any cubit is only 24. The 28 mean digits and their 15 types of mini-intervals are the design grids for creating diverse sets of 24 fingers that form the lengths of diverse types of cubits, including their related feet, arrows, and palms, in the same measurement rod. For Example, (i) cubits of 39.375cm, 41.25cm, 45cm, and 52.5cm, (ii) feet of 26.25cm, 27.5cm, 30cm, and 35cm, (iii) arrows of 19.6875cm, 20.625cm, 22.5cm, and 26.25cm, (iv) palms of 6.5625cm, 6.875cm, 7.5cm, and 8.75cm. In this context, the cubit of 45cm long is the mean cubit[14], which is composed of two arrows of 22.5cm. In some rods, the arrow symbol takes the digit number 12, to mark its mean length of 22.5cm; and in other types, it takes the digit number 13, to become 24.375cm long, or ½ a cubit of 48.75cm.

1	2	3	4	5	6	7	8	9	10	11	12	13	14	15	16	17	18	19	20	21	22	23	24	25	26	27	28
2.1875*4= 8.75cm				1.71875*8= 13.75cm								1.875*16 = 30 cm															
larg palm				small palm				short foot			arrow	arrow	foot	cubit							short cubit				royal cubit		
2.1875	2.1875	2.1875	2.1875	1.71875	1.71875	1.71875	1.71875	1.71875	1.71875	1.71875	1.71875	1.875	1.875	1.875	1.875	1.875	1.875	1.875	1.875	1.875	1.875	1.875	1.875	1.875	1.875	1.875	1.875
													16s	15s	14s	13s	12s	11s	10s	9s	8s	7s	6s	5s	4s	3s	2s
			8.75	6.875							22.5	24.375	26.25							39.375	41.25		45		48.75		52.5
			1/6 royal cubit				1/4 foot of 27.5cm				One arrow of 22.5cm	One foot or one arrow	One foot of 26.25cm							1.5 foot of 26.25cm (one cubit)	1.5 foot of 27.5 (one cubit)		2 arrows of 22.5cm (mean cubit)		2 arrows/feet of 24.375cm		One royal cubit of 52.5cm

Figure 3: The architectonic design of the Egyptian measurement rod that contains diverse types of cubits, foots, arrows, palms, and fingers; its cubits starts from the cubit of 39.375cm up to the royal cubit of 52.5cm, where each cubit contains only 6 palms. Its design direction is from left to right, and for geometrical use, it will be turned upside-down; in order to make the first subdivided digit of 2s in the left side of the architect/engineer, and below the intended drawing line. This integrated design of the Egyptian measurement rod of diverse cubits implies that it might was used in many parts of the known world in their days, similar to the rod of diverse-scale ratios that we use today.

Correlating the design principles for producing the Egyptian cubit rod that mentioned above with reliable historic reference, might reveal some ancient measuring systems. Herodotus (430BCE, p.559, 2-149) correlated the measurement units that he used as follows: (i) one hundred fathoms is exactly a stadium (stade) of six hundred feet; (ii) the fathom is six feet in length, or four cubits; (iii) a cubit measures six palms, and a foot is four palms; (iv) parasang is thirty stadia; and (v) schoene is sixty stadia. Using these correlations between measurement-units, table.1 shows the possible metric values of each, based on using four types of feet that are included in the design of the Egyptian measurement rod. Besides, Herodotus (430BCE, p.285, 2-9) said, "the distance between Heliopolis and the sea is fifteen hundred stadia, and sailing up the river from Heliopolis to Thebes, the distance is eighty-one schoenes". Using Google Earth Data, of 2015, these two distances, along the course of the Nile River are about 250km and 670km, respectively. This implies that Herodotus used the measurement system based on the foot of 0.275m, i.e., he used the stade of 165m, and the schoene of 8,250.0m.

Moreover, Diodorus (60BC-30AD, p.24, 47-1) said, based on citing Hecataeus, "ten stadia from the first tomb of concubines of Zeus (i.e. entrance of the valley of the Queens), stands a monument of the king known as Osymandyas (i.e. Ramesseum temple), and its first pylon is two plethra in breadth". Using Google Earth data (2015), this walking distance, along the road, is about ~1.89Km, and the breadth of the first pylon of the Ramesseum temple is ~63m. Besides, Diodorus (60BC-30AD, p.26, 51-5) said, "ten schoene above Memphis king Moeris (Seti-I) excavated a lake". Using Google Earth data, the distance from Memphis to the old entrance of Lake Moeris (beside Lahun Pyramid) is ~78km. It implies that Diodorus used the measurement system based on the foot of 0.2625m, i.e., he used the stade of 189m, the plethra of 31.5m, and the schoene of 7,875.0m. In this regard, Nallino (1910, pp.268-275) said[15], "Eratosthenes used the stade of Alexandria of 157.5m long", which equals 600 feet, each is 0.2625m long.

Table 1: The possible metric values of diverse measurement units mentioned by Herodotus.

Name	(mean) Length in meter	(large-royal) Length in meter	(small) Length in meter	(short) Length in meter	Design intervals
Finger (δάκτυλος)	0.01875 (mean δ)	0.021875 (7/6 of δ)	0.0171875 (11/12 of δ)	0.01640625 (7/8 of δ)	1/16 foot (1/24 cubit)
Palm (παλάμη)	0.075	0.0875	0.06875	0.065625	1/4 foot (1/6 cubit)
Arrow	0.225	0.2625	0.20625	0.196875	3 palms (1/2 cubit)
Foot (ποδες)	0.30	0.35	0.275	0.2625	4 palms (2/3 cubit)
Cubit (πήχηείς)	0.45	0.525	0.4125	0.39375	6 palms
Fathoma (οργιά)	1.8	2.1	1.65	1.575	6 feet
Plethra (πλέθρον)	30	35	27.5	26.25	100 feet
	36	42	33	31.5	120 feet
	39	45.5	35.75	34.125	130 feet
Stade (στάδιον)	150	175	137.5	131.25	500 feet
	180	210	165	157.5	600 feet
	216	252	198	189	720 feet
Parasang (παρασάγγες)	4,500.0	5,250.0	4,125.0	3,937.5	30 Stadia of 500 feet
Schoene (σχοίνον)	9,000.0	10,500.0	8,250.0	7,875.0	60 Stadia of 500 feet

Furthermore, Herodotus (430BCE, pp.426-437) mentioned the dimensions of the pyramids of Memphis[16], that Strabo (1932-1967, p.89) said, "they lie about forty stadia (6.30 Km.) south of Memphis". This location is the plateau that includes the pyramids of Dahshur, where its elevation is about a hundred feet high[17], or 27.5m. Despite Herodotus did not mention the Sphinx and the remaining small pyramids in the Giza plateau; scholars of Egyptology have thought that he describes the Giza Pyramids. Herodotus called the bent pyramid, as the pyramid of Cheops[18]. Concerning the dimensions of its existing causeway (of 690m in Google Earth data 2015), he said, "it is five stadia in length (137.5m*5), ten fathoms wide (16.5m), at the highest part, eight fathoms; and the pyramid is a square, eight plethra[19] each way, and the height (along its inclined edge) the same"[20]. The pyramid of Cheops's daughter (satellite pyramid), which lies directly south of, and close to, the bent pyramid, and each side of it measures one and half plethra[21]. The red pyramid, he called the pyramid of Chephren, and said, "it was built of the same bigness as that of Cheops, save that it falls forty feet (40*0.275=11m) short of it in height (along its inclined edge)"; and he added, "I have myself measured it". It implies that the length that Herodotus assigned to the side of the pyramid of Cheops (bent pyramid) is the result of measuring the base of the red pyramid, assuming that their square-bases are equal in size. The black pyramid (now called pyramid of Amenmhat-III) he called the pyramid of Mycerinus or Rhodopis[22] and said, "the length of each side of its square base falls short of three plethra by twenty feet"[23]. Diodorus (60BC-30AD, pp.32-33, 63-3 & 64-2,7) said, "the lengths on each side of the bases of pyramids of Chemmis (Cheops), Cephres (Chephren), and Cherinus (Mycerinus) are: seven plethra, a stade, and three plethra, respectively". Concerning the names of builders of pyramids of Memphis, Diodorus (60BC-30AD, 64-13) said, "with regard to the pyramids there is no complete agreement among either the inhabitants of the country or the historians; for according to some the kings mentioned above (in his text) were the builders, according to others they were different kings". Herodotus (440BCE, p.431, 2-128) also said, "people call these pyramids after the shepherd Philitis". Table.2 shows the dimensions of the pyramids of Mimphes (in Dahshur), by Herodotus and Diodorus, using different types of plethra, in comparison with the available data by Petrie and other scholars.

Table 2: Bases dimensions of Dahshur Pyramids by ancient and modern scholars.

Pyramids of Memphis, in Dahshur	Herodotus		Diodorus		Petrie[24] & Rossi[25]
	Plethra (πλέθρων)	meter	Plethra (πλέθρων)	meter	meter
Red Pyramid	**8** * (27.5m)	220	**7** * (31.5m)	220.5	220.5
Bent Pyramid	Assumed = to Red pyramid	--	**6** * (31.5m) or (stade)	189	189.56
Satellite Pyramid	**1.5** * (35.75m)	53.625	--	--	52.42
Black Pyramid	(3*35.75m) – (20 *0.275m)	101.75	**3** * (34.125m)	102.375	105

5. THE LARGEST DIVIDER OF PYRAMIDS MODULES

Based on the above discussion, the common divider of pyramids of the fourth dynasty could be reckoned by dividing h of each pyramid by 7 and then finding out what is the size of the palm that divides the intended design modules of the five pyramids. Trying out diverse types of palms, the result was the mean-palm of 7.5 cm. This imply that each module in the five pyramids is composed of a number of mean-palms; if it is not a prime number, there might be a divider larger than the size of the mean-palm. For example, each module in the great pyramid is composed of 279 mean-palms, where 3 is the only integer from 2 to 7 that divides it. That is, the Egyptian arrow of 0.225m is the largest module divider d for designing the great pyramid, where it is 2/3 of the mean-foot of 0.30m and ½ of the mean-cubit of 0.45m. The case is also similar in the other four pyramids. Table.3 shows that the Egyptian arrow of 22.5cm is the largest common divider d of

Table 3: The intended design dimensions and astronomical parameters of pyramids of the fourth dynasty.

Pyramid	Modules of h and l (m)		Divisions in one module (n_d)	Intended design length ($m*n_d*d$) d=0.225m in meter	Astronomical parameters for O_i=21.672°; where O_t, and φ are in degrees; and κ and κ_φ are in meter			
					O_t	φ	κ	κ_φ
Great Pyramid	h	7	93	146.475	24.10°	25.93°	50.9	71.6
	l	11		230.175				
2nd Pyramid	h	7	91	143.325	22.986°	19.46°	51.8	53.2
	l	10+1/2		214.9875				
3rd Pyramid	h	7	42	66.15	24.30°	26.89°	22.9	33.8
	l	11+1/10		104.895				
Red Pyramid	h	7	66	103.95	22.986°	19.46°	9.66	54.2
	l	14+1/2+1/4+1/10		220.5225				
Combined or Bent Pyramid	$h_{\beta 1}$	7	85	133.875	21.672°	0.0°	50.4	**46.5**
	$l_{\beta 1}$	9+1/2+1/3+1/15		**189.3375**				
	$h_{\beta 2}$	7	66	**103.95**	23.3205°	21.672°	36.8	42.1
	$l_{\beta 2}$	15+1/20+1/63		223.7282143				

the design modules of pyramids of the fourth dynasty. It shows also, the design modules m, the number of module divisions n_d, the related intended design lengths (h, l, k and κ_φ), and the astronomical design parameters (O_i, O_t and φ) of each pyramid. Similarly, one can identify d in each of the four pyramids problems in RMP, using the royal cubit of 0.525m, in order to identify the metric lengths of pyramids in these problems that are mentioned in section-1 above, taking into consideration that in RMP#56-58, h_m=7, and in RMP#59, l_m=7. In RMP#56, d=0.15cm (1/2 mean-foot), n_d=125, and h= 0.15*125*7= 131.25m. In RMP#57-58, d=0.35m (royal foot), n_d=20, and h=0.35*20*7=49m. In RMP#59, d=0.30m (mean foot), n_d=2, and l= 0.30*2*7= 4.2m.

6. CONCLUSION

This paper showed four astronomical design algorithms for quantifying the slopes of pyramids, in relation to the pyramids design modules of RMP; and that could be used on earth or elsewhere, using diverse obliquity limits. It proved that Herodotus never talked about, and defiantly never saw, the Giza pyramids; and he meant only the pyramids of Memphis, i.e., Dahshur pyramids. It decoded the design of the ancient Egyptian cubit rod of 52.5cm, and showed that it includes other types of cubits, feet and arrows. It showed that Herodotus used the measuring system based on the foot of 27.5cm, and Diodorus used other system based on the foot of 26.25cm. It also, showed that the arrow of 22.5cm (half cubit of 45cm) is the largest common divider of the design modules of pyramids of the fourth Egyptian dynasty. In general, this paper showed an additional proof to support that defiantly there was consistent astro-mathematical theory in pyramids design (the Horizon theory), which was formulated and architectonically encoded by designers of pyramids of the fourth dynasty in ancient Egypt.

REFERENCES

Aboulfotouh, Hossam (2014), The Architectonic Encoding of the Minor Lunar Standstills in the Horizon of the Giza Pyramids, *Mediterranean Archaeology and Archaeometery*, Vol. 14, No. 1, pp.343-352.

Aboulfotouh, Hossam (2012), The Geometric Grids of the Hieratic Numeral Signs, *Mediterranean Archaeology and Archaeometery*, Vol. 12, No. 1, pp.129-138.

Aboulfotouh, Hossam (2007), the Relativistic Tilts of Giza Pyramids' Entrance Passages, *Mediterranean Archaeology and Archaeometery*, Vol. 7, No. 1, pp.23-37.

Aboulfotouh, Hossam (2005), Horizon Theory-Part-II: Internal Design Concept of the Great Pyramid, 1st international Conference of the Department of Restortion, Faculty of Fine Arts at Minia University, March 2005, http://www.geocities.com/fotouh28/H-Aboulfotouh-Horizon-II.pdf.

Aboulfotouh, Hossam (2002), The Horizon Theory, Part-I: Original Concept Plan of the Pyramids Plateau, 1st International Conference of the UIA Work Programme "Architecture and Heritage," Region-V, on "Architecture and heritage as a paradigm for knowledge and development: lessons of the past, new inventions and future challenges", Bibliotheca Alexandrina, March 2002. http://www.geocities.com/fotouh28/H-Aboulfotouh-Horizon-I.pdf

Al-Maqrizi (1364–1442AD) *Al Mawaes Wal A'atebar Bezeker Al-khetat Wal Asar* (Lessons and Consideration by Mentioning Areas and Old Buildings) Dar Al-Tahrier, Bolaque Edition, Cairo, 1849.

Belmonte; Juan Antonio & Magli; Giulio (2015) Astronomy, Architecture, and Symbolism: The Global Project of Sneferu at Dahshur, *Journal for the History of Astronomy*, Vol. 46(2) pp. 173–205.

Nallino, Carlo (1911) *Arabian Astronomy: its History During the Medieval Times*, Oriental Papers, Arabian House Bookshop, First Ed. Rome, Second Ed. Beirut, 1993 (in Arabic).

Diodorus the Sicilian (60BC-30AD), *Library of History*, Book-I, made English by G. Booth, Printed by Edw. Jones, London, MDCC (1700 AD), p.1-52, Ch-63.

Eyre, C. J. (1991) Book Review "Herodotus Book II Commentary 99-182 by Alan B. Lloyd", *The Journal of Egyptian Archaeology*, Vol. 77, pp. 223-225.

Gillings, R. J. (1982) *Mathematics in the Time of the Pharaohs*, Dover Publications Inc, New York.

Herodotus (484 - 425 BC) *The Histories:* Books I-II, with English Translation by A. D. Godley, Harvard University Press, London, William Heinemann LTD., 1920-1975.

How, W. W.; Wells, J., (1913–1927) *A Commentary on Herodotus*, Oxford University Press, ii. 124-36, p.249, (Electronic Edition).

Laskar, J., (1986) Secular Terms of Classical Planetary Theories Using the Results of General Theory, *Astronomy and Astrophysics*, Vol. 157, p.59-70.

Lepsius K. R. (1866) *Die altägyptische Elle und ihre Eintheilung; Philol. Abhandle. der Akademie der Wissenschaften zu Berlin*, 67 p., 5 pl. Berlin.

Manetho, (1940-1964), Aegyptiaca Αἰγυπτιακα (Epitome), with an English Translation by W. G. Waddell, Cambridge, Massachusetts, Harvard University Press, London, William Heinemann Ltd, MCMLXIV.

Meeus, Jean, (1991), *Astronomical Algorithms*, Willmann-Bell Publications Inc., USA.
Petrie, W.M.F., (1883), *Pyramids and Temples of Gizeh*, London.
Petrie, W.M.F., (1887) *A Season in Egypt*, the Leadenhall Press, London.
Rossi, Corinna (2003), *Architecture and Mathematics in Ancient Egypt*, Cambridge University Press, New York.
Sarton, George (1936), On the Curious Subdivision of the Egyptian Cubit, *ISIS*, Vol. 25, No. 2, pp.399-402.
Scott, Nora E. (1942), Egyptian Cubit Rods, *The Metropolitan Museum of Art, New Series*, Vol. 1, No.1, pp.70-75.
Scott, R. B. Y. (1958), The Hebrew Cubit, *Journal of the Biblical Literature*, Vol. 77, No. 3, pp.205-214.
Shaltout, M. & Belmonte, J. A., (2005) On the Orientation of Ancient Egyptian Temples I: Upper Egypt and Lower Nubia" *JHA*, Pre-print series: PP 03/2005.
Spence, Kate (2002) Book Review "Roger Herz Fischler: The Shape of the Great Pyramid", *ISIS*, Vol. 93, No.1, pp.83-84.
Strabo (36BC-24AD), *Geography*, Book XVII, with an English translation by Horace Leonard Jones, Harvard University Press, London, William Heinemann Ltd, 1932-1967 (Pyramids 17-1-33).
The Penny Cyclopedia of the Society for the Diffusion of Useful Knowledge (1840), Charles Knight and Co., London, MDCCCXL, Vol. XVIII and Vol. XIX.
Wittmann, A. (1979) the Obliquity of the Ecliptic, *Astronomy and Astrophysics*, 73, pp.129-131.

NOTES

---

[1] Without some basic knowledge in spherical astronomy, it is hard to retrieve any mathematical method for designing the slopes (angles of inclination) of the Egyptian pyramids. RMP includes mere numerical ratios without the design method. In her review, Spence (2002) disputed the strange conclusion by the mathematician Herz-Fischler who said in his book, "there may have been no true theoretical basis for the angles of inclination of pyramids".

[2] Strabo said, "on proceeding forty stadia from the city (from Memphis heading south as he walks), one comes to a kind of mountain-brow; on it are numerous pyramids the tombs of kings, of which three are noteworthy; and two of these are even numbered among the Seven Wonders of the World, for they are a stadium in height."

[3] Eyre, C. J. (1991, pp.223-225) said Herodotus's text mercilessly plagiarized by Diodorus Siculus.

[4] How *et al* (1913-1927, p. 250, ii 124) called it «height along sloping side»; in the pyramids chapter, Al-Maqrizi (1849) said, based on citing Abdulrhim Alquise, "the perimeter of the Great Pyramid's base is 2000 black cubits and its height is 500 black cubits", implying the height along its inclined edge.

[5] In the opinion of scholars, e.g., Gillings (1982, p.186) and Clagett (1999, pp.166-168), a cubit measures 7 palms, and thus they thought it represents the rise of the pyramids slopes in RMP. That is despite, Sarton (1936, pp.399-402) had noted out that the 7 palms shown in the cubit rods are not equal, and Herodotus (430BCE, p.559, 2-149) said, "a cubit measures only 6 palms".

[6] Gillings (1972, p.185) pronounced it as *Seked*, and said, "it means the slope of the sides of a pyramid".

[7] Using flat mutation of 0.47" per year from the epoch 2000 AD (23.45°), the obliquity 24.10° meets ~ 3055-3065 BCE; and using, M. Bessel's Eq.: ε = 23° 28' 18" + 0.48368" T – 0.00000272295" T^2 (Nallino, 1911, p.270) & (The Penny Cyclopedia, 1840, Vol. XVIII, p.495), 3065BCE meets ε= 24.109°; where ε is O_t, and T is the years elapsed from the epoch 1750AD. Bessel's equation is the only polynomial fit that gives value close to the record of Chou Li in 1100BC, ε=23.881° (data in: Wittmann 1978). The polynomial fits that are set to finite limits, e.g., between 22.5° and 24.5° (Meeus, 1991, p.135) or between 22.61° and 24.23° (Laskar, 1986, p.86), give ε=~24.027° for 3070BCE; and fail to give ε = ~23.922° for the azimuth 296.75° of winter solstice of the axis of the main Karnak temple, at λ=25.72°, in ~ 1620BCE (data in: Shaltout *et al*, 2005, p.15).

[8] Its recorded slope is 56° 18' (Rossi, 2003, p.249; based on citing, Lauer, Observations, p.95).

[9] Al-Maqrizi (1849), based on citing Abdelatif Al-Bugdady, said, "the demolition of the granite casing of the small pyramid was by the order of king Al-Aziz Osman in 593H, whose illiteracy made him imagine that there are treasures beneath it, but the lack of finance obliged the workers to discontinue after 8 months of demolition work".

[10] Since, its h= ~104.61m that is composed of 154 steps (The Penny Cyclopedia, 1840, Vol. XIX, p.152), and its l=220.5m (Rossi, 2003, p253), any record above that of Vyse 43° 36' 11" is invalid.

[11] Petrie (1887, p.32) reckoned it as follows: *tan* 55° 1' * 33m= 47.15m.

[12] Over time, shrinking might cause slight reduction in the length of wooden rods; thus, any royal cubit length other than 52.5cm, e.g., 52.375cm or 52.3cm (Scott, 1942, p.70), is incorrect and misleading.

[13] Further, 13/12 of the mean δ produces the cubit of 48.75cm; and 9/8 of δ produces the cubit of 50.625cm; 10/9 of δ produces the cubit of 50.0cm; and 61/60 of δ produces the cubit of 45.75cm that 2/3 of it is the English foot of today.

14 Scott (1942, p.70) called it a short cubit, and Rossi (2003, p60) called its half (arrow) of three mean palms as small-hands. Scott (1958, pp.205-214) also mentioned the lengths of other cubits, e.g., 39.6cm and 50.0cm.

15 Carlo Nallino (p.274) referred to the works of the German scholar Fr. Hultsch.

16 In Greek: πυραμίδας κατατμηείσαι τας ἐν Μέμφι (Herodotus, 430BCE, p.282, 2-8).

17 Herodotus (430 BCE, p.430-431, 2-127) perhaps mentioned the elevation from the plain of Memphis or from the water surface of the Nile River; and considering the metric value of the foot he used, Google Earth data of 2015 show the same elevation.

18 Despite, Africanus had cited Manetho and said, "the fourth dynasty kings are: Soris, Suphis (i), Suphis (ii), Mencheres, etc." (Manetho, 1940-1964, p.47), Egyptologists thought that Cheops (or Chmmes) is the 2nd king of the 4th dynasty. Herodotus (430BCE) put Cheops the 4th king after Sesostris, and Diodorus (60BC-30AD, p.32) put him the 5th king after Sesostris, who most probably is the so-called Ramses-III, where Sesostris is a royal or military title before the king's cartouche. The names: Chemmis, Cephres, and Cherinus, match the cartouches of the so-called Ramses-VIII, Ramses-IX, and Ramses-XI, respectively.

19 In Greek: ὀκτώ πλέθρα (Herodotus, 440 BC, p.426, 2-124).

20 Strabo (36BC-24AD, p.89, 17-I, 33) said, "high up approximately midway between the sides, it has a movable stone, and when this is raised up there is a sloping passage to the vault"; related to this, see Fig-XI: Door of the South Pyramid of Dahshur, by Petrie (1883).

21 In Greek: ὁλον καί ἡμίσεος πλέθρον (Herodotus, 440 BC, p428, 2-128).

22 Most probably, Rhodopis is the so-called Nubheteti-Khered that her tomb is found under this pyramid; the correct pronunciation of the last word 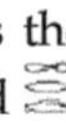 in her name is Radophi (or Pharida) without the Greek "ς"; the upper letter is *Ph* not *Kh*, which becomes the last letter if we follow the ancient Egyptian spiral sequence of writing the letters of words.

23 In Greek: εἰκοσι ποδών καταδέουσαν κώλον έκαστον τριών πλέθρων (Herodotus, 440 BC, pp.436-437, 2-134).

24 Data of the bent and satellite pyramids are by Petrie (1887, p.28).

25 Data of the red and black pyramids are by Rossi (2003, p.253, based on citing, Stadelmann, MDAIK 39, p.253, and Arnold, Amenemhet III., p.9, respectively).

Permissions

All chapters in this book were first published in MAA, by University of the Aegean; hereby published with permission under the Creative Commons Attribution License or equivalent. Every chapter published in this book has been scrutinized by our experts. Their significance has been extensively debated. The topics covered herein carry significant findings which will fuel the growth of the discipline. They may even be implemented as practical applications or may be referred to as a beginning point for another development.

The contributors of this book come from diverse backgrounds, making this book a truly international effort. This book will bring forth new frontiers with its revolutionizing research information and detailed analysis of the nascent developments around the world.

We would like to thank all the contributing authors for lending their expertise to make the book truly unique. They have played a crucial role in the development of this book. Without their invaluable contributions this book wouldn't have been possible. They have made vital efforts to compile up to date information on the varied aspects of this subject to make this book a valuable addition to the collection of many professionals and students.

This book was conceptualized with the vision of imparting up-to-date information and advanced data in this field. To ensure the same, a matchless editorial board was set up. Every individual on the board went through rigorous rounds of assessment to prove their worth. After which they invested a large part of their time researching and compiling the most relevant data for our readers.

The editorial board has been involved in producing this book since its inception. They have spent rigorous hours researching and exploring the diverse topics which have resulted in the successful publishing of this book. They have passed on their knowledge of decades through this book. To expedite this challenging task, the publisher supported the team at every step. A small team of assistant editors was also appointed to further simplify the editing procedure and attain best results for the readers.

Apart from the editorial board, the designing team has also invested a significant amount of their time in understanding the subject and creating the most relevant covers. They scrutinized every image to scout for the most suitable representation of the subject and create an appropriate cover for the book.

The publishing team has been an ardent support to the editorial, designing and production team. Their endless efforts to recruit the best for this project, has resulted in the accomplishment of this book. They are a veteran in the field of academics and their pool of knowledge is as vast as their experience in printing. Their expertise and guidance has proved useful at every step. Their uncompromising quality standards have made this book an exceptional effort. Their encouragement from time to time has been an inspiration for everyone.

The publisher and the editorial board hope that this book will prove to be a valuable piece of knowledge for researchers, students, practitioners and scholars across the globe.

List of Contributors

Munirah A. Almushawh
King Saud University, Department of Archaeology, Riyadh, Saudi Arabia

Silvia Motta and Adriano Gaspani
I.N.A.F. Istituto Nazionale di Astrofisica Osservatorio Astronomico di Brera- Milano Italy

Alexey Stoev and Penka Maglova
Space Research and Technology Institute, Bulgarian Academy of Sciences, Stara Zagora Department

Mina Spasova
Institute for the Study of Societies and Knowledge, Bulgarian Academy of Sciences, Sofia

David A. Fisher
Independent Researcher

Stanisław Iwaniszewski
Posgrado en Arqueología, Escuela Nacional de Antropología e Historia, Instituto Nacional de Antropología e Historia, Mexico City

David Espinosa-Espinosa and Marco V. García-Quintela
University of Santiago de Compostela

A. César González-García
Institute of Heritage Sciences, Incipit-CSIC, Santiago de Compostela, Spain

Andrea Orlando
Institute of Sicilian Archaeoastronomy, Italy
Fondazione Floresta Longo, Italy

Sebastiano Tusa
Regional Department of Cultural Heritage and Sicilian Identity, Italy

Davide Gori
Wood plc, Italy
Institute of Sicilian Archaeoastronomy, Italy

M. Incerti, S. D'Amico and S. Giannetti
Architecture Department of Ferrara (Italy)

G. Lavoratti
Architecture Department of Florence (Italy)

Hans Martz de la Vega
National School of Anthropology and History, Mexico

Vance R. Tiede
Astro-Archaeology Surveys

Julio Cuenca Sanabria and Carlos Gil Sarmiento
PROPAC and Cabildo de Gran Canaria, Spain

José Carlos Gil
PROPAC and Cabildo de Gran Canaria, Spain
Instituto Canario de Estadística (ISTAC), Gran Canaria, Spain

José de León
Heritage Service, Cabildo de Gran Canaria, Spain

Cipriano Marín
Unesco Center of the Canary Islands and Cabildo de Gran Canaria, Spain

Juan Antonio Belmonte
Instituto de Astrofísica de Canarias and Universidad de La Laguna, Tenerife, Spain

José Miguel Márquez Zárate
Architect, Tenerife, Spain

Eva Spinazzè
Ca' Foscari University of Venice, IUAV University of Architecture Venice, University of Zurich, Independent Researcher

Alexandra Edwards and Edmundo Edwards
Pacific Islands Research Institute
Rapa Nui Planetarium

Barthelemy d'Ans
Rapa Nui Planetarium
Universidad Peruana de Arte Orval

Marc Frincu
Romanian Society for Cultural Astronomy and West University of Timisoara Faculty of Mathematics and Computer Science

J. McKim Malville
Department of Astrophysical and Planetary Sciences, University of Colorado

Andrea Orlando
Institute of Sicilian Archaeoastronomy, Italy
Fondazione Floresta Longo, Italy

Davide Gori
Wood plc, Italy
Institute of Sicilian Archaeoastronomy, Italy

Carlo Veca
Archaeologist, Superintendence of Cultural Heritage, Malta

Walmir Thomazi Cardoso
Pontifícia Universidade Católica de São Paulo; Grupo de História, Teoria e Ensino de Ciências, Universidade de São Paulo

Roberto de Andrade Martins
Universidade Federal de São Paulo; Fundação Municipal de Ensino Superior de Bragança Paulista; Grupo de História, Teoria e Ensino de Ciências, Universidade de São Paulo

Andrew M. Munro and Steven R. Gullberg
The University of Oklahoma, College of Professional and Continuing Studies, 1700 Asp Avenue, Norman, OK 73072, USA

Susanne M. Hoffmann
Independent Scholar, formerly: Excellence Cluster TOPOI Hannoversche Str. 6, 10115 Berlin, Germany

Giuseppe Severini
Casa della Musica di Randazzo, Italy
Istituto di Archeoastronomia Siciliana, Italy

Andrea Orlando
Institute of Sicilian Archaeoastronomy, Italy
Fondazione Floresta Longo, Italy

Stefano Giovanardi, Gabriele Catanzaro, Giangiacomo Gandolfi and Gianluca Masi
Planetarium and Astronomical Museum of Rome

Giangiacomo Gandolfi
Planetario e Museo Astronomico di Roma Capitale

Silvia Sclavi, Marzia Monaco, Flavio Carnevale and Claudia Scatigno
La Sapienza University of Rome

Silvia Gaudenzi
La Sapienza University of Rome
INAF-IAPS Rome

Marcello Ranieri
INAF-IAPS Rome

Francesco Vito Polcaro
INAF-IAPS Rome
ACHe, University of Ferrara

Christiaan Sterken
University of Brussels, Pleinlaan 2, 1050 Brussels, Belgium

George F. Steiner
Negev Rock Art Center, Israel

M. Incerti, G. Lavoratti and S. Iurilli
Architecture Department of Ferrara (Italy)

Hossam M. K. Aboulfotouh
Faculty of Environmental Design, King Abdulaziz University, Jeddah, KSA
Faculty of Fine Arts, Minia University, Minia, Egypt

Index

www.ingramcontent.com/pod-product-compliance
Lightning Source LLC
Chambersburg PA
CBHW081416220926
51613CB00010B/252
9781639873906